I0047199

David Peck Todd, William Thynne Lynn

Stars and Telescopes

A Hand-Book of Popular Astronomy

David Peck Todd, William Thynne Lynn

Stars and Telescopes
A Hand-Book of Popular Astronomy

ISBN/EAN: 9783337407674

Printed in Europe, USA, Canada, Australia, Japan

Cover: Foto ©berggeist007 / pixelio.de

More available books at **www.hansebooks.com**

Stars and Telescopes

A Hand-book of Popular Astronomy

Founded on the 9th Edition of
Lynn's Celestial Motions

BY

DAVID P. TODD

M.A., PH.D.

*Professor of Astronomy and Director of the Observatory, Amherst
College. Author of "A New Astronomy," etc., etc.*

BOSTON
LITTLE, BROWN, AND COMPANY
1899

DEDICATED

TO

Mʳˢ D. WILLIS JAMES

PREFACE

STARS AND TELESCOPES is intended to meet an American demand for a plain, unrhetorical statement of the astronomy of to-day: and it has been based upon M^r L^YNN'S *Celestial Motions*, because his little book had already covered excellently part of the ground, and had passed through nine English editions. Permission for this use of *Celestial Motions* was most courteously granted by its author, formerly of the Royal Observatory at Greenwich. In accord with M^r L^YNN'S wish, I have indicated

WILLIAM THYNNE LYNN

which is his work and which my own, both *in situ*, and on the page of contents. For the greater part of Chapter XVI I am indebted to the kindness of D^r SEE.

With a single exception, the ample illustrations are wholly new additions. For excellent photographs I acknowledge gratefully my indebtedness —

To Mrs HENRY DRAPER of New York,
To MM. LOEWY, DESLANDRES, and HENRY of Paris,
To Mr CHRISTIE, Astronomer Royal of Greenwich,
To Dr GILL, H. M. Astronomer at the Cape,
To Sir HOWARD GRUBB of Dublin,
To Dr ROBERTS of Crowborough Hill, Sussex,
To Professor YOUNG of Princeton University,
To Professor LANGLEY, Secretary of the Smithsonian Institution,
To Professor ROWLAND of the Johns Hopkins University,
To Professor PICKERING, Director of Harvard College Observatory,
To Commodore PHYTHIAN, U. S. Naval Observatory,
To Professor HALE, Director of the Yerkes Observatory,
To Mr BRASHEAR, Director Allegheny Observatory,
To Messrs WARNER and SWASEY of Cleveland.

For the use of illustrations in astronomical periodicals also —

To the Editors of the *Vierteljahrsschrift der Astronomischen Gesellschaft* (Leipzig),
To the Editors of *Bulletin Astronomique* (Paris),
To the Editor of *Nature* (London),
To the Editors of *The Observatory* (London),
To M. FLAMMARION, Editor of *L'Astronomie*,
To Dr SCHWAHN, Editor of *Himmel und Erde* (Berlin),
To Mr MAUNDER, Editor of *Knowledge* (London),
To Dr CHANDLER, Editor of *The Astronomical Journal*,
To Professor PAYNE, Editor of *Popular Astronomy*,
To Professor HALE, Editor *The Astrophysical Journal*.

For the courteous loan of engraved blocks also —

To Dr BARNARD of the Yerkes Observatory,

To Professor PAYNE, Director Northfield Observatory,

To the American Book Company of New York,

To the Messrs D. APPLETON & Company of New York,

To the G. & C. MERRIAM Company of Springfield.

For the use of illustrations in printed volumes likewise —

To MM. GAUTHIER VILLARS et Fils of Paris (Father SECCHI's *Le Soleil*, and M. FLAMMARION's *La Planète Mars*),

To W. ENGELMANN's Verlag of Leipzig (Dr VOGEL's *Populäre Astronomie*, and Dr SCHEINER's *Die Photographie der Gestirne*),

To HARTLEBEN's Verlag of Vienna (Baron v. SCHWEIGER-LERCHENFELD's *Atlas der Himmelskunde*),

To Messrs A. and C. BLACK of Edinburgh (Dr DREYER's *Tycho Brahe*),

To Messrs TAYLOR and FRANCIS of London (Mr DENNING's *Telescopic Work for Starlight Evenings*),

To EDWARD STANFORD of London (M^r McCLEAN's *Photographic Spectra of the Sun and Metals*),

To Messrs HOUGHTON, MIFFLIN & Company of Boston (Mr LOWELL's *Annals*, vol. i., and Mrs TODD's *Corona and Coronet*),

To Messrs LONGMANS, GREEN & Company of New York (Mr PROCTOR's *Old and New Astronomy*),

To Messrs CHARLES SCRIBNER's SONS of New York (*Scribner's Magazine*).

Many of the portraits of astronomers are taken from the published volumes of their lives and works, to the editors and publishers of which due acknowledgment is here returned. To the courtesy of Mrs ADAMS I am indebted for the appropriate design of ADAMS'S *Ex Libris*. Other portraits have been made available through the kindness of relatives and executors.

Stars and Telescopes has advantaged greatly from the criticism of several astronomers who have examined chapters in proofs, though no one will hold them responsible for any errors that may have escaped notice : —

The late Professor NEWTON of Yale University, *Meteors and Meteoric Bodies*,
M. CHARLOIS of Nice, *The Small Planets*,
Professor NEWCOMB of Washington, *The Planets*,
Professor YOUNG of Princeton, *The Sun*,
Professor PICKERING of Cambridge, *The Fixed Stars*,
Mr LOWELL of Boston, *The Ruddy Planet*,
Professor BARNARD of Williams Bay, *The Comets*.

To all I am glad to accord my obligation in fullest measure.

DAVID P. TODD.

OBSERVATORY HOUSE,
AMHERST: *March* 1899.

** While this work is passing through the press, announcement is made of an important discovery by Professor W. H. PICKERING at the Harvard Observatory, of a ninth moon of Saturn (page 129), with a period of about 2½ years, and five times more distant from Saturn than Japetus, the outermost satellite hitherto known. The new moon has been named Phœbe, and it is probably about 200 miles in diameter. — *29th March* 1899.

CONTENTS

LIST OF ILLUSTRATIONS

INTRODUCTION

A STRONOMY may be styled a very aristocrat among the sciences; but, while its cosmic conceptions never fail to arouse the broadest general interest, it is an interest that has by no means led to a knowledge of this engrossing science proportionately widespread.

Antecedent to the 19th century, astronomy was purely a science of celestial motions. Its mystical parent, astrology, had taught that planetary places influence men's destinies: how important then, to know the place of every planet in the past, and to be able to divine their conjunctions in the future. Astrology was utilitarian to this end, and her offspring was hardly less so; for upon a knowledge of celestial motions, still farther refined and perfected, were founded the indispensable and very practical arts of conducting merchant ships from port to port in safety, and of surveying accurately coast lines and other national boundaries.

But with the beginning of the 19th century came the brilliant and imaginative conceptions of the elder HERSCHEL, concerning the constitution and physical peculiarities of the heavenly bodies, based upon his minute and painstaking scrutiny ; for his thought was acute, as his observation was thorough. Still it did not so much concern him *where* these bodies were as *what* they were. Sir WILLIAM HERSCHEL, indeed, was the first to build monster telescopes of modern excellence, and he vastly augmented our knowledge of the heavens by faithfully using them. Also he prepared the foundation upon which the imposing edifice of physical astronomy has been erected by his followers.

As the marvellous revelations of the heavens have come to us mainly through the telescope in the hands of highly skilful astronomers, mention of their labors is associated with the telescopes that made the researches possible. Inevitably these are intimately related : we cannot neglect the human element, though we may affect to do so. Thereby may human interest lead to wider diffusion of astronomical knowledge.

So the book is embellished also by a profusion of portraits of astronomers, among those not living, — of many whose important work was done so unostentatiously that their names have long since passed from merely popular recognition. Noteworthy and impersonal achievement is everywhere associated with the distinguished lives that have passed in unselfish discovery of astronomical truth — lives of unswerving devotion, of persistent self-sacrifice, of unwearying toil, and here and there of sad disappointment, and even pathetic disaster.

Next in our abiding concern with astronomers and their labors are the tools with which their work is accomplished — telescopes not only, but all those modern adjuncts introduced by the physicist, and called photometers, spectroscopes, and bolometers, as well as the multifold adaptations of the photographic camera which have revolutionized nearly all departments of the science, and afforded a welcome and unquestioned service. Nor have I neglected either the men who made the telescopes or the buildings in which all such instruments are housed.

Stars and Telescopes begins with a running commentary or outline of astronomical discovery, with a rigid exclusion of all detail. Second and third chapters deal almost as briefly with the Earth, and the Moon, our nearest celestial neighbor in good and regular standing. Due prominence is given that occult phenomenon of recent discovery, called technically the variation of latitude; so persistently has research upon it been prosecuted that the tiny oscillations of the Earth's axis within the globe itself can be predicted for a decade to come. So signally has photography aided in delineation of the surface of our satellite that photographs, reproduced by the best modern processes, have been made to serve instead of word descriptions of lunar scenery. Of less popular interest is the Calendar. But a chapter on the astronomical relations of light is intended to emphasize the prime sig-

nificance of this phenomenon in all celestial
enquiry. Without it the heavens would be as a
book forever sealed.

The Sun, monarch of the planetary system,
and source of all our light and life, could not
be dismissed without the fuller treatment demanded by his importance. The ever new
problem of his distance from us, providing
as it does our foot-rule of the universe; the
continually changing appearances of his surface; the power of his light, and the intensity
and maintenance of his heat — all are set forth,
not only the most recent results, but the instruments and methods by which in part they have
been derived.

Solar eclipses, too, are introduced, but only
in the most general fashion. Their prediction is,
however, exciting sufficiently wide interest to
warrant a table and chart of these engrossing
phenomena in the future.

The chapter embodying a general outline of
the solar system necessarily brings to the read-
er's attention the labors of the most celebrated
of all the early astronomers — TYCHO, KEPLER,
and NEWTON. But an account of the individual
planets of our system called for still farther expansion, if the present state of our knowledge of
those bodies was to be abundantly set forth.
Particularly have the most recent results been
presented, with a view of exhibiting in a popular
way the state of technical literature down to the
present day. Wherever the desired verification

of theory or observation is still lacking, I have been careful to say so.

In the summer of 1898 was discovered a new planet, probably a genuine member of the asteroidal group, although periodically it approaches nearer to the Earth by one-half than Venus ever does, thereby outclassing not only this planet when in transit over the Sun's disk, but Mars also at his close oppositions, as a medium for finding the Sun's true distance from us. Fresh interest attaching to the small planet group as a whole, I have appended in tabular form a complete list of these tiny members of the solar system, but without amplifying it to include particularities of their paths round the Sun.

The chapter on Comets deals especially with historic detail of these filmy voyagers; and the great meteoric showers of 1899 and 1900 must be justification for extending the space allotted to meteorites — the sole material messengers from outer space ever known to come within human reach.

Even more amplified is the chapter on the fixed stars and nebulæ, because the methods of the new astronomy have achieved unparalleled revelations in the stellar realm. So fertile are the starry fields, and so persistently have they been tilled by a band of tireless workers for the last quarter of a century, that the compacting of results into a single chapter was found impossible. The cosmogony was therefore allotted a chapter by itself, and due space accorded the

newest research in our knowledge of the building
and development of worlds.

Both to encourage and to gratify a desire to
consult original sources, there is added to each
chapter a bibliographic note, never intended to
be complete, and with both popular and scientific
papers purposely intermingled.

But nearly equal in interest with the stars
themselves are the telescopes by which alone we
ascertain their distances, inconceivably vast, and
the spectroscopes that tell the tale of stellar
motion and constitution. So *Stars and Telescopes*
may be said to culminate in an account of the
famous instruments, their construction and mounting and use. To the gradual development of
skill in the arts and to present-day perfection
of mechanical and photographic methods is the
progress of astronomy vastly indebted in all its
ramifications. Just as the precision of the 'old
astronomy' could never have been attained independently of the skill of workers in metals
who provided accurate clocks and meridian
circles, so the 'new astronomy' would have
been wholly handicapped in its development
but for the splendid prisms and perfect gratings
of the modern optician. Evolution, too, of the
process of making great lumps of glass has been
mainly mechanical, not to mention the methods
of fashioning disks into lens and prism. During
the first half of the 19th century, there was little
chance to forget the optician and telescope
builder, because the greatest instruments of

those days were constructed by the astronomers themselves, who used them so successfully. Too often, however, in our own day we omit to mention the optician and instrument maker, unaided by whose consummate skill original research in physics and astronomy would have made but slender progress. Not only in the title of this book, therefore, are the significant labors of the mechanician recognized, but its last chapter is lengthened to a degree fitting a highly mechanical age.

By the most conservative estimate, the threshold of the edifice of stellar investigation is but barely crossed; its corridors have been subjected to the merest reconnaissance; nothing but the crudest plan of the intricate temple of the stars can yet be sketched with confidence. Here comes an insistent demand for more light and greater telescopes still — more light, not for the expected discovery of new bodies, but for studying in greater detail the constitution of the component bodies of the stellar universe already known, and for investigating with greater precision their motions relatively to our solar orb and his tributary family of planetary globes.

Greater successes for the optician of the future seem safely predicated upon the rapid advances of the recent past. No step can, however, be taken except at infinite pains and large expenditure. Mechanician and optician are ready to do their faithful part, and opportunity to go forward must not be too long denied them.

Continuity of effort will alone guarantee success ; each generation must build upon the platform of its immediate predecessor. Otherwise gaps oc-. cur. With the passing of the elder HERSCHEL, nearly a generation elapsed before Lord ROSSE'S great telescope was finished; and he could then profit relatively but little by HERSCHEL'S unparalleled experience. Already the CLARKS have passed from earth, father and both sons. But we have worthy successors, still in the prime of brilliant achievement: they await to-day their keenly coveted opportunity.

D. P. T.

AMHERST, *February* 1899.

STARS AND TELESCOPES

CHAPTER I

OUTLINE HISTORY OF ASTRONOMICAL DISCOVERY

IT is not proposed here to enter into any discussion of the knowledge of astronomy possessed by the ancient Egyptians, Babylonians, or other Eastern nations. Among the Greeks, EUDOXUS, born at Cnidus in Caria about B. C. 409, is the first astronomer that need be named. His works have not come down to us ; but it is known that one of them was a description of the constellations, which is versified in the *Phainomena* of ARATUS of Soli in Cilicia, and is the work quoted by PAUL in his speech before the Areopagus at Athens.

The greatest of the ancient Greek astronomers was HIPPARCHUS, who, though a native of Nicæa in Bithynia, made most of his observations at Rhodes. He was the first to form a systematic catalogue of stars, induced to do so by the appearance of a new star in the heavens, which is thought to be the same as one recorded in the Chinese annals as having appeared in the constellation Scorpio in the year corresponding to B. C. 134. The most probable date of the death of HIPPARCHUS is B. C. 120. His catalogue is known to us only by its incorporation, with some modifications, into the great work of the Alexandrian astronomer PTOLEMY, entitled 'Η Μαθηματικὴ Σύνταξις, which, after the Arabians, is usually called *The Almagest.* PTOLEMY, principally known from the system of the world which bears his name, died early in the reign of the Emperor AURELIUS, which commenced in A. D. 161.

The greatest of Arabian astronomers was ALBATENIUS (so called from his birthplace, Baten, in Mesopotamia), who flourished in the ninth century of our era. About a century later the Persian astronomer AL-SÛFI formed a catalogue of stars from his own observations. Another catalogue was made by, or rather under the auspices of, the Mongol prince MIRZA MOHAMMED TARAGHAI, or ULUGH BEGH (as he is generally called), grandson of the famous TIMOUR, from observations at Samarkand about A. D. 1433.

The labors of PEURBACH and MÜLLER (usually called REGIOMONTANUS), toward the end of the same century, prepared the way for COPERNICUS, who showed the simple explanation of the planetary motions which resulted from supposing the Sun to be placed in the centre of the system. His work, *De Revolutionibus Orbium Cœlestium,* was published in 1543, in which year the illustrious author died. The foundation of the true theory of the solar system was thus laid, but the time for its establishment had not arrived. It was rejected by the great Danish astronomer, TYCHO BRAHÉ, in favor of a system called from him the Tychonic, in which the Moon and Sun were supposed to revolve round the Earth, and the planets round the Sun. TYCHO'S observations, however, furnished the means by which, less than 20 years after his death, KEPLER deduced the laws of planetary motion. It was in 1618 that he discovered his third and last law, which shows the mutual correspondence existing between the motions and distances of all the planets, and proves (though it was reserved for the genius of NEWTON to demonstrate this consequence) that they obey the same law of force directed toward the Sun.

COPERNICUS (1473–1543)

Meanwhile GALILEO GALILEI, equally famous, had been discovering a system of moons revolving round Jupiter as the planets do round the Sun, confirming by that and by other discoveries the truth of the Copernican system, and establishing the laws of motion, without a knowledge of which farther progress in astronomy as a science was impossible. GALILEO died in 1642 ; and in the same year was born NEWTON, the prince of philosophers, who discovered the law of gravitation, and demonstrated its competency to explain the most important of the lunar and planetary motions. This was made possible by the French astronomer PICARD's determination of the true size of the Earth, and by FLAMSTEED's observations of the Moon, commenced at Greenwich in 1676, the year after the founding of the Royal Observatory there. The *Philosophiæ Naturalis Principia Mathematica*, NEWTON's great work, was published in 1687.

Of the satellites of Saturn, HUYGENS discovered one, and CASSINI four, in the latter part of the 17th century; and HUYGENS had explained the ring-like nature of the appendage to Saturn, the existence of which had been first noticed by GALILEO. It was not until some time after the death of NEWTON in 1727 that his theory was further developed, and shown to be capable of explaining not only the principal courses of the planets, but the smaller deviations in their movements. This is due to the labors of several eminent mathematicians, but especially to LAGRANGE and LAPLACE. The *Mécanique Analytique* of the former was published in 1787, and the *Mécanique Céleste* of the latter was completed in 1825.

Toward the end of the 18th century arose an astron-

GALILEO (1564–1642)

omer, Hanoverian by birth and English by adoption, who, by the discovery of Uranus in 1781, not only

SIR WILLIAM HERSCHEL'S 40-FT. TELESCOPE
(Inscribed to his Royal Patron, King GEORGE THE THIRD, 1795)

extended what had previously been supposed to be the boundary of the solar system, but who, turning upon the sidereal heavens with unwearied diligence the powerful instruments constructed by himself,

acquired for mankind a knowledge of the distribution and motions of the stars, which inaugurated a
new era in the history of astronomy. Sir WILLIAM
HERSCHEL, elected in 1820 the first president of the
Royal Astronomical Society, died two years subsequently. That Society published in 1827 a compiled

SIR WILLIAM HERSCHEL (1738–1822)

catalogue of stars, which was the standard work of
reference on the subject until superseded by the catalogue of the British Association containing 8,377 stars,
published in 1845.

Soon after this the boundary of the solar system
was still farther enlarged by the famous discovery of
Neptune in 1846 ; and the year following, Sir JOHN
HERSCHEL published the results of his observations at

the Cape of Good Hope, where he had devoted some years to diligent scrutiny of the stars and nebulæ visible only in the southern hemisphere, — a research similar to that in the northern heavens, initiated by his father and extended by himself and others.

Early in the present century four new planets were discovered, the first of a group of many hundred revolving between the orbits of Mars and Jupiter, and so much smaller than all the others that distinct terms — planetoids or asteroids — were suggested for them. But as they differ from the other planets mainly in size, it is now more usual to call them small or minor planets. They remained four in number until 1845, when a fifth was found. Since then, discovery has been so rapid that members of that group are now approaching 500 in number, and probably there are hundreds more. The later discoveries cannot be seen without a powerful telescope, and nearly all the most recent were first recognized on photographic plates by Dr MAX WOLF of Heidelberg and M. CHARLOIS of Nice. The number of known bodies of the solar system has been farther increased by the discovery in 1877 of two satellites revolving round Mars, and of the fifth satellite of Jupiter in 1892.

Also, the accurate observations of the last half century have enabled astronomers to determine the approximate distances of some of the fixed stars ; a Centauri, in the southern hemisphere of the sky, and about 275,000 times more distant than the Sun, being, so far as is known, the nearest.

It is now a third of a century since a new branch of astronomy was inaugurated by the analysis of the light of celestial bodies with the spectroscope.

This began with the investigations of KIRCHHOFF and
BUNSEN in 1859, which were brilliantly followed up by
Dr HUGGINS (at first in conjunction with MILLER), and
subsequently by SECCHI, JANSSEN, YOUNG, LOCKYER,
VOGEL, DRAPER, PICKERING, MAUNDER, and many
others. Not only has the spectroscope enabled us
to learn much of the chemical constitution and condition of celestial bodies both in and beyond the
solar system, but its later developments have revealed
motions of several of the stars in the line of sight (in
some cases approaching, in others receding), which
cannot be recognized in any other way. Likewise,
the same instrumental method has led to the important discovery that many stars, although appearing
single under direct scrutiny with the highest telescopic
powers, are really double. One very interesting discovery of spectrum analysis is that the light of many
of the nebulæ (the great nebula in Orion, for example) does not proceed from multitudes of stars, as was
once supposed, but from incandescent or intensely
heated matter in a gaseous condition.

The first person who is known to have noticed
(though it is not unlikely NEWTON had done so
before) a few of the dark lines crossing the prismatic
spectrum at definite distances, was WOLLASTON in 1802,
and he is known as their discoverer ; but as the accurate positions of a large number of them were first
measured in 1815 by FRAUNHOFER, the celebrated
German optician, they are always called Fraunhofer
lines. The measurement of the positions and distances of these lines in the spectra of different celestial objects, and the comparison of the spectra of the
heavenly bodies with the spectra of incandescent terrestrial substances, have led to the important discov-

eries alluded to above. But the spectroscope is not
the only new engine of astronomical research introduced in recent years. The application of photog-

FRAUNHOFER (1787–1826)

raphy to celestial observations, and the instruments
for making exact determination of the comparative
amounts of light from different heavenly bodies have
both inaugurated new fields of study which have borne,
and are bearing, the richest fruit.

S & T — 2

JABLONOW, *De Astronomiæ Ortu ac Progressu, etc.* (Rome 1763).

LALANDE, *Astronomie* (Paris 1792).

LA PLACE, *Exposition du Système du Monde* (Paris 1800).

MONTUCLA, *Histoire des Mathématiques* (Paris 1802).

LALANDE, *Bibliographie Astronomique avec l'Histoire de l'Astronomie* (Paris 1803).

BAILLY, *Histoire de l'Astronomie Ancienne et Moderne* (1805).

DELAMBRE, *Histoire de l'Astronomie Ancienne* (Paris 1817).

DELAMBRE, *Histoire de l'Astronomie du Moyen Age* (Paris 1819).

DELAMBRE, *Histoire de l'Astronomie Moderne* (Paris 1821).

DELAMBRE, *Histoire de l'Astronomie du XVIIIe Siècle* (1827).

LA PLACE, *Précis de l'Histoire de l'Astronomie* (Paris 1821).

BENTLEY, *Hindu Astronomy* (London 1825).

ROTHMANN, *History of Astronomy*, ' Library of Useful Knowledge' (London 1832).

NARRIEN, *Origin and Progress of Astronomy* (London 1833).

JAHN, *Geschichte der Astronomie* (Leipzig 1844).

GRANT, *History of Physical Astronomy* (London 1852).

MAIN, *History of Astronomy*, 8th edition of *Encyclopædia Britannica*, vol. iii. (1853) ; 9th ed. ii. (1875), 744 (PROCTOR).

LOOMIS, *Recent Progress of Astronomy* (New York 1856).

LEWIS, *Astronomy of the Ancients* (London 1862).

BIOT, *Études sur l'Astronomie Indienne et Chinoise* (Paris 1862).

WOLF, *Geschichte der Astronomie* (Munich 1877).

HOUZEAU, *Bibliographie Générale*, i. (1887), Introduction.

YOUNG, 'Ten Years' Progress,' *Nature*, xxxv. (1887), 67, 86, 117.

BERTIN, 'Babylonian Astronomy,' *Nature*, xl. (1889), 237, 261, 285, 360.

EPPING and STRASSMAYER, *Astronomisches aus Babylon* (Freiburg 1889).

WOLF, *Handbuch der Astronomie*, i. (Zurich 1890).

CLERKE, *History of Astronomy during the XIXth Century* (London 1893).

TANNERY, *Recherches sur l'Histoire de l'Astronomie Ancienne* (Paris 1893).

LOCKYER, *The Dawn of Astronomy* (New York 1894).

BERRY, *Short History of Astronomy* (New York 1899).

Excellent chronological tables of salient events are in the *Annuaire* of Brussels Observatory (1877), p. 52, CHAMBERS's *Astronomy*, vol. ii. p. 468, and in Miss CLERKE's *History*, p. 531. Consult, also, the extensive lists of HOUZEAU's *Vade Mecum de l'Astronome*, ch. ii. pp. 34–147, a work of invaluable assistance in all investigation of the origins of astronomy, and its development down to modern times. —*D. P. T.*

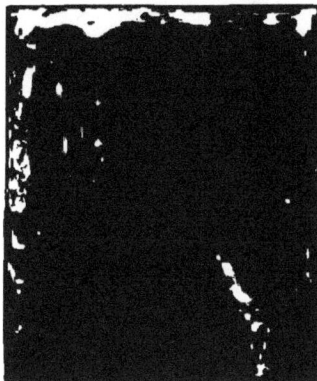

BAILLY (1736–1793)

(JEAN SYLVAIN BAILLY, *a learned astronomical writer and historian, was honored with the presidency of the National Assembly of 1789, and the mayoralty of Paris, in which capacity he served with sterling integrity. Besides his 'Histoire de l'Astronomie,' in five volumes, he published many elaborate treatises, and held the high honor of membership in the three French academies simultaneously*)

DELAMBRE (1749–1822)

(JEAN JOSEPH DELAMBRE *was perhaps the most prolific of all French astronomical writers.* HERSCHEL'S *discovery of Uranus in 1781 first gave him opportunity for distinction by constructing tables of the new planet's motion. He was collaborateur with* MÉCHAIN *in measuring an arc of meridian, and succeeded* LALANDE *in 1807 at the Collège de France.* DELAMBRE *was an officer of the Legion of Honor*)

CHAPTER II

THE EARTH

THE general shape of the Earth on which we live is that of a sphere or globe. Although the irregularities of its surface present in places what appear, to our eyes, enormous elevations, the altitude of the highest mountain does not really amount to nearly the thousandth part of the diameter of the Earth; and the depth of the lowest sea-bottom is about the same as the height of the highest mountain.

But while our powers of locomotion are thus confined within so comparatively small a part of the vast universe of creation, the wonderful sense of sight has enabled us to acquire, through the agency of light, some knowledge of other bodies around us. Many of them are seen to be in motion, though it has been discovered that these movements are not all real, but in part apparent, produced by real motions of the Earth itself. All our knowledge respecting these bodies is embraced under the term astronomy; and as that science has taught us to consider the Earth itself as one of the moving bodies of the universe (one of the planets, in fact, revolving round the Sun, from which they derive their light and heat), a knowledge of its motions, as well as of its shape and size, forms also an important part of astronomy.

The Earth's actual form is not spherical, but slightly flattened at the poles, — the diameter taken anywhere

across the equator measuring 7,926.6 miles, and the
polar diameter, or that taken from pole to pole, 7,901.5
miles, 25 less than the equatorial.[1]

On the polar diameter as an axis the Earth rotates
from west to east, while revolving in the same direction in an orbit round the Sun. The axis of the Earth
is inclined to the plane of this orbit at an angle of 66°
32′ 52″.0 (in 1900), and its rate of increase is nearly
0″.5 annually. The true period of axial rotation is

[1] The Clarke spheroid of 1878, generally adopted, gives
more exactly: —

Earth's
$\begin{cases}
\text{Equatorial semi-diameter} = 20,926,202 \text{ feet} = 6,378\text{-} \\
\quad 301 \text{ metres} = 3,963.296 \text{ miles.} \\
\text{Polar semi-diameter} = 20,854,895 \text{ feet} = 6,356,515 \\
\quad \text{metres} = 3,950.738 \text{ miles,}
\end{cases}$

the polar compression being $\frac{1}{293.46}$. This supposes that the
figure of our globe corresponds to an oblate spheroid, its equator being a circle. But it is found that all existing measures
of the Earth are better represented if the equator is regarded
as slightly elliptical; in other words, the Earth is an ellipsoid
with three unequal axes. The greatest equatorial bulge lies
very nearly coincident with Liberia on one side and the Gilbert
Islands at its antipodes; while the City of Mexico in the western hemisphere and Ceylon in the eastern mark the meridian
of least eccentricity. A good notion of the refinements in
modern geodetic work is conveyed by stating that this investigation places Ceylon only 1,500 feet nearer the Earth's centre
than Liberia; farther observations are, however, necessary to
establish this conclusion beyond dispute. But owing to the
irregular past action of forces operant in moulding the Earth's
crust, it is wholly unlikely that our planet has the general outline of any mathematical solid; so that it becomes the business
of the geodesist to assume the most probable figure, and then
measure the deviations of the actual Earth or geoid therefrom
in every particular. It is to be expected that the extensive
series of pendulum observations now conducted by represen-
tative nations will tentatively determine, not only the Earth's
general figure, but all local deviations, with greater accuracy
than the processes of the ordinary geodesy. — *D. P. T.*

called a sidereal day; it is divided into 24 sidereal hours, and is equal to $23^h 56^m 4^s.09$ of ordinary time.

But, the convenience of life rendering it necessary to regulate our time by the Sun, the day of our ordinary reckoning is a solar day, and this also is subdivided into 24 equal parts called solar hours. The reason for the two kinds of day will be seen on reflecting that the Earth's actual motion round the Sun (or, what is in effect the same thing, the apparent motion of the Sun among the stars) places it a little farther east every day. Evidently, then, the Earth, rotating eastward also on its axis, and having completed an entire rotation referred to the stars, must turn a little farther to complete an entire rotation referred to the Sun. The solar day, therefore, is longer than the sidereal day; and these apparent solar days were employed as a measure of time in France until 1816. But they are inconvenient for this purpose because of their unequal length, which is due partly to variations from day to day in the Sun's apparent motion among the stars. In practice, a mean or average day for the whole year (called the mean solar day) is now used, and the difference between the mean and the apparent solar time at any instant is termed the equation of time.[2] The

---

[2] Coincidence of true Sun and mean Sun occurs four times annually; that is, the equation of time is then zero, and Sun and mean time clock agree. Following are the approximate days of each year when this takes place, also the dates when the equation of time is a maximum, either positive (Sun slow), or negative (Sun fast): —

	m				m	
12 February,	Sun 14.5	slow.	25 July,	Sun	6.2	slow.
15 April,	0.0		31 August,		0.0	
14 May,	Sun 3.9	fast.	1 November,	Sun	16.3	fast.
15 June,	0.0		24 December,		0.0	

Between these dates the equation of time is intermediate in

mean solar day is universally accepted as the elemental
unit of time in all other astronomical measurements of
duration, and for the purposes of civil life as well.

Of these mean solar days the Earth occupies
365.25636 (equal to 365^d 6^h 9^m 9^s.0) in revolving
round the Sun, and this period is called the sidereal
year, because at the end of it the Earth is in the same
position with respect to the Sun and the stars. But,
just as the conveniences of life lead us to regard as
the day not the precise period of time in which the
Earth completes one ro-
tation on its axis rela-
tively to the stars, so do
they not allow us to
adopt as the year the
exact period of time in
which the Earth revolves
in its orbit round the
Sun. For what makes
the observance of the
year necessary to us is
the change produced by
the variations in the seasons ; and in conse-
quence of a slow conical
movement of the Earth's
axis (occupying about
25,800 years to com-
plete a whole round, and

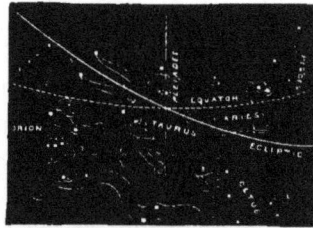

POSITION OF THE VERNAL EQUINOX
B. C. 2170

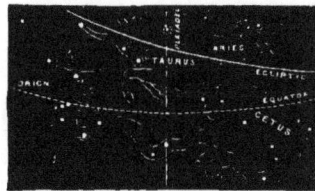

TO INDICATE THE PRESENT POSITION
OF THE VERNAL EQUINOX, NOW
PASSED BEYOND TAURUS AND
ARIES

called precession of the equinoxes, from the effect it
produces upon the equinoctial points), the tropical

value, and *The Ephemeris*, or *Nautical Almanac*, for the particular year must be consulted to find the precise amount at
noon each day. — *D. P. T.*

year, in which all the changes of the seasons are run through, is 20^m 23^s.5 shorter than the sidereal year. In fact, the true length of the ordinary, or tropical year, is 365^d.24219, or 365^d 5^h 48^m 45^s.5.

It is now apparent that the Earth has three principal motions : (1) a rotation of the globe round its own axis, (2) a revolution round the Sun, during which the axis remains nearly parallel to itself, and (3) a slow conical motion of the axis so performed as to retain nearly the same inclination to the plane of the orbit throughout. But it should be mentioned that the axis is subject to an oscillation, in a period of a little less than 19 years, which alternately increases and diminishes the inclination by a small amount. This is called nutation, and was discovered in 1747 by BRADLEY, the third Astronomer Royal.[3]

[3] Recent investigation has brought to light still another motion of the Earth which causes a periodic variation in the latitude of all places on its surface. An early research of EULER demonstrated that if a 'free rigid oblate spheroid' (then supposed to represent what was known about the Earth) be set in rotation round an axis making a small angle with its axis of figure, and be not acted on by any force, the position of the axis of rotation within the spheroid will describe a circular cone round the axis of figure with a uniform motion. This principle applied to the Earth gave a theoretic period of about 306 days, and the eminent German astronomer, C. A. F. PETERS, was the first to discuss the question whether observations were sufficient to reveal a corresponding variation in terrestrial latitudes. Later the investigations of NYRÉN and others showed that if an inequality of this period existed at all, its amplitude could not exceed a few hundredths of a second — that is, a few feet on the surface of the Earth. As, however, the Earth is not a perfectly rigid spheroid, but is partly covered by a fluid envelope, and has perhaps the elasticity of steel (to adopt the conclusion of Lord KELVIN), Professor NEWCOMB finds the theoretic period not 306, but 441 days — a remarkably good

The mean density of the Earth is about five and a half times that of water; which shows that some portion at least of the interior must be much denser than that of the exterior crust. The entire surface area of our planet is about 197,000,000 square miles. Of this about three fourths, or 145,000,000 square miles, are covered by water; and the fact that the area of water

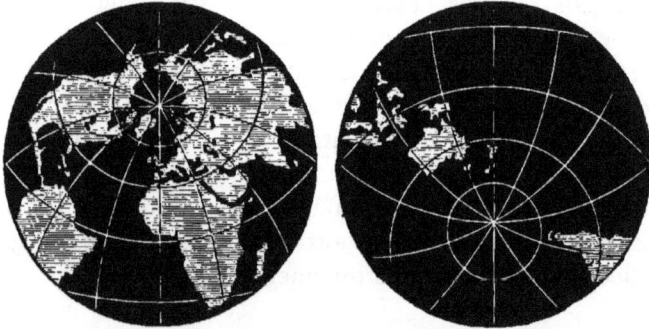

in the southern hemisphere is larger than that in the northern proves that the former portion of the Earth is somewhat denser than the latter, enabling it to retain the larger bulk of water. If the globe be divided into two hemispheres such that the surface of one contains the largest possible amount of land, and the other the

agreement with the period of 427 days determined from observation by Mr CHANDLER in 1892. Professor NEWCOMB regards this agreement as affording conclusive and independent evidence of the rigidity of our globe; and we may, he says, accept the conclusion that the pole of rotation of the Earth describes a circle round its pole of figure in about 427 days, and in a direction from west toward east, as a result of dynamical theory. Observations still in progress (1899) at Tokyo, Kasan, Pulkowa, Prague, Potsdam, Lyons, Philadelphia, and elsewhere, have revealed the exact nature of the polar oscillation since 1890, as in the opposite diagram, and defined its motion in the future. The curve is mainly elliptical rather than circular, and the extent of fluctuation is about 60 feet.— *D. P. T.*

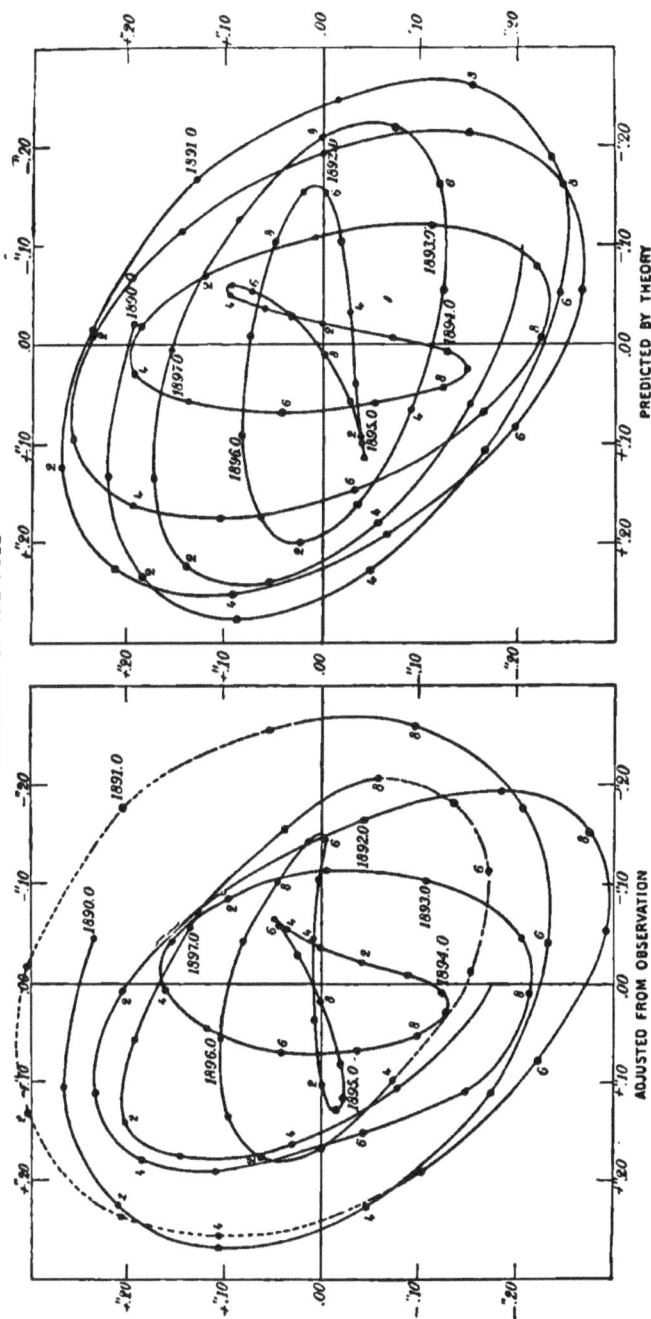

TRAJECTORY OF THE POLE

ADJUSTED FROM OBSERVATION

PREDICTED BY THEORY

(These diagrams represent the wanderings of the pole, from January 1890 to August 1897, as given by the research of Dr. CHANDLER in 1898. On the left is the motion of the pole as derived from actual observations; on the right its motion during the same interval as given by theory; note the close correspondence between the two. Each chart represents a circumpolar area about 60 feet square. Onward from 1898 the predicted motion of the pole carries it round counter-clockwise in an elliptical inwinding spiral, roughly speaking, until 1901. Its motion is then reversed for about a year, during which it describes a rude circle clockwise, and less than 10 feet in diameter. In February 1902, the pole again resumes its usual counter-clockwise motion, and for the next two years it describes an outwinding spiral. During 1890-92 the general sweep of polar motion was much less restricted than in 1895-1904)

largest possible amount of water, the pole of the former portion will be very near the British Islands, and the pole of the other very near New Zealand.

FORM AND SIZE

TODHUNTER, *History of the Mathematical Theories of Attraction and the Figure of the Earth.* 2 vols. (London 1873.)
CLARKE, *Geodesy* (Oxford 1880).
WOODWARD, *American Journal Science*, cxxxviii. (1889), 337.
PRESTON, ' Pendulums,' *Am. Jour. Science*, cxli. (1891), 445.
GORE, J. H., *Geodesy* (Boston and New York 1891).
FOWLER, ' Measurement of Earth,' *Knowledge*, xx. (1897), 148.

INVARIABILITY OF ROTATION AND VARIATION OF LATITUDE

NEWCOMB, *American Journal Science*, cviii. (1874), 161.
NEWCOMB, *Astron. Papers American Ephemeris*, i. (1882), 461.
GLASENAPP, DOWNING, *The Observatory*, xii. (1889), 173, 210.
CHANDLER, *The Astronomical Journal*, xii., xiv. (1891-94).
NEWCOMB, *Month. Not. Royal Astron. Society*, lii. (1892), 336.
DOOLITTLE, *Proc. Am. Assoc. Advancement Science*, xlii. (1893).
BALL, ' Wanderings of North Pole,' *Smithsonian Report* 1893.
WOODWARD, *The Astronomical Journal*, xv. (1895), 65.
ALBRECHT, *Astronomische Nachrichten*, cxlvi. (1898), 129.
CHANDLER, *The Astronomical Journal*, xix. (1898), 105.

AURORA BOREALIS, TERRESTRIAL PHYSICS, ETC.

LOVERING, *Mem. and Trans. Am. Acad. Arts and Sci.*, 1867-73.
HUXLEY, *Physiography* (London 1882).
TROMHOLT, *Under the Rays of Aurora Borealis* (London 1885).
LEMSTRÖM, *L'Aurore Boréale* (Paris 1886).
BISHOP, ' Sky glows,' *Warner Observatory* (Rochester 1887).
SYMONS, *The Eruption of Krakatoa* (London 1888).
BALL, ' Krakatoa,' *In Starry Realms* (London 1892).
WALLACE, ' Our Molten Globe,' *Fortn. Rev.*, lviii. (1892), 572.
MILL, *The Realm of Nature* (New York 1892).
KING, ' Age of the Earth,' *Smithsonian Report* 1893, p. 335.
BECKER, ' Interior of Earth,' *North Am. Rev.*, clvi. (1893), 439.
BONNEY, *The Story of our Planet* (London 1893).
KELVIN, *Popular Lectures and Addresses*, ii. (New York 1894).
DAWSON, *Salient Points in Science of Earth* (New York 1894).
ANGOT, *Les Aurores Polaires* (Paris 1895).

For Earth's density, tides, etc., consult HARKNESS, *Washington Observations* 1885, App. iii. pp. 159 and 161 ; and GORE, 'A bibliography of Geodesy,' *Report U. S. Coast Survey* 1887 ; also POOLE's Indexes, vols. i.-iv., and *Astrophysical Journal*, i.-viii. (1895-98). — *D. P. T.*

CHAPTER III

THE MOON

IN its annual journey round the Sun, the Earth is accompanied by a smaller body called the Moon; her movement relatively to the Earth being in the nature of a motion in an elliptic orbit round the latter, she is considered as a satellite or secondary planet thereto.

The Moon, then, being by far our nearest neighbor among the heavenly bodies, much more has been learned about her than about the others; selenography, in fact, has of recent years become almost a science of itself. The great work of BEER and MÄDLER upon the Moon was published in the year 1837, that of Mr NEISON (now Mr NEVILL) in 1876.

On account of the Moon's proximity to us, an exact knowledge of her apparent motion among the stars is of the greatest practical use in navigation, thereby enabling the mariner to find his longitude at sea. The difficulty of this problem, which has exercised the labors of many distinguished mathematicians, arises from the perturbations to which the Moon's elliptic motion round the Earth is subjected through the gravitating action of the Sun and the nearer or more massive planets.[4] By reason indeed

[4] Foremost among these investigators in the last half-century have been HANSEN in Germany, PONTÉCOULANT and

HANSEN (1795–1874)

(PETER ANDREAS HANSEN *was one of the greatest mathematical astronomers of the 19th century. His tables of Sun and Moon, although now nearly a half century old, are still the basis of computation by which all timepieces are regulated, and the ships of all nations guided. Twice he was honored by the award of the gold medal of the Royal Astronomical Society*)

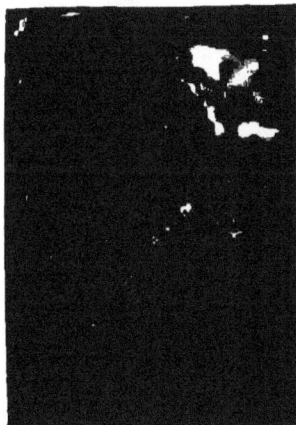

DELAUNAY (1816–1872)

(CHARLES EUGÈNE DELAUNAY *was perhaps the greatest of modern French geometers. He succeeded* ARAGO *and* LE VERRIER *as director of the Paris Observatory.* DELAUNAY'S *masterpiece entitled ' La Théorie du Mouvement de la Lune ' is published in two quarto volumes of the ' Mémoires de l'Académie des Sciences de l'Institut de France ' (Paris* 1860–67)

of the much greater mass of the Sun, he exerts a more powerful attractive force upon the Moon than the Earth does, although the latter is so much nearer. The mathematician might, therefore, take exception to calling the Moon a satellite of the Earth ; and she is certainly not so, in the full sense that the moons of Jupiter and Saturn are satellites of those planets. Still, as we have said, the Moon's motion, relatively to the Earth, is a motion round it ; she is connected with our planet by an indissoluble tie, and, from her great benefit to us, will always be called 'our satellite,' exciting our gratitude and commanding our constant attention.

The actual duration of the Moon's orbital motion round the Earth is 27^d 7^h 43^m $11^s.545$; but here again convenience compels us to regard as a lunar month or lunation not this, the time of her sidereal

---

DELAUNAY in France, AIRY and ADAMS in England, and Professor NEWCOMB and Mr G. W. HILL in America. HANSEN, DELAUNAY, and AIRY spent a goodly portion of their lives upon the intricate mathematics of the 'lunar theory,' as it is termed. However, the formation of final tables for predicting the Moon's motions is a task so prodigious that only one of these investigators (HANSEN) arrived at this complete solution of the problem ; publishing in 1857 his *Tables de la Lune construites d'après le principe Newtonien de gravitation Universelle*. These, in conjunction with Professor NEWCOMB's *Corrections*, are now exclusively employed in calculating future positions of the Moon ; and at the present time the error of prediction does not often exceed 4″, while its average is half that quantity, or about $\frac{1}{1000}$ part of the Moon's breadth (that is to say, about two miles in space). The story of the labors of mathematicians in determining the 'secular acceleration of the Moon' (one out of nearly a hundred perturbations which derange its motion round the Earth) is told by Professor NEWCOMB in the introduction to his masterful work entitled *Researches on the Motion of the Moon* (Washington Observations, 1875). — *D. P. T.*

revolution, but the mean period between successive conjunctions with the Sun. This period is sometimes called a synodic revolution, and amounts to 29^d 12^h 44^m $2^s.684$; upon it of course depend the Moon's phases as presented to the Earth, her illumination being due to the light of the Sun. The Moon also receives solar light reflected from the Earth; and when the illuminated part of our globe turned toward her is the greatest (that is, near our New Moon), this light is sufficiently strong to be reflected back again, enabling us to see the other part of her surface, as at Full Moon, only very faintly. This is popularly called 'the old Moon in the young Moon's arms.'

In consequence of the comparative proximity of our satellite, her distance and size can be determined much more accurately than those of any other heavenly body. Her mean distance is 238,840 miles (the mean horizontal parallax being 57' 2".3); but her actual distance varies, in different parts of her orbit, between 221,610 and 252,970 miles.[5] The eccentricity of the Moon's orbit is 0.0549; and its inclination to the ecliptic, or Earth's orbit, varies between 4° 57' and 5° 19', the mean value being 5° 8' 40".

[5] It is worthy of note that, while the Moon comes nearest to the Earth of all the heavenly bodies, the meteors alone excepted, its distance (about 60¼), in terms of the radius of its primary body, exceeds that of all other known satellites in the solar system. Only Japetus, the outer satellite of Saturn, approaches this maximum, its distance being 59.6 times the equatorial diameter of that planet; while the satellites nearest their primary, when estimated in like manner, are Phobos, the inner moon of Mars (2.8), and the newly-found satellite of Jupiter (2.6). — *D. P. T.*

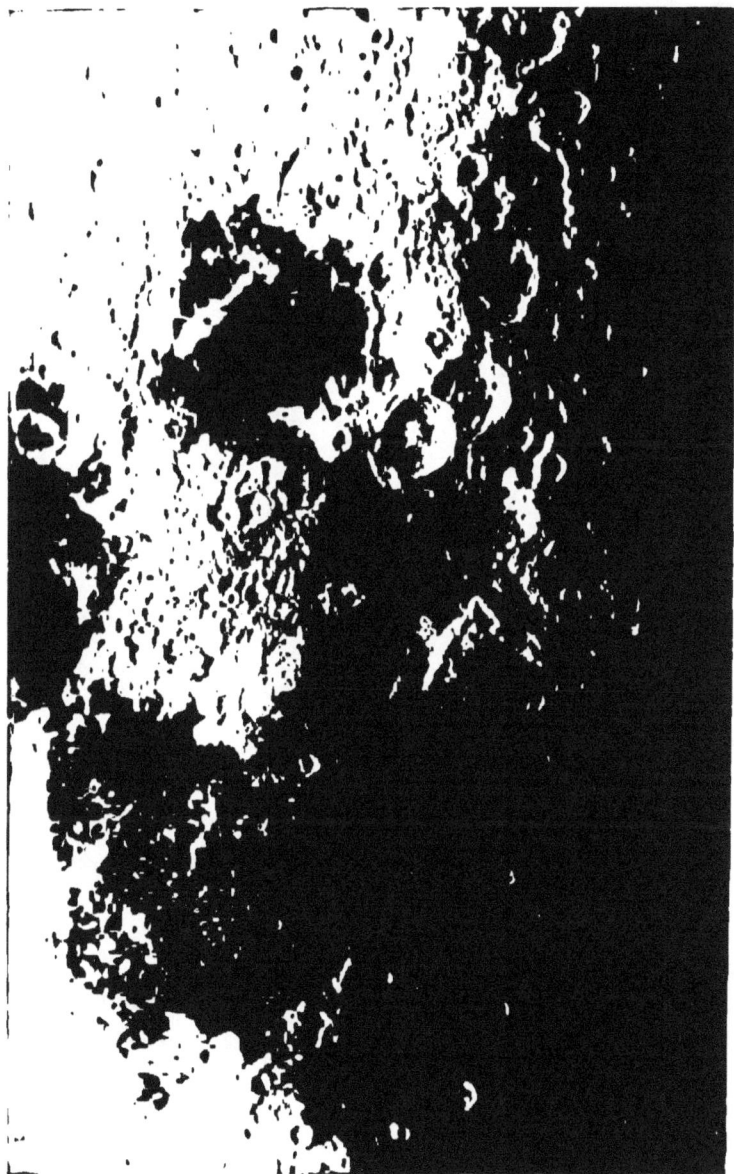

A PART OF THE MOON (WESTERN CENTRAL REGION)

(From a photograph by the Brothers HENRY *at Paris, 23rd March 1893. Age of the Moon, 5 days 16 hours. The three large craters above the centre of the field are Theophilus (lowest one), Cyrillus, and Catharina; and the diameter of each is about 65 miles)*

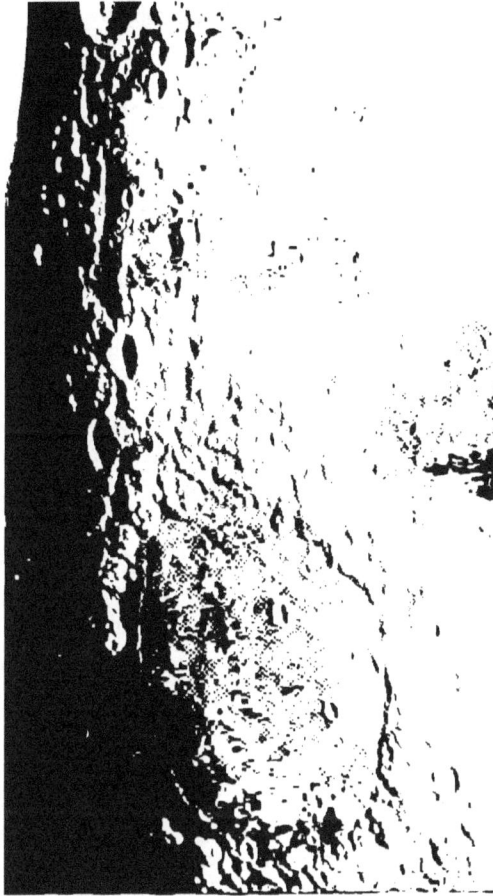

SURFACE OF THE MOON, 1ST SEPTEMBER, 1890

(*Photographed at the Lick Observatory*)

The diameter of the Moon is 2,163 miles, or somewhat less than two sevenths that of the Earth; so that the bulk of our planet is about 50 times as great as that of the Moon. But her density being little more than half that of our own globe, her mass amounts to only about $\frac{1}{80}$ that of the Earth.

The Moon rotates on her axis very much more slowly than our Earth does. In fact, the duration of her rotation is exactly the same as the period of her revolution round the Earth. The consequence is that we always see precisely the same face of the Moon; an entire half of her surface being always concealed from us, excepting such small portions of it near the visible half as come into view from librations (as they are called), mostly produced by the varying velocity of her orbital motion and the inclination of her orbit to that of the Earth.

That portion of the Moon's surface which we are able to see and study presents an appearance very different from that which any portion of the Earth's surface would do if similarly scrutinized. All indications show that there is no atmosphere surrounding the Moon; [6]

[6] The adjoining picture will serve to illustrate one of the critical indications referred to — the planet Jupiter in part hidden behind the limb of the Moon. No distortion of the delicate details of the planet's surface is apparent along the Moon's edge, as would be the case if the Moon had an appreciable atmosphere. Planetary occultations are now excellently photographed. Lunar photography, begun by the elder DRAPER in 1840, and continued by his son HENRY, and by RUTHERFURD, in New York, has culminated in splendid series of

JUPITER REAPPEARING FROM OCCULTATION, 7TH AUGUST 1889 (DENNING)

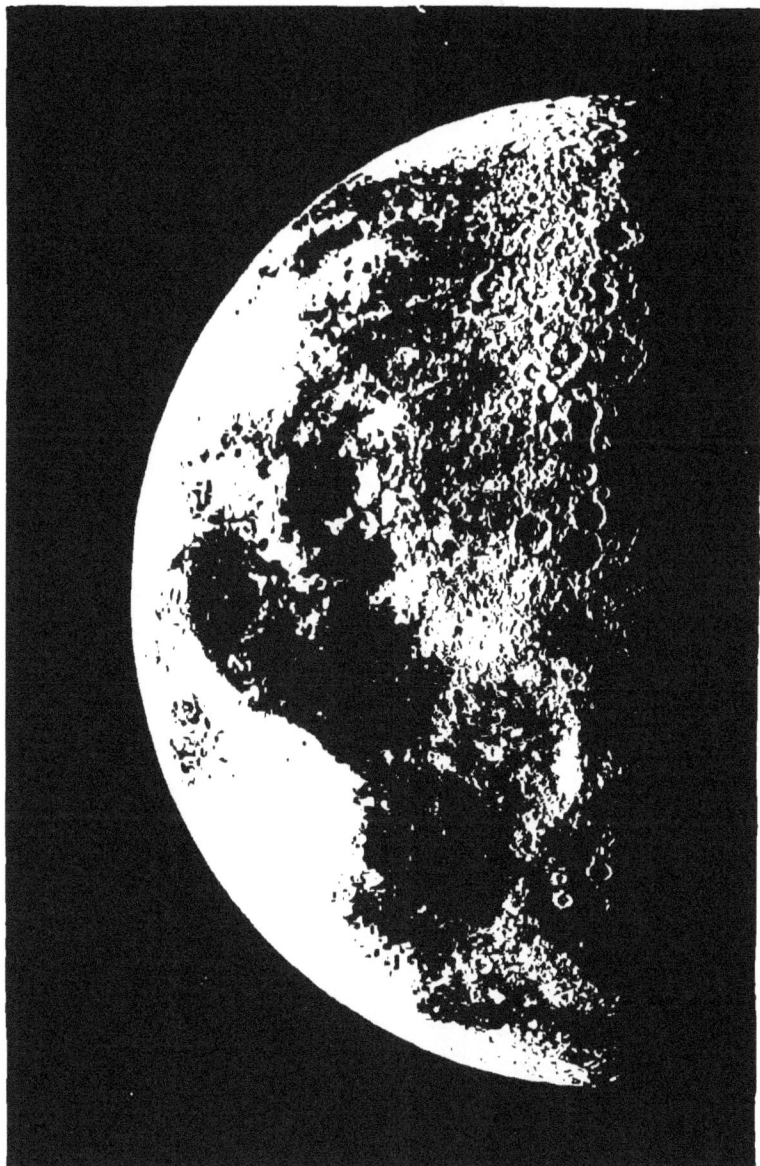

THE MOON NEAR FIRST QUARTER, LOEWY AND PUISEUX

from a photograph in the great photographic atlas of the Paris Observatory

or at least that if there be any, it must be of excessive tenuity, and unable to hold clouds, or any appreciable amount of aqueous vapor, in suspension. Indeed, if there be any water on the surface of our satellite, it can only be very trifling in amount, and confined to the lowest depressions. Yet there is decisive evidence of the action of water on the lunar surface in bygone ages, so that this has probably in the course of time been all absorbed into the interior, or has by degrees become chemically combined with the solid material of the exterior.

The surface is diversified by a large number of crateriform cavities (many of them containing a conical hill at the bottom), and extensive arid plains. The latter retain the name of seas, given them when selenography was in its infancy, and when it was erroneously thought that they were large bodies of water. Markings of a very peculiar character have also been

negatives procured with the 13-inch Harvard telescope on Mount Wilson, California, and with the 36-inch telescope of the Lick Observatory. At the Columbian Exposition, Chicago, 1893, were exhibited many superb enlargements of portions of the moon from both series, some of the Harvard prints being on a scale corresponding to a lunar diameter of 14 feet. Many of the Lick originals (of nearly six inches diameter) are technically perfect enough to bear enlargement to six feet; while M. Prinz of the Royal Belgian Observatory has made photographic amplifications of the lunar crater Copernicus from these negatives to a scale on which the Moon's diameter would exceed 30 feet. The finest photographs rival the classic handiwork of Mädler, Schmidt, and Lohrmann, but are secondary in this, that the very small craters are sometimes indistinctly registered by photography. Also the Lick negatives of the Moon have been industriously studied by Professor Weinek of Prague, who appears to have discovered new rills and craters, and has elaborated a beautiful series of enlarged drawings, which, reproduced by heliogravure, form very perfect detail pictures of lunar scenery. — *D. P. T.*

noticed in comparatively recent times, consisting of long whitish streaks running across other formations. These are called rills, and are usually supposed to be the results of crackings in the surface ; but it has been suggested that they may be dried water-courses.[7]

[7] The ordinary surface structure of the Moon, apparently volcanic in the main, is best seen at or near the time of quadrature, under oblique illumination. But the lunar rills or streaks, difficult to observe when the shadows of the mountains are most conspicuous, and very prominent when these shadows are imperceptible, demand a front illumination for their visibility. Theoretically, the material of the streak-surface must, as pointed out by Dʳ COPELAND, be made up of multitudes of surfaces more or less completely spherical, but either concave or convex; and he therefore regards the streaks as produced by a material pitted with minute cavities of spherical figure, or strewn with minute solid spheres. On critical investigation of the rill systems (with a 13-inch telescope, at an elevation of 8,000 feet, at Arequipa, Peru), Professor W. H. PICKERING finds that the streaks of the system surrounding Tycho and other craters radiate, not from the centre of the ring mountains, but from a multitude of craterlets upon their rims. Also there are not, as heretofore supposed, any streaks hundreds or even thousands of miles long; but their usual length varies from ten to fifty miles, while they seldom exceed one quarter mile in breadth at the crater. In the volcanic region surrounding Arequipa, the roads are, in some places, partially covered with a white pumice-like material, and its behavior under different inclinations of the line of illumination to the line of vision has led Professor PICKERING to a conclusion quite identical with that of Dʳ COPELAND, that the general appearance of the streaks is most readily explained by the hypothesis of a light-colored powder extending away from the craterlets. The only farther step necessary to a satisfactory explanation of these mysterious markings is a reasonable elucidation of the process by which this powder has come to be radially disposed. WÜRDEMANN's theory, cited by Dʳ GILBERT in his address on ' The Moon's Face: a Study of the Origin of its Features ' (*Bulletin Philosophical Society of Washington*, xii. 285), is that meteorites, striking the Moon with great force, have splashed whitish matter in various directions, — which seems plausible. — *D. P. T.*

GUILLEMIN and MITCHELL, *The Moon* (New York 1873).

PROCTOR, *The Moon ; her Motions, Aspects, etc.* (London 1873).

ROSSE, Bakerian Lecture on ' Radiation of Heat from the Moon,' in *Philosophical Transactions* for 1873, p. 587.

NASMYTH and CARPENTER, *The Moon* (London 1874).

NEISON, *The Moon, and . . . its Surface* (London 1876).

LOHRMANN, *Mondcharte* (Leipzig 1878).

SCHMIDT, *Charte der Gebirge des Mondes* (Berlin 1878).

The Selenographical Journal (a London monthly, begun in 1878).

OPELT, *Der Mond* (Leipzig 1879).

HARRISON, *Telescopic Pictures of the Moon* (New York 1882).

LANGLEY, ' Temperature of the Moon,' *Mem. Nat. Acad. Sciences*, iii. (1885) ; iv. (1889) with bibliography.

STRUVE, L., *Total Lunar Eclipses*, 1884 and 1888 (Dorpat 1889 and 1893), giving for the lunar semidiameter 15′ 32″ 65.

ROSSE and BOEDDICKER, ' Lunar Radiant Heat,' *Sci. Trans. Royal Dublin Society*, iv. (1891).

VERY, ' Distribution of Moon's Heat, and its Variation with Phase,' Utrecht Society of Arts and Sciences, 1891.

WEBB and ESPIN, *Celestial Objects* (London 1893).

WEINEK, *Publications of the Lick Observatory*, iii. (1894).

PRINZ, *Agrandissements des Photographies Lunaires* (1894).

PICKERING, W. H., *Ann. Harv. Coll. Obs.*, xxxii. pt. i. (1895).

ELGER, *The Moon :* description and map (London 1895).

LOEWY, PUISEUX, *Atlas Photographique de la Lune* (Paris 1896–98).

HOLDEN, *Lick Observatory Atlas of the Moon* (1897).

WEINEK, *Photographischer Mond-Atlas* (Prague 1897–98).

SCHWEIGER-LERCHENFELD, *Atlas der Himmelskunde* (Vienna 1898).

HERZ, VALENTINER'S *Handwörterbuch der Astronomie* (1898).

KRIEGER, *Mond-Atlas* (Trieste 1898).

VERY, ' Range of temperature,' *Astrophys. Jour.* viii. (1898), 199.

So extensive is the popular literature of the Moon that reference only can be made to the ample lists of POOLE'S *Index to Periodical Literature*, vol. i. (1802–1881), pp. 865–866 ; vol. ii. (1882–1886), p. 296 ; vol. iii. (1887–1891), pp. 286–287 ; vol. iv. (1892–1896), p. 381.

The scientific literature of the Moon occupies sixty-five pages of HOUZEAU and LANCASTER'S *Bibliographie Générale de l'Astronomie*, vol. ii. (Brussels 1882), cols. 1185–1317. Consult, also chap. xiii. of HOUZEAU'S *Vade Mecum de l'Astronome* (Brussels 1882) ; and the lists in *The Astrophysical Journal*, 1895–98. — *D. P. T.*

CHAPTER IV

THE CALENDAR

ALL those astronomical concepts which pertain to the measurement of time having already been given, the relations of this important element to the concerns of ordinary life are now presented, as they obtain in the construction of the calendar.

Originally the words almanac and calendar had the same meaning; but, as often happens in such cases, usage has, in the course of time, given a slightly different meaning to each of them. Almanac, as now understood, means an annual volume containing information of various kinds specially useful in the year to which it applies. Every almanac contains a calendar which gives in tabulated form for the year (divided into its months) the days of the week corresponding to each day of the month, with the dates of the principal anniversary days, and the times of the most important celestial phenomena, such as the rising and setting of the Sun and Moon, the Moon's changes, etc. 'Almanac' comes to us from the Arabic, 'al' being the definite article in that language, and 'manakh' signifying a calendar. The word calendar is from the Latin *calendæ*, by which name the Romans called the first day of each month; and this is probably derived from a word *calare*, to call (cognate with the Greek καλεῖν), because it was customary in very ancient times to summon people together at the beginning of a month to make known the calendar arrangements for that month.

Now it is evident that in distributing time its divisions must be made to correspond to those periodical celestial phenomena which regulate the course and ordinary actions of our lives. Most important of these is the day, or period of time in which the Earth, by revolving on its axis and turning successively toward and from the sun, causes the alternations of light and darkness. There is an unfortunate ambiguity in the word day, as it sometimes means the whole length of this revolution (arbitrarily divided into twenty-four hours, as there are no natural divisions except light and darkness), and sometimes that portion of it between sunrise and sunset. In all parts of the world, except the equatorial regions, the length of this varies considerably throughout the year.

Besides the day there are two longer periods of time marked by celestial motions which are of great importance in the concerns of life. First of these is the year; and the relation of the ordinary or tropical year to the sidereal year has already been given in Chapter II. An astronomical year does not contain any exact number of days; and the year in question, amounting to 365^d 5^h 48^m $45^s.5$, is called the tropical year. As it would be awkward to make a year consist of a number of days and a fraction of a day, the difficulty is got over by adopting in the calendar two years of differing lengths: one called a common year, consisting of 365 days, and an occasional one called a bissextile or leap year, embracing 366 days. This simple expedient keeps the designation of any part of the year in closer correspondence to the same state of the seasons. By the old Julian reckoning, which erroneously supposed the year to contain 365¼ days exactly, each year which was divisible by four without remainder was considered a leap year, the extra day being added to the month of February. But by the Gregorian or present reckoning, adopted in England in 1752, the leap-year day is dropped at the end of each century, unless the centurial year is divisible by 400 without remainder — in which case the bissextile day is retained. For example, the years 1800 and 1900 are not leap years, while the years 1600 and 2000 are.[8] This further adjustment is equivalent to considering the year as consisting of 365.24250 days, thus making it differ by only 0.00031 of a day (or about 27^s) from its true value, 365.24219 days. The accumulation of this small difference will not amount to an entire day for more than 3,000 years.

The other natural division of time (called a month)

[8] Generally speaking, the number of days by which the Julian calendar differs from the Gregorian is as follows : —

In the 15th century (1400-1500),   9 days.
  16th century (1500-1600),   10 days.
  17th century (1600-1700), also 10 days, as 1600
   was a bissextile year.
  18th century (1700-1800),   11 days.
  19th century (1800-1900),   12 days.
  20th century (1900-2000),   13 days.

As the discovery of Columbus occurred in the 15th century, the day of the year on which it took place (12th October, Julian reckoning, or Old Style) falls nine days later in the Gregorian, or present reckoning (that is, 21st October, New Style). Russia still retains the Julian calendar ; and to avoid confusion, Russian dates are customarily written in fractional form, with the Old Style date as numerator: for example, 14th May (of any year in the 19th century) would be written May $\frac{2}{14}$; and 5th September (Gregorian) becomes in Russia $\frac{\text{August 24}}{\text{September 5}}$. — *D. P. T.*

was suggested by the revolution of the Moon in its orbit; and, as already pointed out, it corresponds, not to the actual period of the Moon's revolution round us referred to the stars, but to her revolution round the Earth as seen from the Sun. Evidently then the latter, called the synodic month, is the longer, because the Earth has been moving through a considerable portion of its orbit while the Moon has been going round it. The length of the synodic or lunar month, or a lunation simply, is 29^d 12^h 44^m 2^s.684; so that about 12$\frac{1}{2}$ lunations are completed in a year. But as it is desirable to divide the year into nearly equal portions, an integral number of which are completed at the end of each year, the months adopted in our calendar have no real relation to the period of the Moon's revolution, but are purely artificial divisions, each of which, excepting February, is a little longer than a lunar month. Some nations, the Hebrew peoples, for example, who attribute greater comparative importance to the lunar month than we do, actually employ it in their calendar, so that their year consists sometimes of twelve and sometimes of thirteen months.

The fact is said to have been first noticed by METON, a Greek astronomer, that 235 lunar months are very nearly equal to 19 tropical years. The former, in fact, contain 6,939^d 16^h 31^m; the latter, 6,939^d 14^h 27^m. From this it results that during each 19th year of our reckoning the position of the Moon with reference to the Sun will be very nearly the same on the same day of the year. Owing to the importance of this fact, the years were distributed into cycles of 19, each of which is called a Metonic cycle, and the number of any year in this cycle is its Golden Number.

The year called by us B. C. 1 corresponded to the number one, and A. D. 1 to the number two in this cycle. As a cycle will be completed in 1899, the golden number of 1894, for example, is 14 ; and that of 1900 is 1, in the next period or cycle.

The Sunday Letter, also termed the Dominical Letter, primarily of use in determining the date of Easter, is employed in ascertaining the day of the week on which any day of the year falls. To find the Sunday Letter of any common year, affix the first seven letters of the alphabet in their usual order to the first seven days of January; then the letter adjoining Sunday is the Dominical or Sunday Letter of that year. If, for example, New Year's Day falls on Sunday, the Dominical Letter of the year is A ; if on Saturday, B ; if on Friday, C ; and so on. Manifestly, now, if every year contained only 364 days $(364 = 52 \times 7)$, all years would have the same Sunday Letter perpetually ; that is to say, a given day of any month would always fall on the same day of the week. But as the common year contains 365 days, it must always end on the same day of the week on which it began ; so that any year (whether common or bissextile) immediately following a common year must begin one day later in the week. Therefore its Sunday Letter falls one letter earlier in the alphabetic cycle. But the addition of the intercalary day at the end of February, and not at the end of the year, makes it usual to assign two Dominical Letters to each bissextile year; the reason for which may be illustrated by the following example : The year 1892 beginning on Friday, its Sunday Letter is C. But the intercalary day, 29th February, disturbing the regular periodic succession of the common year, the

THE ROYAL OBSERVATORY, GREENWICH — W. H. M. CHRISTIE, C. B., F. R. S., *Astronomer Royal*

(*Of all the observatories of the world, this one, founded by* CHARLES *the Second in 1675, has, by its centuries of observations of the heavenly bodies, made the greatest practical contributions to that precise knowledge of the fundamental elements of exact astronomy upon which the measurement of time and the accuracy of the Calendar depend*)

1st March falls one day later in the week on that account; so that the Dominical Letter for the remaining ten months of the bissextile year evidently drops back one letter in the alphabetic cycle, — or *leaps over* one letter, whence the name ' leap year.' [9]

[9] Easter day is given in the following table, together with the Golden Number, Dominical Letter, and the corresponding years of the Jewish and Mahometan Eras, with the day of the month in each ordinary year when they begin : —

Year	Easter Day	Golden Number	Dominical Letter	Jewish Era		Mahometan Era	
				Year	Begins	Year	Begins
1890	6th April	10	E	5651	15th September	1308	17th August
1891	29th March	11	D	5652	3d October	1309	7th August
1892 B	17th April	12	CB	5653	22d September	1310	26th July
1893	2d April	13	A	5654	11th September	1311	15th July
1894	25th March	14	G	5655	1st October	1312	5th July
1895	14th April	15	F	5656	19th September	1313	24th June
1896 B	5th April	16	ED	5657	8th September	1314	12th June
1897	18th April	17	C	5658	27th September	1315	2d June
1898	10th April	18	B	5659	17th September	1316	22d May
1899	2d April	19	A	5660	5th September	1317	12th May
1900	15th April	1	G	5661	24th September	1318	1st May

Other chronological data are printed each year in *The American Ephemeris and Nautical Almanac*. For farther information on calendars and chronology, consult the articles on these subjects in *The Encyclopædia Britannica* (9th edition), IDELER'S *Handbuch der Mathematischen und Technischen Chronologie*, CHAMBERS'S *Astronomy* (4th edition),vol. ii. bk.10,and SCHRAM'S *Hilfstafeln für Chronologie*. Also, MAHLER'S *Chronologische Vergleichungs-tabellen* will be found helpful. More accessible doubtless is BOND'S *Handy-Book of Rules and Tables for Verifying Dates with the Christian Era*, etc. (London 1869); also Dʳ WISLICENUS'S *Astronomische Chronologie* (Leipzig 1895); and the article by the same author in VALENTINER'S *Handwörterbuch der Astronomie*, i. (Breslau 1897). LEWIS has a popular paper on *Almanacs* in *The Observatory*, xxi. (1898). — *D. P. T.*

CHAPTER V

I T is essential to make early reference to the means by which the heavenly bodies become known to us through the agency of light, and to the effects of the motion of light upon their apparent places.

The undulations, or light waves, are transmitted through a very rare but material substance, diffused through all space, to which the name ether has been given; and to distinguish it from the chemical substance with the same name, it is called luminiferous (that is, light-bearing) ether. The vibratory motions of the particles of this ether, called waves of light, when once originated by a luminous body acting as a source of light, are transmitted through the atmosphere in its ordinary state, because it is permeated by these particles, and through some other substances called for this reason transparent. In passing from one transparent medium into another of different density, the rays of light, propagated originally in straight lines, are bent or turned in a direction inclined to that previously followed. This bending is called refraction, and is greater, the greater is the difference of density of the different media through which the light is successively transmitted. It also varies according to the angle of inclination to the surface of a medium at which a ray of light enters it;

the law being, that if the media be the same, the sine of the angle of incidence (whatever that angle be) is always in a constant ratio to the sine of the angle of refraction. This law was discovered about 1621 by WILLEBRORD SNELL of Leyden. The ratio in any particular case can only be determined by experiment.

Rays of light coming to us from one of the heavenly bodies suffer a constantly increasing refraction in passing through the successive strata of the Earth's atmosphere, which are more and more dense the lower down they are. The general result can only be ascertained by astronomical observations, the celestial object appearing in the direction in which the ray of light proceeding from it enters the eye. The law just mentioned shows that a body exactly overhead, or in the zenith of any place, is seen in its true position, its rays having undergone no refraction at all; and that the refraction is greater the farther the object is from the zenith, or the nearer it is to the horizon. Half-way between the two, or at an altitude of 45°, the refraction amounts to nearly 1′; at 20° to 2½′; at 10° to 5⅓′; at 5° to 10′; close to the horizon it is as much as 33′ or 34′, varying considerably with the temperature and pressure of the atmosphere. As this arc is a little more than the apparent diameter of the Sun or Moon, it follows that when either of those bodies seems to be just above the horizon, it is really just below it, refraction always making a heavenly body appear higher above the horizon than it really is.

That the propagation of light, whether it be a substance, as was formerly supposed,[10] or an undulatory

[10] 'It is possible,' says CLERK MAXWELL, 'to produce darkness by the addition of two portions of light. If light is a sub-

motion in a substance, as it is now known to be, should occupy time, seems *à priori* what might be expected. But so extremely rapid is this propagation that the ancients, having no means of recognizing it by astronomical observations, appear to have thought that it was instantaneous, which it practically is between places on the Earth's surface. The first person to discover that the velocity of its transmission is measurable was the illustrious RÖMER of Copenhagen in 1675, while discussing the observations of Jupiter's first satellite made by himself, and at Paris by CASSINI. He found that the intervals of time between successive immersions of the satellite into the shadow of the planet and its emersions therefrom were different according to whether the Earth in its orbital motion was approaching Jupiter or receding from it; and he concluded that the cause of this was the Earth's motion combined with that of light, which was thus·recognized to be a sensible and measurable quantity. RÖMER's explanation of the differences observed was contested by CASSINI; but it nevertheless gradually obtained general acceptance among astronomers, and about 50 years after its publication a most remarkable confirmation of its truth was obtained by the discovery of the aberration of light. RÖMER's approximate determination of the time required by light in passing from the Sun to the Earth was rather more than eleven minutes; but it is now known that when the Sun is at its mean distance, this interval is 498^s (8^m 18^s), the best determinations of the velocity of light by the refined

stance, there cannot be another substance which, when added to it, shall produce darkness. We are, therefore, compelled to admit that light is not a substance.'

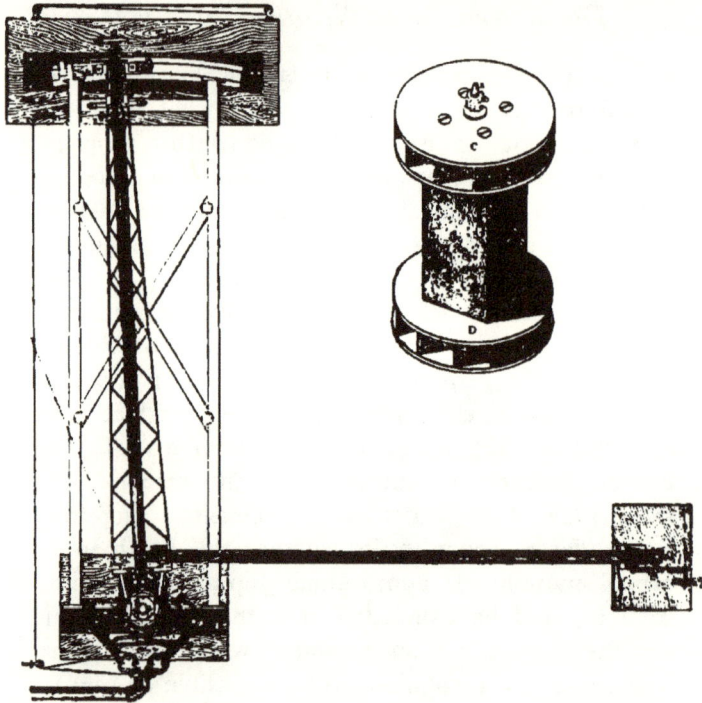

PHOTOTACHOMETER FOR MEASURING THE VELOCITY OF LIGHT

(*View from above*)

(*Only in the latter half of the 19th century was the velocity of light first measured with accuracy. The favorite method is that of the 'Revolving Mirror,'
inaugurated in 1862 by* FOUCAULT, *the eminent French physicist (page 47).
Professor* MICHELSON, *now of the University of Chicago, improved upon his
methods in 1878, and at Washington, in 1880–82, Professor* NEWCOMB *employed
the greatly enlarged apparatus above shown, embodying farther important
modifications. The speed of the revolving mirror,* C D *— a rectangular prism
of polished steel about 1½ inches square — was 250 turns in a second. The three
piers supporting the instrument were about 7 feet apart; and the stationary
mirror reflecting the return ray was nearly 2½ miles distant. For description
of the apparatus and its use, and a full discussion of the problem, see 'Astronomical Papers of the American Ephemeris,' vol. ii. (Washington 1891). Also*
CORNU, *'Annales de l'Observatoire de Paris,' Mémoires, xiii.* (1876); YOUNG
and FORBES, *' Philosophical Transactions,'* 1882. *For bibliography, consult
'Washington Observations' for* 1885, *Appendix iii. p.* 153.)

and accurate methods of modern physicists and astronomers (*vide* opposite page) giving 186,330 miles a second. So that when, for example, we are observing Neptune (a planet 30 times more distant than the Sun), we see it by light which began its journey earthward somewhat more than four hours previously.[11]

A few words in conclusion on the aberration of light. In 1725–27, while making a number of careful observations of γ Draconis (a star which crosses the meridian in the south of England very near the zenith), in the hope of determining its parallax and distance, BRADLEY was surprised to notice a regular change in its apparent position. Its period, like that produced by a measurable parallax, was exactly a year; but the amount was much greater than he had expected, and the direction was the reverse of that which would be caused by parallax. Speculating on the source of

RÖMER (1644–1710)

---

[11] This quantity is technically termed the 'aberration-time,' and is equal (in seconds) to 498 times the planet's distance from the Earth expressed in astronomical units. The mean value of the aberration-time (between Sun and Earth) is the constant term in the 'equation of light.' Also there is an aberration of the planets due to their orbital motions in space. Suppose our globe to be at rest in its orbit ; the aberration of light would then vanish. But as a planet is always seen from the Earth in the direction where it actually was when light left it, obviously its absolute position at the time of observation must differ from the apparent position, because of (1) the aberration-time, and (2) the planet's orbital motion athwart the line of

this, it occurred to him that the Earth's motion round the Sun would lead to an effect of this kind when combined with the gradual propagation of light as discovered by Römer. Calculation confirmed this conjecture ; and the aberration of light (as this phenomenon is called) affords one of the methods by which the velocity of light was subsequently determined.

The nature of the apparent change in a star's place produced by aberration depends upon its position with respect to the plane of the ecliptic. If that position be in the plane of the Earth's orbit, the star will, in consequence, appear to change its place in a straight line only ; if it be nearly in the pole of the ecliptic, the star will appear to describe a circle round its true place ; while, if situate anywhere between the ecliptic and its pole, the change of apparent place will form an ellipse whose eccentricity is greater the nearer the star is to the plane of the ecliptic. The angular

JAMES BRADLEY (1692–1762)

sight. This is called planetary aberration. A similar but very small effect obtains for the Moon, amounting to about $0''.5$, the linear value of which, at the distance of our satellite, is 2,650 feet; that is to say, while light is travelling from the Moon to the Earth (nearly $1^s.3$), the Moon advances about one-half mile eastward in its orbit. — *D. P. T.*

semi-diameter of the circle in the former case, or the semi-major axis of the ellipse in the latter case, is practically a constant quantity, called the constant of aberration ; for the only circumstance that can make it vary is the change of the Earth's orbital velocity at its different distances from the Sun, and the proportion of this variation to the whole velocity is very small. The best modern value of the constant of aberration is 20".492, determined by M. NYRÉN.

The Earth's rotation on its axis, combined with the motion of light, also causes a phenomenon of a similar kind, called the diurnal aberration. Its amount, of course, varies according to the latitude of the place of observation, being greatest at the equator, where the velocity produced by the Earth's rotation is greatest, and smaller the nearer the poles are approached, at which points the diurnal aberration vanishes. At latitude 40°, the average for places in the United States, the constant of diurnal aberration (for an equatorial star at meridian passage) amounts to 0".24 ; the velocity of rotational translation at latitude 40° being about the eighty-fifth part of the Earth's velocity in its orbit.

LÉON FOUCAULT (1819–1868)

CHAPTER VI

THE SUN

O N the Sun's distance depends our knowledge of all absolute magnitudes in the solar system ; for by KEPLER'S third law (page 93), the proportions of the distances of all the bodies moving round the Sun are known exactly from their periodic times of revolution: so that if the Earth's distance from the Sun be known, the distances of all the planets from the Sun and from each other can be deduced by a very simple piece of arithmetic. Also, the Sun's distance is of the first order of significance, because it is the elemental unit in our measures of the distances of the fixed stars.[12] To measure the Sun's parallax with pre-

.

[12] The Sun, cosmically speaking, is simply a star, but the nearest fixed star is 275,000 times more remote ; so that the Sun's vastly greater brightness is, for the most part, due to mere proximity. Still, the distance of the Sun is by no means easy to conceive or illustrate. Recalling that the distance round the Earth's equator is about 24,000 miles, ten times this gives the distance of the Moon, which is practically inconceivable ; but the Sun is 390 times more remote. As the two bodies are about the same in apparent size, it follows that the Sun's actual diameter is about 390 (accurately 400) times greater than the Moon's. The diagram on the opposite page will not only convey a true idea of the relative size of Sun, Earth, and Moon, but by imagining spheres of the given proportions set at the distances indicated, the actual relations of these bodies in space may, in some sense, be comprehended. On the same

RELATIVE SIZE OF SUN, EARTH & MOON

(On scale of 18,000 miles = 1 inch.)

ARC OF A CIRCLE 4 FEET IN DIAMETER TO REPRESENT THE SUN.

THE EARTH

MEAN DISTANCE FROM
THE SUN, 429 FEET.

{ On 1st Jan. 422 ft.
{ On 1st July 436 ft.

THE MOON

MEAN DISTANCE FROM THE EARTH,

13¼ INCHES.

{ At perigee, ⅞ in. nearer.
{ At apogee, ⅞ in. farther.

cision is difficult on account of the great distance of that body compared with the size of the Earth; a small error in the measured parallax produces a large error in the resulting distance.

From the time of FLAMSTEED (Astronomer Royal from 1675 to 1719) it has been known that the Sun's parallax does not exceed 10″. HALLEY pointed out the advantage that might be gained by observing transits of Venus, those rare occasions when she passes at inferior conjunction over the Sun's disk. Accordingly the transits of 1761 and 1769 were observed by a large number of parties sent to different regions of the world. But the solution of the problem in this manner is encompassed by practical difficulties. Many results were obtained by calculation from the observations of the transits in those years; but for a long time it was agreed among astronomers that the value of the Sun's parallax obtained by ENCKE in 1835, which amounted to 8″.57, was the best. More recently many determinations, made by observing Mars at its closer oppositions, and by other methods, have shown

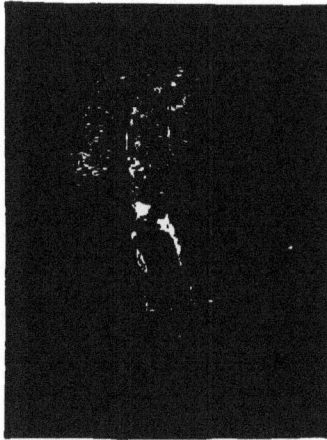

EDMUND HALLEY (1656–1742)

reduced scale, the nearest fixed star would be 16,000 miles distant, equal to a journey from New York westerly to Japan and back. — *D. P. T.*

that the Sun's parallax is greater by at least 0".2
than ENCKE'S value. Moreover, Mr E. J. STONE,
late of Oxford, pointed out that ENCKE'S discussion
was affected by erroneous interpretations of some of
the observations, and that, when these are rightly
explained, they also give a value considerably larger.
The observations of the more recent transits in 1874
and 1882 likewise prove to be most consistent with a
value of about 8".84 ; but the method is now somewhat discredited, as compared with other methods.[13]
From his skilful observations of Mars at Ascension

[13] The available methods of ascertaining the Sun's distance, more than a dozen in number, may be divided into three
classes : (1) by geometry or trigonometry; (2) by gravitational
effects of Sun, Moon, and planets ; (3) by the velocity of
transmission of light. The first includes transits of Venus,
and near approaches of the Earth to Mars, or to small planets
exterior thereto, — at which times the distances of these bodies
from the Earth are not difficult to measure. Adopting, with
Professor YOUNG, the number 100 as indicating a method
which would insure absolute accuracy, this class of determinations will range all the way from 20 to 90. The second class
of methods, too mathematical for explanation here, depends on
the Earth's mass, and their present value may be expressed as
40 to 70; but the peculiar nature of one of them (utilizing the
disturbances which the Earth produces in the motions of Venus
and Mars) offers an accuracy continually increasing, so that
two hundred years hence it alone will have settled the Sun's
distance with a precision entitled to the number 95. But the
best methods now available are embraced in the third class,
which employ the velocity of light (determined by actual physical experiment, as related in the last chapter) ; and their present
worth is about 80 or 90. The problem of the Sun's distance is
one of the noblest ever grappled by the mind of man ; and no
one of the numerous elements with which it is complexly inter-
woven can yet be said to have been determined with the highest
attainable precision. (HARKNESS, *The American Journal of
Science*, cxxii. (1881), 375 ; YOUNG, *General Astronomy* (Boston
1898), chapter xvii.) — *D. P. T.*

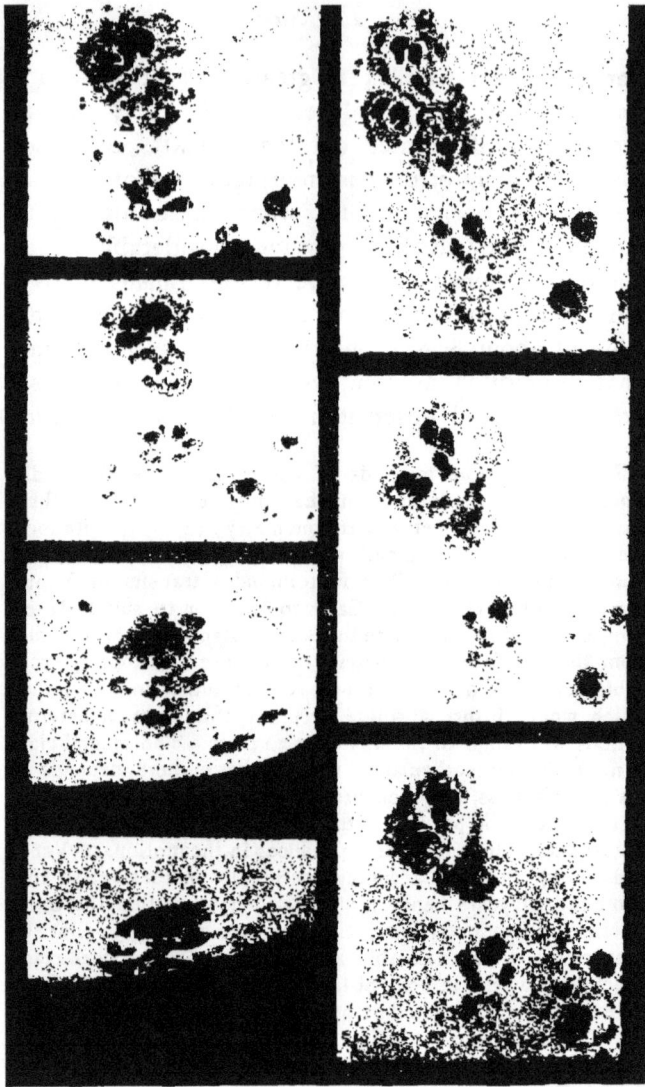

A GROUP OF SUN SPOTS, SHOWING CHANGES IN EIGHT DAYS

(*Photographed by* RUTHERFURD)

Island during that planet's favorable ·opposition in 1877, Dr GILL obtained a solar parallax of 8″.78, which gives for the Sun's mean distance very nearly 93,000,000 miles, — quite certainly within a quarter of a million miles of its true value.[14]

[14] An admirable summary of investigation of the Sun's distance is given by Dr GILL as an introduction to M^n GILL'S *Six Months in Ascension* (London, 1880), — an account of the expedition to that island three years previously. The value of the Sun's parallax, 8″.848 ± 0″.013, determined by Professor NEWCOMB (*Washington Observations*, 1865), and now become classic, is adopted in all the national astronomical ephemerides except the French, which uses LE VERRIER'S slightly larger value. Independent determinations of this constant here given show the measure of modern precision in this important field of research; and the relations of the values to each other will be apparent on recalling that the addition of 0″.01 to the Sun's parallax is equivalent to diminishing his distance about 105,000 miles : —

(1880)	TODD	Velocity of Light	8.808 ± 0.006
(1881)	PUISEUX	Contact and Micrometer Observations, Transit of Venus, 1874	8.8
(1881)	TODD	American Photographs, Transit of Venus, 1874	8.883 ± 0.034
(1885)	NEWCOMB	Velocity of Light	8.794
(1885)	OBRECHT	French Photographs, Transit of Venus, 1874	8.81 ± 0.06
(1887)	CRULS	Brazilian Observations, Transit of Venus, 1882	8.808
(1887)	E. J. STONE	British Contact-Observations, Transit of Venus, 1882	8.832 ± 0.024
(1888)	HARKNESS	American Photographs, Transit of Venus, 1882	8.842 ± 0.012
(1889)	HARKNESS	Planetary Masses	8.795 ± 0.016
(1890)	BATTERMANN	Lunar Occultations	8.794 ± 0.016
(1890)	NEWCOMB	Transits of Venus, 1761 and 1769	8.79 ± 0.034
(1891)	AUWERS	Transits of Venus, 1874 and 1882	8.896 ± 0.022
(1892)	GILL	Opposition of Small Planets { (12) Victoria	8.801 ± 0.006
(1893)	GILL	(80) Sappho	8.798 ± 0.011
(1898)	ELKIN	(7) Iris	8.812 ± 0.009

A nearly complete list of the more important earlier papers is given in NEWCOMB'S *Popular Astronomy* (Appendix). The newly discovered small planet 1898 *DQ*, at its near approach to the earth in 1900, is expected to yield a value of the Sun's distance far exceeding all others in precision (page 408).

It must be noted, that in consequence of the eccentricity of the Earth's orbit (which amounts to 0.01677), the Sun's actual distance varies between a million and a half miles less, and a million and a half miles more than this. It is least about 1st January, when the Earth is at that end of the major axis of its orbit which is nearer the focus occupied by the Sun, and greatest about 1st July, when the Earth is at the opposite end of that axis. These two points are called respectively the perihelion and the aphelion of the Earth's orbit.

Accepting, then, 93,000,000 miles as the Sun's mean distance from us, it is easy to find, by observing his apparent diameter, that his real diameter is 865,350 miles. This is nearly 110 times as great as the Earth's, so that the Sun's volume is able to contain the Earth's 1,300,000 times over. But the comparative effects of their attractive forces show that the Sun's mass is only 331,100 times as great as that of the Earth. Consequently his density must amount to only about one

Professor HARKNESS published in 1891 a laborious paper entitled *The Solar Parallax and its Related Constants* (Washington Observations, 1885), in which this quantity is treated, not as an independent constant, but as 'entangled with the lunar parallax, the constants of precession and nutation, the parallactic inequality of the Moon, the lunar inequality of the Earth, the masses of the Earth and Moon, the ratio of the solar and lunar tides, the constant of aberration, the velocity of light, and the light-equation.' Collating the great mass of astronomical, geodetic, gravitational, and tidal results which have been accumulating for the past two centuries, and applying the mathematical process known as a 'least square adjustment,' he derives the value $8''.809 \pm 0''.006$, giving for the mean distance between the centres of Sun and Earth 92,797,000 miles. A valuable bibliography of the entire subject concludes Professor HARKNESS's paper. — *D. P. T.*

fourth that of the Earth, or rather more than 1⅓ times the density of water.[15]

The shape of the Sun appears to be that of a perfect

[15] Possible changes of the Sun's diameter from time to time have been critically investigated by Dʳ AUWERS of Berlin, and Professor NEWCOMB, with negative results; nor are the observations yet made sufficient to disclose any difference between equatorial and polar diameters. The heliometer (p. 277) affords the best means of measuring the Sun's apparent diameter, or the angle subtended by its disk. The orbit of the Earth being elliptical, this diameter changes in the inverse proportion of the Earth's varying distance from the Sun; at the beginning of the year it is 32′ 32″, and 31′ 28″ early in July, the mean value being 32′ 0″. Supposing the form of the Earth's orbit unknown, daily measures of the Sun's varying diameter would alone, in the course of a year, enable the precise determination of the figure of that orbit, so accurately can these measures now be made. The linear equivalent of one second of arc at the Sun is 450 miles. The present uncertainty in the solar diameter does not much exceed 2″; that is to say, about 900 miles, or approximately $\frac{1}{1000}$ of the entire diameter. Dʳ AUWERS' recent value of the semi-diameter is 15′ 59.″63.

A simple relation between the Sun's mass and its dimensions relatively to the Earth enables us to determine that the force of gravity at the Sun's surface is 27⅗ times greater than it is here; so that while a body on the Earth falls only 16.2 feet in the first second of time, at the Sun its fall in the corresponding interval would be no less than 444 feet. If a hall clock were transported to the Sun, its leisurely pendulum would vibrate more than five times in every second. So great is the Sun's mass that a body falling freely toward it from a distance indefinitely great would, on reaching the Sun, have acquired a velocity of 383 miles per second. The great Krupp gun exhibited at the World's Fair in 1893, if fired from Chamounix in the direction of Mont Blanc, at an elevation of 44°, would propel its projectile of 475 pounds in a curve meeting the earth at Pré-Saint-Didier, 12¼ miles from Chamounix, and whose highest point would be more than a mile above the summit of Mont Blanc. If we could suppose the same gun to be fired similarly on the Sun, so great is the force of gravity there that the projectile would be brought down to rest about half a mile from the muzzle. — *D. P. T.*

sphere. As soon as he was observed through a telescope, it was seen that his surface is usually diversified by a number of black spots of varying dimensions and configurations. In 1611, JOHN, son of DAVID FABRICIUS, of East Friesland, first noticed their apparent motions across the disk, from which it became evident that the Sun is endued with a stately rotation on his axis. These motions are such as would carry an equatorial spot from first appearing on the Sun's disk to appearing there again (if persistent enough to do so) in about 27 days; hence, taking into account the simultaneous motion of the Earth in its orbit round the Sun, it was inferred that the Sun turns on his axis in about 25^d 7^h. It should be stated, however, that recent observations have shown that the spots nearer the Sun's equator move somewhat more rapidly than those farther from it. Spots are very seldom seen at a greater distance from the Sun's equator than about 30° of solar latitude. These regions of the solar surface take as much as $26\frac{1}{2}$ days to make a complete revolution.[16]

[16] From groups of the faculæ (pages 57–58) Dr WILSING has found that the Sun's equator revolves in $25^d.23$; but these observations are exceedingly difficult, and a repetition of the work is desirable. Professor YOUNG and Dr CREW have determined the period of rotation of the Sun's equator by means of the spectroscope, utilizing that technicality called DOPPLER'S principle. This means that the spectra from opposite sides of the Sun (the east side coming toward the Earth, and the west receding from us) are brought into juxtaposition; then, careful measurement of the difference in position of a given line in the two spectra forms the basis for calculating the rapidity of rotation. M. DUNÉR of Lund, Sweden, carrying this research still farther, into high solar latitudes, finds for the equatorial regions a period of sidereal rotation equal to $25^d.46$, in close correspondence with the determinations of CARRINGTON and

The periodicity of the sun spots was discovered by SCHWABE of Dessau in 1838. The subject has since attracted much attention, and the period, so far as it is constant, has been pretty accurately determined by WOLF of Zürich to be a little more than eleven years ; but the physical cause of the periodicity is not yet satisfactorily explained. A recent epoch of maximum abundance and frequency, more than usually protracted, was in 1883–84. The diminution being generally less rapid than the increase, a minimum followed in 1889 ; the next maximum passed during

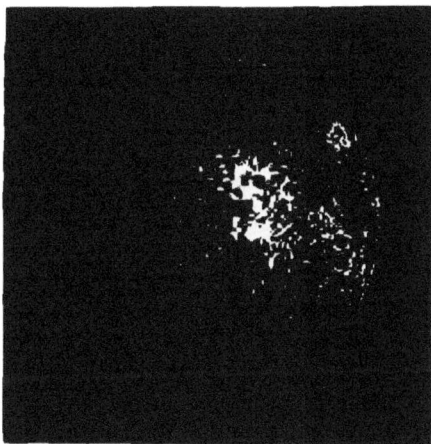

THE SUN (*from a photograph by* RUTHERFURD)

(*Faculae are shown adjacent to the spots near the Sun's limb*)

SPOERER from the spots alone. The slowing down as the poles are approached is remarkably verified, his results giving, for the rotation period at latitude 75°, no less than 38ᵈ.54. M. DUNÉR's observations were made near the time of minimum spots, and it would be interesting to repeat the determination near the epoch of maximum spottedness — *D. P. T.*

the latter half of 1894, and the next minimum may be expected early in 1900.

The solar spots are thought to be produced by dislocations in portions of the photospherical envelopes which surround the Sun, so that we see in them to a depth below that of the ordinary surface.[17] In their immediate neighborhood there are often to be seen patches of more than usual brightness (heapings up, so to speak, of luminous matter), which are called *faculæ*, the Latin word for *torches*. Both phenomena indicate activity and commotion in the outer part (and to some unknown depth below) of the Sun's surface.

[17] The Sun, as seen with telescopes of low magnifying power, is shown on the preceding page; and page 52 illustrates admirably the changes taking place from day to day in an average

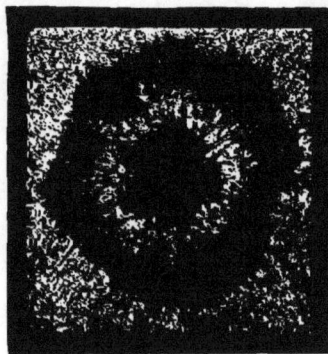

SUN SPOT HIGHLY MAGNIFIED (SECCHI)
(Observe that the penumbra is darker nearer its outer edge)

group of spots as they transit the apparent face of the Sun. Above is a characteristic sun-spot, from a drawing with much detail, by SECCHI. Also, the opposite illustration indicates the regions of greatest spottedness (where the zones are dark-

Other manifestations of this have been recognized in tremendous rushes of gaseous matter, moving with enormous velocities, to great heights above the solar surface. Modern instruments and methods of investigation have enabled astronomers to trace the action and course of these at any time when they are in progress; but the appearances afterward found to be due to them were first perceived on the occurrence of total eclipses of the Sun. These phenomena, now known to be produced by glowing hydrogen, were long called

est), and the apparent positions of these zones at the different seasons. The Sun's axis is inclined 83° to the plane of the Earth's orbit; and if prolonged northward to the celestial sphere, that axis would intersect it near the third-magnitude

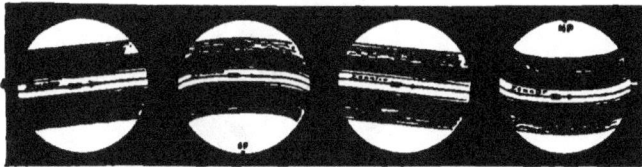

December 6   March 6   June 5   September 5

APPARENT POSITION OF THE SUN SPOT ZONES AT DIFFERENT
TIMES OF THE YEAR (PROCTOR)

(*The darker belts show where the spots are more numerous, and indicate
the apparent direction of spot motions across the disk by solar rotation*)

star δ Draconis: so that in March the Sun's north pole is turned farthest from the Earth; in September it is inclined 7° toward us. Spectroscopic study of the sun-spots shows that their inferior brilliance is due in part to a greater selective absorption than obtains in the photosphere generally. Continuous and systematic records of the solar spots are now kept at Greenwich (in connection with Dehra Dun, India), at Potsdam near Berlin, at Chicago, and elsewhere. Excellent photographs of sun-spots and the solar surface have been obtained at Potsdam (*Himmel und Erde*, ii. (1890) 24; iv, 484). This

red flames, or rose-colored protuberances, attention
having been first attracted to them during the eclipse

famous observatory is illustrated on the opposite page; also
its peculiarly mounted telescope below, with its oblong tube
and duplicate object-glass, which makes it possible to keep the
instrument pointed with the greatest accuracy while exposures
are making with the photographic objective, mounted in a twin
tube. The advantages of the overhanging pier commend themselves at once to the practical observer.

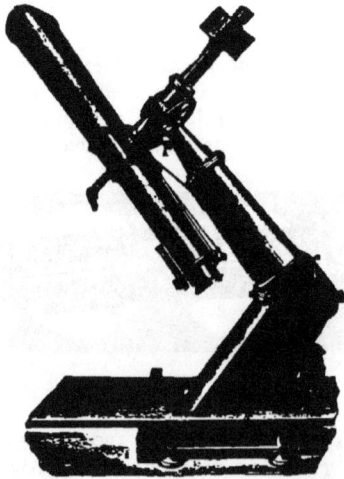

THE 12-INCH EQUATORIAL AT POTSDAM

(*The bent pier affords exceptional convenience of observation*) Scale, 1 inch = 12 feet

Also at Meudon, Paris, M. JANSSEN has had extraordinary
success in photographing the Sun's surface in detail; its granulation, sharply defined in his originals, is somewhat blurred
in the reproduction on page 62. In viewing the Sun with a
telescope this granulation can be satisfactorily seen with a
magnifying power of about 400 or 500, under good atmospheric
conditions.

While the 42 years' faithful work of SCHWABE, as revised by
WOLF and collated with other and scattering results, gives an

THE POTSDAM OBSERVATORY, NEAR BERLIN — Dr H. C. VOGEL, *Director*

which was total in Switzerland in 1706.[18]   It was long
contested whether they belonged to the Sun or to the

average sun-spot period of 11⅑ years, there are great irrregu-
larities: the intervals between maxima have varied from 8 to
15½ years, and between minima from 9 to 14 years.   True interpretation of this indicates with an approach to certainty that

THE SUN'S SURFACE, OR PHOTOSPHERE (JANSSEN)
(Showing its granulation and a large circular spot)

On the same scale the Sun's diameter would be 4 feet, and the Earth like
this: —

the cause of the periodicity does not lie in planetary or any
exterior agency, but that it is seated in the Sun itself. — *D. P. T.*

[18] Professor YOUNG was the first to photograph a solar
prominence, in 1870.   Of the three prominences here shown,

Moon, the dark body of which concealed the Sun during a total eclipse ; and it was then thought that she

GREAT PROTUBERANCE
29th August 1886
(*Height*, 150,000 *miles*)

ERUPTIVE PROTUBERANCE
3d May 1892
(TROUVELOT)

those of 1868 and 1886 were observed during total eclipses; while that of 1892 was drawn in full sun-light, by means of a spectroscope adjusted delicately on the edge of the Sun, this instrument reducing the sky glare, without dispersing very much the light of the prominence itself. This method has now been in common use more than a quarter-century. In October, 1878, Professor YOUNG observed the highest prominence ever recorded, which reached an elevation of nearly 400,000 miles above the Sun's limb. TACCHINI and RICCÒ in Italy, TROUVE-LOT and DESLANDRES of Paris, MAUN-DER and SIDGREAVES in England, and VON KONKOLY and FÉNYI in Hungary, are the most faithful European observers of these wonderful phenomena. By means of the spectro-heliograph devised by Professor HALE of the University of Chicago, the hindering effects of our atmosphere are in considerable part evaded. In April, 1891, he obtained the first photograph of the spectrum of

SOLAR PROMINENCE, —
'GREAT HORN' OF 1868
(*Height*, 100,000 *miles*)

might be surrounded by an atmosphere which occasioned these phenomena. But the careful observations made during later eclipses (particularly in Norway and Sweden in 1851 and in Spain in 1860,

a prominence ever taken without an eclipse; and he is enabled to secure on a single plate (with a single exposure) not only the photosphere and sun-spots, but the chromosphere and protuberances. Also the same instrument (which utilizes monochromatic light, or light of a single color only) has demonstrated

SUN SPOTS AND THE ZONES OF FACULÆ
(*From a single exposure with Professor* HALE'S *spectro-heliograph*)

that the faculæ, which to the eye are ordinarily seen only near the Sun's limb, actually extend all the way across its disk, in approximately the regions of greatest spot-frequency. By the courtesy of Professor HALE, both these results are here illustrated. The bright zones of faculæ are related to the prominences — perhaps identical with them; and they indicate an abundance of glowing calcium vapor. His progressive methods of solar research will soon afford large accumulations of facular observations, from which the laws of their appearance may

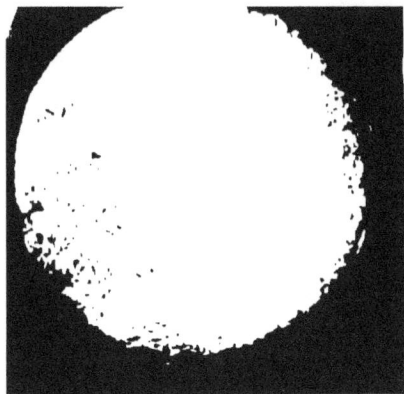
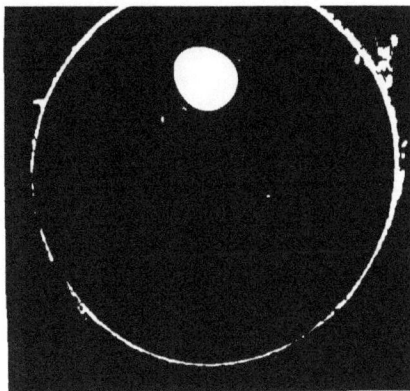

SUN — ACTUAL SURFACE PHENOMENA
(Photograph of Sun's surface showing granulation attributed to sunlight.)

SOLAR ATMOSPHERE — PROTUBERANCES
(The protuberance at right is a height of about 30,000 miles.)

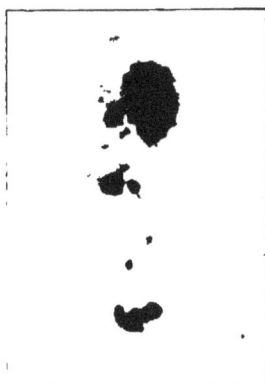

SEPT. 8, 10 A. M.   SEPT. 9, 3 P. M.   SEPT. 10, 11 A. M.

GREAT GROUP OF SPOTS ON THE SUN IN 1898

(From photographs at the Royal Observatory, Greenwich. Sun's diameter on the same scale is about 9 inches. The largest spot is about 35,000 miles long.)

by AIRY, then Astronomer Royal, and by many other
astronomers) showed clearly, from the way in which

be finally determined, and their connection with the formation
of spots and prominences satisfactorily made out. A vast
advantage has been secured through the recent erection of the
Maharajah Takhtasingji Observatory at Poona, India, where a
spectro-heliograph similar to the one at Williams Bay is already
in operation. Professor HALE's spectro-heliograph is illustrated on page 352.

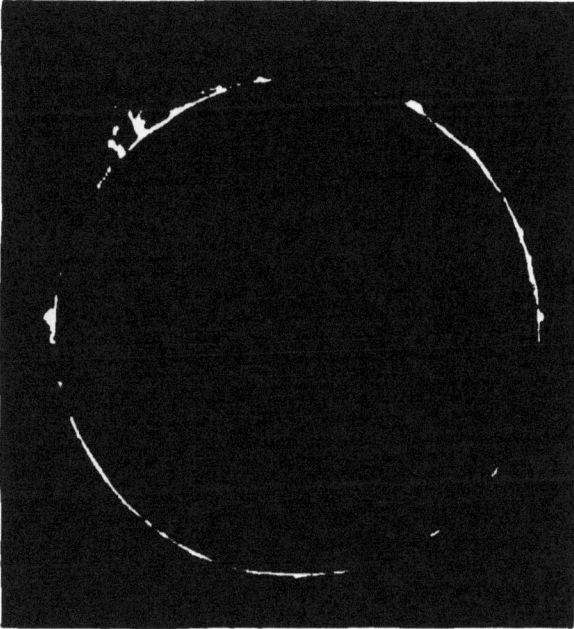

THE SUN'S CHROMOSPHERE (25th July 1892)
(*From a negative obtained with the spectro-heliograph*)

Both spots and prominences have a well-recognized variation in heliographic, or solar latitude ; the former has been investigated by Dr SPOERER of Potsdam, and the latter by
M. RICCÒ of Palermo. Just before the epoch of a minimum

they were first covered and then uncovered by the
Moon, that they were appendages of the Sun.

(1888, for example), the spots are seen nearest the Sun's equator; coincidently with the minimum, these circum-equatorial
spots cease, and a series breaks out afresh in high solar latitudes. Thenceforward to the time of the next minimum, the
mean latitude of the spots tends to decline continuously, as
shown in the adjacent diagram. This fluctuation is called 'the

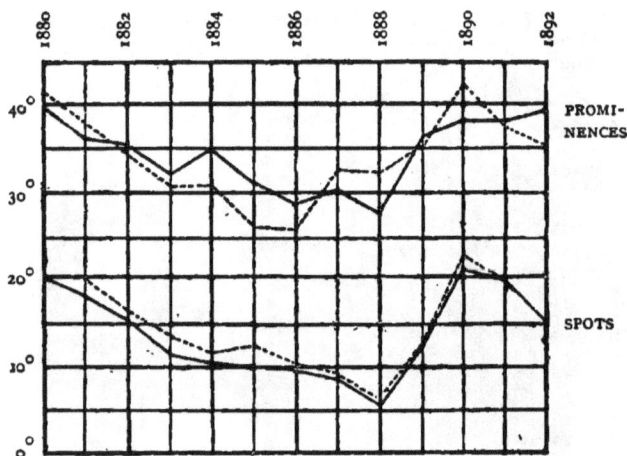

VARIATIONS IN LATITUDE OF PROTUBERANCES AND SPOTS

(*Prominences according to* RICCÒ, *spots according to* SPÖRER)
(*Continuous line, solar latitude north; dotted, south*)

law of zones.' SPÖRER'S careful research farther shows an
occasional predominance of spots in the Sun's southern hemisphere not counterbalanced by a corresponding appearance in
the northern. Also, during the last half of the 17th century
and the early years of the 18th, there seems to have been a remarkable interruption of the ordinary course of the spot cycle,
and the law of zones, too, was apparently in abeyance. M.
GUILLAUME of Lyons recorded in 1896 two instances of small
and short-lived spots in solar latitude 44° and 47°. On approach
of the minimum, there is often a transient outburst of very large
spots, as in September 1898, well exemplified in the photographs.

opposite page 64. An unexplained disturbance of the magnetic
needle, often violent if the spots are large, takes place coincidently with the appearance of spot areas; and there is a
fairly accurate correspondence between the number of spots
and the fluctuations of the needle, as exhibited in Dr WOLFER'S
diagram below. Brilliant auroras, too, are to be expected.

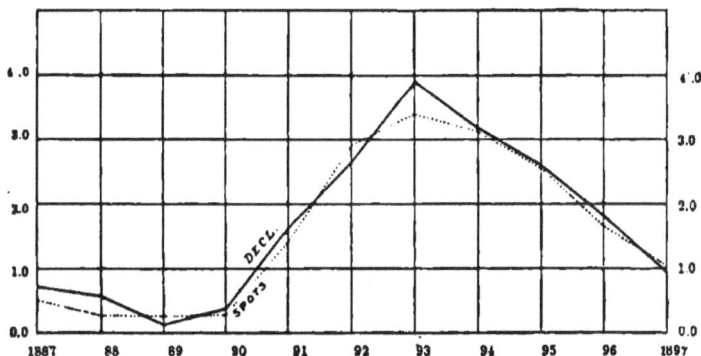

VARIATIONS OF MAGNETIC DECLINATION AND NUMBER OF SPOTS

(*Compared by Dr WOLFER of Zürich*)

As the upper half of the opposite diagram shows, latitude
variations of prominences follow closely fluctuations of spots,
although exhibiting greater divergence between the two hemispheres than the spots do. Professor YOUNG has classified these
phenomena of the solar limb into eruptive and quiescent prominences. While the former are metallic and their distribution
follows nearly the same law of zones as the spots, the latter are
apparently more cloud-like, and are found in all latitudes, even
about the solar poles. — *D. P. T.*

SESTINI, 'Observations of Spots,' *Washington Obs.*, 1847.
CARRINGTON, *Observations of Spots on the Sun* (1853-61) at
  Redhill (London 1863).
TACCHINI and others, *Mem. Soc. Spettroscopisti Italiani* (Palermo
  and Rome, since 1872).
SPÖRER, *Beob. der Sonnenflecken zu Anclam* (Leipzig 1874).
SECCHI, *Le Soleil* (2 volumes and atlas), Paris 1875-77.
PROCTOR, *The Sun: Ruler, Fire, Light* (London 1876).

NEWCOMB, *Popular Astronomy* (New York 1877).
FIÉVEZ, *Annuaire Obs. Bruxelles*, 1879, 255. Bibliography.
ZÖLLNER, *Wissenschaftliche Abhandlungen*, iv. (Leipzig 1881).
DEWAR, McLEOD and others, *Reports Brit. Assoc. Adv. Sci.*
    1881, p. 370 ; 1884, p. 323; 1889, p. 387 ; 1894, p. 201.
PERRY, 'Solar Surface,' *Proc. Roy. Institution*, xii. (1888), 498.
LANGLEY, *The New Astronomy* (Boston 1888).
UNTERWEGER, 'Lesser Spot-periods,' *Denk. kais. Akad. Wissenschaften Wien*, lviii. (1891).
DUNÉR, *Recherches sur la Rotation du Soleil* (Upsala 1891).
FÉNYI, 'Prominences,' *Publ. Haynald Obs.* vi. (1892).
MAUNDER, 'Sunspots and their Influence,' *Knowledge*, xv. (1892), 128 ; xxi. (1898), 228.
BALL, *The Story of the Sun* (New York 1893).
v. OPPOLZER, E., 'Sunspots,' *Astron. and Astro-Phys.* xii. (1893), 419, 736.
SPÖRER, 'Sonnenflecken,' *Publ. Obs. Potsdam*, x. (1895).
SAMPSON, 'Rotation and Mechanical State,' *Mem. Roy. Astr. Soc.* li. (1895), 123.
JANSSEN, *Ann. Obs. Astron. Physique Meudon*, i. (Paris 1896).
YOUNG, *The Sun* — newly revised edition (New York 1896).
MEYERS *Konversations-Lexikon*, xvi. p. 95 (Leipzig 1897).
WOLFER, 'Oberfläche,' *Publ. Stern. Polytech.* (Zürich 1897).
VERY, 'Heliographic Positions,' *Astrophys. Jour.* vi., vii. (1897–98).
WILCZYNSKI, *Hydrodynamische Untersuchungen mit Anwendungen auf die Sonnenrotation* (Berlin 1898).
FONTSERÉ Y RIBA, *Sobre la Rotacion del Sol* (Barcelona 1898).

Dʳ ALFRED TUCKERMAN'S *Index to the Literature of the Spectroscope* ('Smithsonian Miscellaneous Collections,' No. 658, Washington 1888), contains at pp. 88–132 an exceedingly full bibliography, conveniently subdivided under sunspots, rotation, protuberances, eruptions, chromosphere, corona, solar spectrum in general, solar atmosphere, etc.

CHAPTER VII

MORE ABOUT THE SUN—SOLAR PHYSICS

'IT is because the secrets of the Sun,' writes Sir NORMAN LOCKYER, 'include the cipher in which the light messages from external Nature in all its vastness are written, that those interested in the "new learning," as the chemistry of space may certainly be considered, are so anxious to get at and possess them.' But even more significant to us are the heat radiations of the Sun, because they are determinant in all animal and vegetable life, and are the original source of nearly every form of terrestrial energy recognized by mankind. Through the action of the solar heat-rays the forests of palæozoic ages were enabled to wrest carbon from the atmosphere and store it in forms afterward converted by Nature's chemistry into peat and coal ; through processes incompletely understood, the varying forms of vegetable life are empowered to conserve, from air and soil, nitrogen and other substances suitable for and essential to the life-maintenance of animal creatures. Breezes operant in the production of rain and in keeping the air from hurtful contamination; the energy of water, in stream and dam and fall ; trade-winds facilitating commerce between the continents ; oceanic currents modifying coast climates (and no less the tornado, the waterspout, the typhoon, and other manifestations of natural forces, excepting earthquakes, frequently destructive to the works of man), — all are traceable primarily to the heating power of the Sun's rays acting upon those readily movable substances of which the Earth's exterior is in part composed.

As the Sun shines with inconceivably greater power than any terrestrial source, an idea of its total light is difficult to convey intelligibly in terms of the ordinary standards of the physicist. Its intrinsic brightness, or amount of light per square unit of luminous surface, exceeds the glowing carbon of the electric arc light about $3\frac{1}{2}$ times, or the glowing lime of the calcium light about 150 times. 'Even the darkest part of a sun spot outshines the lime light' (YOUNG). Some rude notion of the total quantity of light received from the Sun is perhaps obtainable on comparison with the average full Moon, whose radiance the Sun exceeds 600,000 times. In consequence of absorption of the Sun's light by its own atmosphere, the Earth receives very much less than it otherwise would; while if the absorbing property of that atmosphere were entirely removed, the Sun would (according to Professor LANGLEY) shine with a color decidedly blue, resembling the electric arc. As a farther effect of this absorption, the intrinsic brightness at the edge or limb is $\frac{2}{3}$ that of the centre of the disk (according to Professor PICKERING); and Dr VOGEL makes the actinic or photographic intensity only $\frac{1}{2}$ for the same region. While this shading off toward the edge is at once apparent to the eye, when the entire Sun is projected on a screen, the rapid actinic gradation is more marked in photographs of the Sun, which strongly show the effect of under-exposure near the limb, if the central regions of the disk have been rightly timed.

KIRCHHOFF of Berlin (whose portrait is given on the following page) in 1859 formulated the following principles of spectrum analysis: (1) Solid and liquid bodies (also gases under high pressure) give, when incandescent, a continuous spectrum ; (2) Gases under low pressure give a discontinuous but characteristic bright-line spectrum; (3) When white light passes through a gas, this medium absorbs rays of identical wavelength with those composing its own bright-line spectrum. These principles fully account for the discontinuous spectrum of the Sun, crossed as it is by the multitude of Fraunhofer lines. But it must be observed that the relative position of these lines will vary with the nature of the spectroscope used: with a prism spectroscope the relative dispersion in different parts of the spectrum varies with the material of the prism ; with a grating spectroscope (in which the dispersion is produced by reflection from a gitter, or grating, ruled upon polished speculum metal with many thousand lines to the inch), the dispersion is wholly independent of the material of the gitter, thereby giving the normal solar spectrum. Compared with this a prismatic spectrum has the red end unduly compressed, and the violet end as unduly expanded.

RUTHERFURD, assisted by CHAPMAN, ruled excellent gratings mechanically; but the last degree of success has been attained by Professor ROWLAND of Baltimore, whose ruling engine covers specular surfaces, either plane or concave, six inches in diameter with accurate lines, up to 20,000 to the inch. The

GUSTAVE ROBERT KIRCHHOFF (1824-1887)

concavity of the gratings vastly simplifies the accessories of the spectroscope, for researches in which they are applicable. So great is the dispersion obtainable that the solar spectrum, as photographed by ROWLAND with one of these gratings and enlarged three fold, is about forty feet in length. An illustration of his ruling engine is given on p. 72, and its superiority consists primarily in the accurate construction and perfect mounting of the screw, which has 20 threads to the inch, and is a solid cylinder of steel, about 15 inches long, and 1¼ inches in diameter. (Article ' Screw,' *Encyclopædia Britannica*, 9th edition.) The perfect gratings ruled with this engine are now supplied to physicists all over the world.

By means of a spectroscope properly arranged with suitable accessories, the Sun's spectrum has been both delineated and photographed alongside of the spectra of numerous terrestrial substances. Foremost among recent investigators in this field, and in mapping the solar spectrum, are THOLLON in France, LOCKYER and HIGGS in England, THALÉN in Sweden, SMYTH

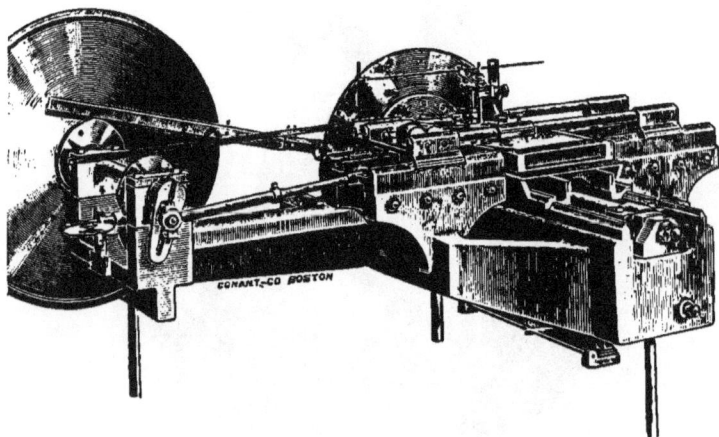

PROFESSOR ROWLAND'S RULING ENGINE

in Scotland, and in America ROWLAND, YOUNG, TROWBRIDGE, and HUTCHINS. Their research, together with that of previous investigators, principally KIRCHHOFF and ANGSTRÖM, VOGEL and FIÉVEZ, has led to the certain detection of at least 35 elemental substances in the Sun, among which are : —

(Al) Aluminium,	(Cr) Chromium,	(Mg) Magnesium,	(Ag) Silver,
(Ba) Barium,	(Co) Cobalt,	(Mn) Manganese,	(Na) Sodium,
(Cd) Cadmium,	(Cu) Copper,	(Ni) Nickel,	(Ti) Titanium,
(Ca) Calcium,	(H) Hydrogen,	(Sc) Scandium,	(V) Vanadium,
(C) Carbon,	(Fe) Iron,	(Si) Silicon,	(Zn) Zinc.

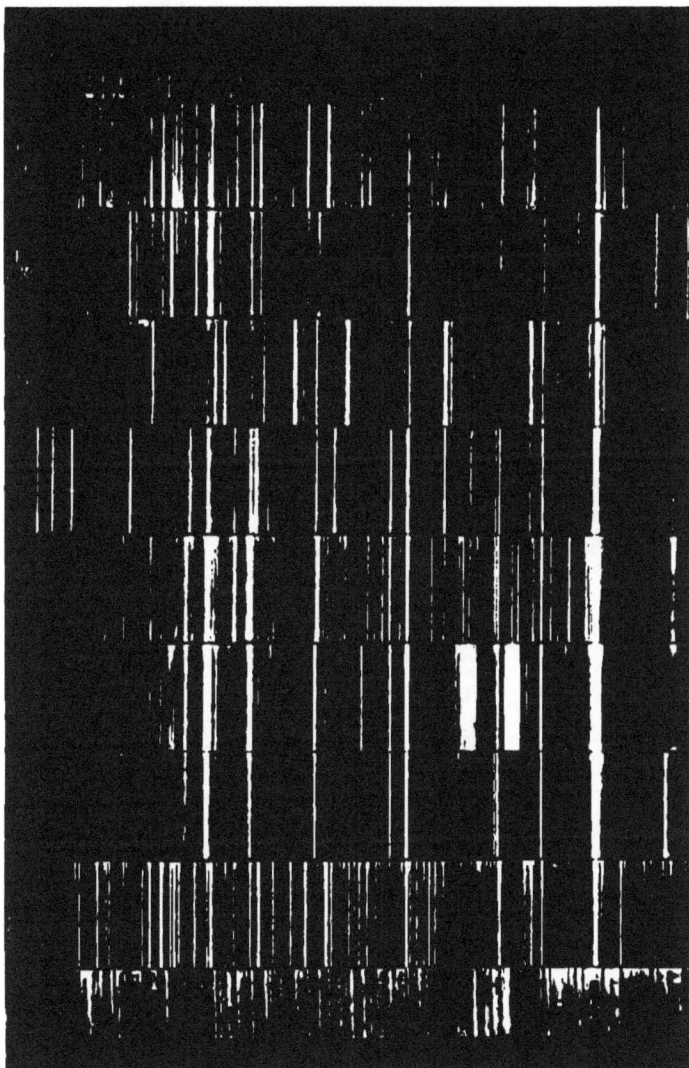

COMPARATIVE PHOTOGRAPHIC SPECTRA OF THE SUN AND THE
IRON-COPPER GROUP OF METALS (McCLEAN)

(Spark spectra of the metals in air — Metals not all pure)

Hydrogen, iron, nickel, titanium, calcium, and manganese are the most strongly marked. RUNGE's researches in 1897 definitively established the existence of oxygen in the Sun. Chlorine and nitrogen, so abundant on the Earth, and gold, mercury, phosphorus, and sulphur, are as yet undiscovered. Also the solar spectrum appears to indicate the existence of many metals in the Sun not now recognized upon the Earth; but it must be remembered that our globe is known only superficially, and there is every reason for believing that the Earth, if heated to incandescence, would afford a spectrum very like that of the Sun itself.

On the opposite page is a reproduction from one of M^r McCLEAN's large charts of the *Comparative Photographic Spectra of the Sun and the Metals* (London 1891). The region is at the violet end of the spectrum, near the Fraunhofer lines H and K, which are due to calcium. The solar spectrum is at the extreme top and bottom, with ANGSTRÖM's scale, and next to it is the spectrum of iron : mark the coincidence of the bright lines of the latter with dark lines of the former. In all, more than 2,000 coincidences with iron lines have been recognized. The symbols of other chemical elements are given at the left side, opposite their characteristic spectra. The chemical spectra of many metallic elements freed from impurities are not yet fully known, but these are in the process of thorough investigation by ROWLAND, and KAYSER and RUNGE of Hanover. On the completion of these researches a farther and more searching comparison will be made with the solar spectrum, hundreds of the dark lines in which are due to absorption by the Earth's atmosphere, and consequently called telluric lines. Especial studies of these have been made by MM. JANSSEN, THOLLON and CORNU, BECKER, and McCLEAN. Mr HIGGS, studying those strikingly marked bands in the solar spectrum due to absorption by the oxygen in our atmosphere, and known as 'great B' and 'great A,' finds that the double lines are in rhythmic groups, in harmonious sequence capable of representation by a simple geometric construction. Whether the solar spectrum is constant in character is not known ; with a view to the determination of this question in the future, Professor PIAZZI SMYTH conducted a series of observations for fixing the absolute spectrum in the year 1884.

Regarding the solar spectrum (prismatic) as a band of color merely, the maximum intensity of heat-rays falls just below the red (at some distance inferior to the dark Fraunhofer line

A); that of light falls in the yellow (between D and E); and that of chemical or photographic activity, in the violet (between G and H); but in the normal spectrum, these three maxima are brought more closely together, approaching the middle of the spectrum which nearly coincides with the yellow D lines of sodium.

Beyond the red in the solar spectrum is a vast region wholly invisible to the human eye; but modern physicists have devised methods for mapping it with certainty. Sir JOHN HERSCHEL, J. W. DRAPER, and BECQUEREL were the pioneers in this research, the last utilizing various phosphorescent substances upon which an intense spectrum had been projected for a long time. Direct photographic maps of the infra-red region are very difficult, because the actinic intensity is exceedingly feeble; and ABNEY, by means of collodion plates specially prepared with bromide of silver, has made an extended catalogue of the invisible dark bands. But Professor LANGLEY has pushed the mapping of the infra-red spectrum to an unexpected limit by means of the bolometer, a marvellously sensitive energy-measure of his own invention. In order to understand in outline the operation of the bolometer, or spectro-bolometer, it is necessary to recall that, as the temperature of a metal rises, it becomes a poorer conductor of electricity; as it falls, its conductivity increases, iron at $300°$ below centigrade zero being, as Professor DEWAR has shown, nearly as perfect an electrical conductor as copper. The characteristic feature of the bolometer is a minute strip of platinum leaf, looking much like an exceedingly fine hair or a coarse spider web. It is about $\frac{1}{4}$ inch long, $\frac{1}{100}$ inch broad, and so thin that a pile of 25,000 strips would be only an inch high. This bolometer film, then, having been connected into a galvanometer circuit, is placed in the solar spectrum formed either by a grating or through the agency of rock salt prisms; and as it is carried along the region of the infra-red, parallel to the Fraunhofer lines, the fluctuations of the needle may be accurately recorded.

In this manner he first represented the Sun's invisible heat spectrum in an energy-curve; but his recent application of an ingenious automatic method, accessory to the bolometer, has enabled him to photograph its indications in a form precisely comparable with the normal spectrum. Bolography is the name given by Professor LANGLEY to these processes which, by the joint use of the bolometer and photography, have automatically produced a complete chart of the invisible heat spec-

trum equal in length to ten times the entire luminous spectrum
of the Sun, though indications of heat extend still farther. Below, on this page, is an illustration of this extraordinary instrument, reproduced from a photograph kindly sent me by Professor LANGLEY. It shows only the bolometer portion of the
entire apparatus, in its case and with the connections as in
actual use. The bolometer was built by GRUNOW of New York,
and forms part of the equipment of the astro-physical observatory of the Smithsonian Institution at Washington. So sensitive
is this delicate instrument that it is competent to detect a tem-

BOLOMETER IN WATER-JACKET, WITH CYLINDRIC LENS

perature fluctuation as minute as the millionth part of a degree
centigrade. It is proper to add that the researches conducted
with such an instrument, often appearing remote and meaning-
less to a layman, are eminently practical in their bearing, as
they pertain directly to the way in which the Sun affects the
Earth, and man in his relations to it, and to the method of distribution of solar heat, forming thus, among other things, a
scientific basis for meteorology.

At the end of the solar spectrum remote from the red is the
ultra-violet region, ordinarily invisible; a portion of which may,
however, be seen by receiving it upon uranium glass or other

fluorescent substances. Glass being nearly opaque to the short wave-lengths of violet and ultra-violet, the optical parts of in-struments for this research are made of quartz or calc-spar, or the necessary dispersion is obtained by using the diffraction grating. The superior intensity of the chemical or actinic rays in this region renders photography of especial service; and sensitive films stained with various dyes have been effectively employed. The painstaking investigations of RUTHERFURD, CORNU, H. DRAPER, ROWLAND, and VOGEL have provided splendid maps of the invisible ultra-violet spectrum, exceeding many times the length of the visible spectrum. The farther region of the ultra-violet is pretty abruptly cut off by the ab-sorptive action of our atmosphere.

The constant of solar heat, first investigated by HERSCHEL and POUILLET in 1837-38, was re-determined by Professor LANGLEY in 1881. He adopts *three calories* (small) as the solar constant, — which signifies that 'at the Earth's mean distance, in the absence of its absorbing atmosphere, the solar rays would raise one gram of water three degrees centigrade per minute for each normally exposed square centimetre of its surface. . . . Expressed in terms of melting ice, it implies a solar radiation capable of melting an ice-shell 54.45 metres deep annually over the whole surface of the Earth.' Professor LANGLEY'S *Researches on Solar Heat and its Absorption by the Earth's Atmosphere: A Report of the Mount Whitney Expedi-tion,* form No. xv of the *Professional Papers of the Signal Service* (Washington 1884). SCHEINER (1898) adopts 3.75 calories.

To express the solar heat in terms of energy: When the Sun is overhead, each square metre of the Earth's surface receives (deducting for atmospheric absorption) an amount of heat equivalent to $1\frac{1}{2}$ horse-power continuously. In solar engines like those of ERICSSON and MOUCHOT, about $\frac{7}{8}$ of this is vir-tually wasted. Of heat-radiation emitted from the Sun and passing along its radius, Professor FROST finds that about $\frac{1}{4}$ part is absorbed in the solar atmosphere, which, were it re-moved, would allow the Earth to receive from the Sun 1.7 times the present amount. Imagine that hemisphere of our globe turned toward the Sun to be covered with horses, ar-ranged as closely together as possible, no horse standing in the shadow of any other; then cover the opposite hemisphere with an equal number of horses : the solar energy intercepted by the Earth is more than equivalent to the power of all these animals exerting themselves to the utmost and continuously.

It is easy to show that 'the amount of heat emitted in a minute by a square metre of the Sun's surface is about 46,000 times as great as that received by a square metre at the Earth, . . . that is, over 100,000 horse-power per square metre acting continuously.' . . . (YOUNG). If the Sun were solid coal, this rate of expenditure would imply its entire combustion in about 6,000 years. The effective temperature of the Sun's surface is difficult to determine, and has been variously evaluated, from the enormously high estimates of SECCHI, ERICSSON, and ZÖLLNER, to the more moderate figures of SPOERER and LANE, who deduced temperatures of 80,000° to 50,000° Fahrenheit. According to ROSSETTI, it is not less than 18,000° Fahrenheit, an estimate probably not far wrong. M. LE CHATELIER, however, in 1892, found the temperature at little short of 14,000°, and WILSON and GRAY, about 12,000°. Dr SCHEINER's recent observations upon the peculiar behavior of two lines in the spectrum of magnesium confirm these lower values in a remarkable way, apparently showing that the Sun's temperature lies between that of the electric arc (about 6000°), and that of the electric spark (probably as high as 20,000°).

VON HELMHOLTZ (1821–1894)

The maintenance of this stupendous outlay of solar energy is explainable on the theory advanced by VON HELMHOLTZ in 1856, who calculated that an annual contraction of 250 feet in the Sun's diameter will account for its entire radiation in a year, — a rate of shrinkage so slow that many centuries must elapse before it will become detectable with our best instruments. Accepting this theory, Lord KELVIN estimates that the Earth cannot have been receiving the Sun's light and heat longer than 20,000.000 years in the past; and Professor NEWCOMB calculates that in 5,000,000 years the Sun will have contracted to one half its present diameter, and it is unlikely that it can continue to radiate sufficient heat to maintain life of types now present on the Earth longer than 10,000,000 years in the future. Assuming that solar heat is radiated uniformly in all directions, a simple computation shows that all the known planets receive almost a two-hundred-millionth part of the entire heat given out by the Sun, the Earth's share being about $\frac{1}{15}$ of this. The vast remainder seems to us essentially wasted, and its ultimate destination is unknown.

To epitomize Professor YOUNG's statement of the theory of the Sun's constitution, generally accepted : —

(*a*) The Sun is made up of concentric layers or shells, its main body or nucleus being probably composed of gases, but under conditions very unlike any laboratory state with which we are acquainted, on account of the intense heat and the extreme compression by the enormous force of solar gravity. These gases would be denser than water, and viscous, in consistency possibly resembling tar or pitch.

(*b*) Surrounding the main body of the Sun is a shell of incandescent clouds, formed by condensation of the vapors which are exposed to the cold of space, and called the photosphere. Telescopic scrutiny shows that the photosphere is composed of myriad 'granules,' about 500 miles in diameter, excessively brilliant, and apparently floating in a darker medium.

(*c*) The shallow, vapor-laden atmosphere in which the photospheric clouds appear to float is called the 'reversing layer,' because its selective absorption produces the Fraunhofer lines in the solar spectrum. During the India eclipse of 1898 it was shown to be about 700 miles in thickness (pp. 364–65). The reversing layer contains a considerable quantity of those vapors which have given rise to the brilliant clouds of the photosphere, just as air adjacent to clouds is itself saturated with the vapor of water.

(*d*) The chromosphere and prominences are permanent gases, mainly hydrogen and helium, mingled with the vapors of the reversing layer, but rising to far greater elevations. Jets of incandescent hydrogen appear to ascend between the photospheric clouds, much like flames playing over a coal fire. Calcium vapor is the most intensely marked, even more so than that of iron, which has over 2,000 line-coincidences, while calcium has only about 80. In 1895 Professor RAMSAY first identified helium as an earthy element.

(*e*) Still above photosphere and prominences is the corona, hitherto observable only during total eclipses, and extending to elevations far greater than any truly solar atmosphere possibly could. The characteristic green line of its spectrum, due to a substance called ' coronium,' is brightest close to the Sun's limb, and during the eclipse of 1st January 1889 it was traced outward by Professor KEELER to a distance of 325,000 miles. Professor NASINI of Padua in 1898 announced that, in the spectrum of volcanic gases of the Solfatara di Pozzuoli, he finds a line corresponding in position with this coronium line, 1474 K̄ as it is often called. Coronium may therefore be a terrestrial substance, as well as solar. But much of the coronal light originates in something other than coronium, because of the dark lines in its spectrum. These indicate solar light, reflected probably from small meteoric particles, possibly the débris of comets circulating about the Sun in orbits of their own. Calcium, hydrogen and helium do not appear in the corona.

Sir WILLIAM HUGGINS and Dr SCHUSTER maintain the view that the coronal streamers are in part due to electric discharges. The corona is a very complex phenomenon, by no means fully understood; and no theory has yet been shown competent to undergo the ultimate test, — that of predicting the general configuration of coronal streamers at future eclipses.

Among modern solar theories may be mentioned that of SCHMIDT, an optical theory of the solar disk, making the Sun wholly gaseous, in fact a planetary nebula; and that of Dr BRESTER of Delft, published in 1892, a theory of the Sun characterized by much novelty. Rejecting the hypothesis of eruptional translation of solar matter, he conceives the Sun to be a relatively tranquil gaseous body, of essentially the same elementary composition as our Earth; and he attempts to show, in accordance with well-known properties of matter, that the same cause which would keep the mass in repose must produce also ' chemical luminescence,' as he terms it. Great material eruptions, then, are merely deceptive appearances, being simply moving flashes in tranquil, incandescent gases.

The surface of the Sun (photosphere, spots, faculæ, and prominences) is now a subject of daily study at many observatories, particularly at Potsdam, Meudon, Rome, and the Yerkes Observatory of the University of Chicago; and observations are rapidly accumulating, the complete discussion of which ought soon to settle many points in the solar theory, now disputed. But as the Sun's corona is visible only a few hours in a century, our knowledge of that object makes haste very slowly, and must continue to do so, unless the photographic method of Sir WILLIAM HUGGINS (apparently successful in 1883, though later not), shall make it possible to study the brighter streamers of the corona without an eclipse. Recent results, however, are not encouraging. Failing to detect even a trace of the corona by its light, Professors HALE and WADSWORTH have attempted to chart its principal streamers by their heat radiations, by means of the bolometer, though as yet without success.

For sources of the early and historic papers on the solar spectrum by ANGSTRÖM, BREWSTER, the HERSCHELS, father and son, KIRCHHOFF, MASCART, VAN DER WILLIGEN and many others, consult HOUZEAU and LANCASTER, *Bibliographie Générale*, ii. (1882), 800.

CORNU, 'Sur le spectre normal du Soleil,' *Ann. de l'École Normale*, iii. and ix. (Paris 1874 and 1882).

TAIT, *Recent Advances in Physical Science* (London 1876).

DRAPER, J. W., 'Distribution of Heat in the Spectrum,' in his *Scientific Memoirs* (New York 1878), p. 383.

LANGLEY, *Proc. Am. Assoc. Adv. Science*, xxviii. (1879), 51.

VOGEL, H. C., 'Sonnenspectrum,' *Publ. Obs. Potsdam*, i. (1879).

PEIRCE, 'The Cooling of the Earth and the Sun,' in his *Ideality in the Physical Sciences* (Boston 1881).

HAUGHTON, 'Sun-heat and Terrestrial Radiation,' Royal Irish Academy (Dublin 1881 and 1886).

SMYTH, *Madeira Spectroscopic* (Edinburgh 1882).

KIRCHHOFF, *Gesammelte Abhandlungen* (Leipzig 1882).

FIÉVEZ, 'Étude du spectre solaire,' *Annales de l'Observatoire Royal*, iv. and v. (Brussels 1882 and 1883).

SIEMENS, *On the Conservation of Solar Energy* (London 1883).

SIEMENS, 'Solar Temperature,' *Proc. Roy. Institution*, x. (1883), 315.

LANGLEY, 'Invisible Prismatic Spectrum,' *Memoirs National Acad. Sciences*, ii. (1883), 147.

SPÖRER, 'Physikalische Beschaffenheit,' *Viertel. Astron. Gesell.* xx. (1885), 243.

LOCKYER, *The Chemistry of the Sun* (London 1887).

KELVIN, 'Sun's Heat,' *Proc. Roy. Institution*, xii. (1887), 1.

TROWBRIDGE, 'Oxygen in Sun,' *Proc. Am. Acad.* xxiii. (1888), 1.

v. FRAUNHOFER, *Gesammelte Schriften* (Munich 1888).

WATTS, *Index of Spectra* (Manchester 1889).

BECKER, 'The Solar Spectrum at Medium and Low Altitudes,' *Trans. Royal Society Edinburgh*, xxxvi. (1890), 99.

MAUNDER, *Spectroscopic Astronomy*, in CHAMBERS'S *Astronomy*, vol. ii. (Oxford 1890), bk. viii.

McCLEAN, *Comparative Photographic Spectra of the High Sun and the Low Sun* (London 1890).

THOLLON, 'Nouveau dessin du spectre solaire,' *Annales de l'Observatoire de Nice*, iii. (Paris 1890), with Atlas.

SCHMIDT, *Die Strahlenbrechung auf der Sonne* (Stuttgart 1891).

YOUNG, 'Solar Physics,' *Sidereal Messenger*, x. (1891), 312.

ROWLAND, *Johns Hopkins University Circular*, February 1891.

KELVIN, 'On the Age of the Sun's Heat,' and 'On the Sun's Heat,' *Popular Lectures and Addresses*, i. (London 1891).

BRESTER, *Théorie du Soleil* (Amsterdam 1892).

LE CHATELIER, 'Temperature of the Sun,' *Astronomy and Astrophysics*, xi. (1892), 517.

KNOPF, *Die Schmidt'sche Sonnentheorie* (Jena 1893).

GORE, 'Fuel of the Sun,' *The Visible Universe* (New York 1893).

HIGGS, *Photographic Atlas of the Normal Solar Spectrum*, in 3 series (Liverpool 1893).

HALE, 'Corona without Eclipse,' *Astron. and Astrophys.* xiii. (1894), 662.

WILSON and GRAY, 'Temperature of Sun,' *Astron. and Astrophys.*, xiii. (1894), 382 ; *Phil. Trans.* clxxxv. (A 1894), 361.

SCHEINER and FROST, *Astronomical Spectroscopy* (Boston 1894), with bibliographies of numerous papers.

LANGLEY, 'Infra-red Spectrum,' *Nature*, li. (1894), 12 ; *Report Brit. Assoc. Adv. Sci.* 1894, p. 465.

RAMSAY, 'Helium,' *Proc. Roy. Soc.* lviii. (1895), 65, 81.

YOUNG, 'Helium,' *Pop. Sci. Mo.* xlviii. (1896), 339.

LOCKYER, 'Helium,' *Science Progress*, v. (1896). 249.

MAUNDER, 'Helium,' *Knowledge*, xix. (1896), 86, 284.

ROWLAND, *Solar Spectrum Wave-lengths* (Chicago 1897).

KEELER, 'Astrophysics,' *Astrophys. Jour.* vi. (1897), 271.

CLERKE, 'Coronium,' *The Observatory*, xxi. (1898), 325.

VOGEL, 'Kirchhoff'schen Spectralapparat,' *Sitz. Kön. Preuss. Akad. Wiss. Berlin* 1898, 141.

SCHEINER, 'Temperature of Sun,' *Himmel und Erde*, x. (1898), 433 ; *Publ. Astron. Soc. Pacific*, x. (1898), 167.

For a nearly complete bibliography of recent literature, consult POOLE's Indexes under *Sun*, and *Spectroscope* (page 395 of this book) ; *Astronomy and Astro-Physics*, xi.-xiii. (1892-94) ; also *The Astrophysical Journal* (i.-viii.) 1895-1898.

CHAPTER VIII

TOTAL SOLAR ECLIPSES

TOTAL eclipses of the Sun, the most impressive of natural phenomena, and formerly serviceable only to the mathematical astronomer in correcting the tables of solar and lunar motions, and in the determination of longitudes, are now observed by the astrophysicist chiefly, for the knowledge afforded as to the Sun's constitution and radiations.

Nearly·70 of these phenomena happen every century. In order that an eclipse may be total, the apex of the conical lunar shadow must at least reach the Earth. But when the Moon is near apogee, its shadow does not extend to the Earth, and an eclipse happening at that time is of the more frequent type known as annular, seven of which take place on the average every eight years; in some regions of our globe the Sun may then be seen as a ring of light surrounding the Moon's disk. Sir ROBERT BALL's diagram on the opposite page makes clear the relations of Sun, Earth, and Moon in eclipses of the various types.

On the occasion of a total eclipse, the Moon's orbital advance, 2,100 miles hourly, causes her narrow shadow to trail easterly over the surface of the Earth. Total obscuration is visible only within this trail, a region which may exceed 8,000 miles in length, but whose average breadth near the equator is less than 100 miles. Obviously, then, a total eclipse at a given place must be an exceedingly infrequent occurrence ; indeed, every spot on the globe is likely to come within the range of the Moon's shadow but once in about three and one half centuries.

While the lunar shadow is sweeping over land and sea from west to east, it is to be noted that the axial

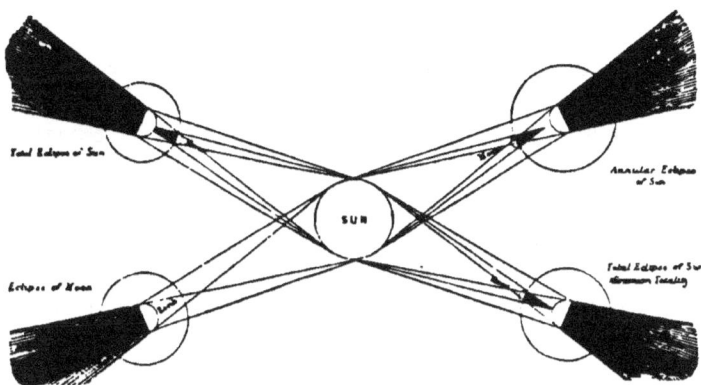

(1) MOON'S SHADOW CUT OFF BY EARTH (TOTAL SOLAR ECLIPSE)
(2) MOON'S SHADOW DOES NOT REACH EARTH (ANNULAR ECLIPSE)
(3) MOON IN EARTH'S SHADOW (TOTAL LUNAR ECLIPSE)
(4) MOON'S SHADOW JUST REACHES EARTH (*total solar eclipse in middle of path, but annular at both ends*)

rotation of our globe is carrying the observer eastward also, 1,040 miles hourly at the equator, so that the Moon's shadow as it sweeps past him has a minimum speed of 1,060 miles per hour (the difference of the two velocities). Swift as this motion is, the lunar shadow has repeatedly been seen advancing and receding with appalling speed. Total eclipse lasts only while the observer remains within the Moon's shadow. It is apparent, then, that the longest total eclipses must take place near the equator, because the observer located there is transported most rapidly in the same general direction as the moving lunar shadow. Also, the longest eclipses happen in summer, for the Earth is then farther from the Sun, thus making its diameter appear smaller, so that our satellite can cover it longer. With the Moon near perigee, and other necessary and favorable conditions, the calculation has been made that it is possible for the Sun to be totally obscured nearly eight minutes. Total eclipses have every range of duration, from this limit down to a single second; but no totality is known to have been observed longer than 5^m 36^s (1868, in India), while the average duration is about three minutes.

Eclipses may be approximately predicted by means of the Saros, a period of 18 years, 11⅓ days (or 18 years, 10⅓ days, if five leap years have intervened). At the end of this cycle another and similar eclipse will occur, only 120° of longitude farther west. The eclipse of 1896, for example, is a return of the famous ones of 1878, 1860, and 1842; and there will be other repetitions in 1914 and 1932. When, however, three periods have passed, an eclipse may return to the same general region on the Earth, but the displacement in latitude will ordinarily amount to several hundred miles; for example, the eclipse of 1842 was total in Austria, but its third return in 1896 fell visible in Norway. By utilizing the mathematical data of the *Astronomical Ephemeris*, published by the government three years in advance, a brief computation will give the time of every phase of an eclipse, exact to a small fraction of a second, for any place on the

Earth where it may be visible. The subject of the remarkable recurrence of eclipses, and the great precision with which they may be predicted, is treated more fully in Number 1 of the 'Columbian Knowledge Series.'

To select a single salient result from each recent eclipse whose observation has important bearing on our knowledge of solar physics: 18th August 1868 (India), when M. JANSSEN, using a high-dispersion spectroscope, succeeded for the first time in observing the solar prominences after (as well as during) the eclipse; 7th August 1869 (Iowa, U. S.), when Professor YOUNG at Burlington observed and accurately identified the green line (1474 on KIRCHHOFF'S scale) in the spectrum of the Sun's corona, which is regarded as due to the vapor of a solar element not yet found on the Earth, and hence called 'coronium'; 22nd December 1870 (Spain), when, just as totality was coming on, Professor YOUNG observed a multitude of the Fraunhofer lines in the Sun's spectrum instantly reversed into bright lines, — a phenomenon repeatedly confirmed, and which indicates the existence of a stratum in the Sun's atmosphere known as the 'reversing layer'; 12th December 1871 (India), when M. JANSSEN saw Fraunhofer lines superposed upon the faint continuous spectrum of the corona; 29th July 1878 (Western United States), when Professor NEWCOMB in Wyoming and Professor LANGLEY on Pike's Peak, Colorado, observed a vast ecliptic extension of the coronal streamers to a distance of eleven millions of miles, east and west of the Sun; 17th May 1882 (Egypt), when Dʳ SCHUSTER photographed the spectrum of the corona for the first time, getting no less than 30 measurable lines; 6th

May 1883 (Caroline Island), when, with a large increase of knowledge relating to the prominences, the
coronal structure and its spectrum, the non-existence
of intramercurian planets was settled to the satisfac-
tion of most astronomers by MM. TROUVELOT and
PALISA; 29th August 1886 (Grenada), when Dr
SCHUSTER found the maximum actinic intensity in the
continuous spectrum of the corona displaced consid-

THE CORONA OF 17TH MAY 1882
(WESLEY, *from* SCHUSTER'S *photographs*)

erably toward the red (in comparison with the spectrum of sunlight), proving that the scattering light
from small particles is feeble; 1st January 1889
(California), when Mr BARNARD at Bartlett Springs
and Professor W. H. PICKERING at Willows obtained
exceedingly fine detail photographs of the Sun's
corona; 16th April 1893 (Senegal), when M. DESLAN-
DRES, by photographs of the coronal spectrum on

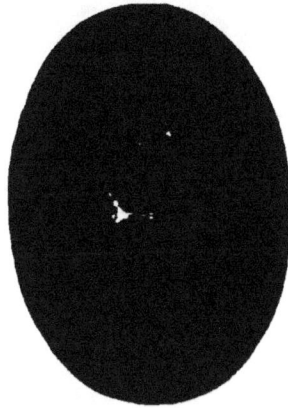

VON OPPOLZER (1841–1886)

(THEODOR VON OPPOLZER, *at first a student of medicine, was for the last thirteen years of his life connected with the European geodetic survey. His chief astronomical works are : a comprehensive treatise in two volumes on the determination of planetary and cometary orbits ; and his ' Canon der Finsternisse,' giving the approximate elements of 5,200 lunar and 8,000 solar eclipses, from the earliest times down to A.D. 2163)*

PERRY (1833–1889)

(STEPHEN JOSEPH PERRY, *director of the Stonyhurst Observatory, conducted two government expeditions to observe the transits of Venus, in 1874 at Kerguelen Island, and 1882 in Madagascar ; also four eclipse expeditions, in 1870, 1886, 1887, and the second total obscuration of 1889. The last cost him his life. He was a Fellow of the Royal Society, and President of the Liverpool Astronomical Society)*

both sides of the Sun, found evidence, on comparing them, that the corona rotates on its axis bodily with the Sun; 9th August 1896 (Nova Zembla), when M' SHACKLETON verified the existence of the reversing layer by photographing its spectrum; 22d January 1898 (India), when Mⁿ MAUNDER secured the first fine photographs of the outermost streamers of the corona.

A fact of much significance in solar research, and now generally regarded as established, is the periodical fluctuation of the streamers of the corona coincidently, or nearly so, with the eleven year cycle of the spots on the Sun. The total eclipse of 1882, on the opposite page, shows the type of corona near the times of maximum spots, — a corona rather fully developed all around the Sun, with an abundance of relatively short, bright streamers, often complexly interlaced; while the eclipse of 1878, in the adjacent figure, exhibits the type occurring at the sunspot minimum, — a corona with uneven radiance, but with a vast extension of its streamers east and west, and beautifully developed in delicately curving filaments around the solar poles.

THE CORONA OF 29TH JULY 1878
(HARKNESS, *from photographs*)

The physical cause underlying this cycle of the corona is not yet made out; and the investigation of this problem, with a host of others arising in connection with total eclipses, leads astronomers to anticipate their occurrence with absorbing interest.

The total eclipses of the next quarter-century most favorable for observation are (map opposite) : —

DATE	REGIONS OF GENERAL VISIBILITY	DURATION OF TOTALITY
1900, May 28	Mexico, United States (from New Orleans to Norfolk), Spain, and Algeria.	2 m.
1901, May 18	Sumatra, Borneo, and Celebes.	6 m.
1905, August 30	Labrador, Spain, and Egypt.	4 m.
1907, January 14	Russia, Turkestan, and China.	2 m.
1912, October 10	Colombia and Brazil.	1 m.
1914, August 21	Norway, Sweden, and Russia	2 m.
1916, February 3	N. extremity of S. America.	2 m.
1918, June 8	U. S., from Oregon to Florida.	2 m.
1919, May 29	Brazil, Liberia, and Congo.	6 m.
1922, September 21	Northern Australia.	6 m.
1923, September 10	California to Texas.	4 m.
1925, January 24	Canada and Maine.	2 m.
1926, January 14	Africa, Sumatra, and Borneo.	4 m.

The totality-path of the great eclipse of 9th September 1904, and those of 3rd January 1908 and 28th April 1911, unfortunately lie for the most part over the unavailable wastes of the Pacific Ocean. No total eclipse will be visible in New England or the Middle States till 24th January 1925. On 20th June 1955 occurs the longest eclipse for many centuries, totality lasting more than seven minutes in the island of Luzon, at or very near Manila.

A full list of popular articles on eclipses, and the researches undertaken by eclipse expeditions to all parts of the world, will be found in POOLE'S *Index to Periodical Literature*, vol. i. (1802–81), pp. 381–2; vol. ii. (1882–86), p. 129; vol. iii. (1887–91), p. 127; vol. iv. (1892–96), p. 167. Also, FLETCHER'S *Index to General Literature* (Boston 1893), p. 90.

JOHNSON, *Eclipses, Past and Future* (London 1874).

BESSEL, *Analyse der Finsternisse* (Leipzig 1876).

RANYARD, *Memoirs Royal Astronomical Society*, xli. (1879).

NEWCOMB, *On the Recurrence of Solar Eclipses, with Tables of Eclipses from* B. C. 700 *to* A. D. 2300 (Washington 1879).

HUGGINS, 'Corona,' *Proc. Royal Society*, xxxix. (1884), 108.

HASTINGS, *Memoirs National Academy Sciences*, ii. (1884), 107.

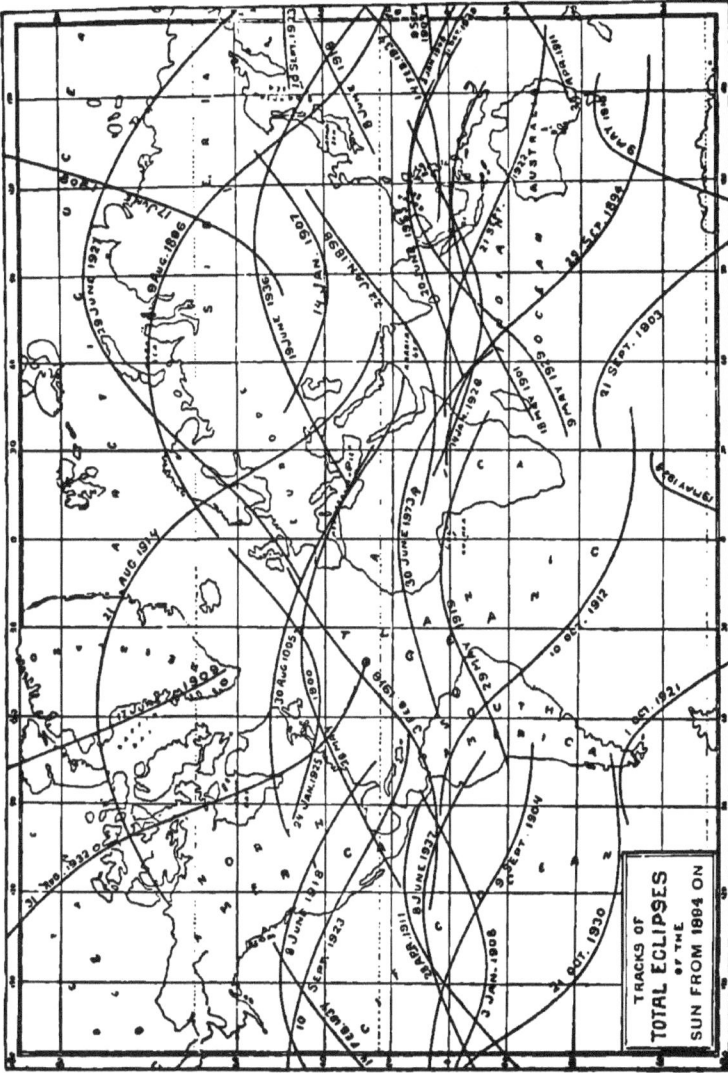

TRACKS OF TOTAL ECLIPSES OF THE SUN FROM 1864 ON

v. OPPOLZER, T., *Canon der Finsternisse* (Vienna 1887).

WESLEY, in CHAMBERS' *Astronomy* (London 1889), i. 311.

PROCTOR, *Old and New Astronomy* (New York 1892).

CLERKE, *History of Astronomy* (London 1893).

TODD, *Total Eclipses of the Sun* (Boston 1894). Bibliography.

GINZEL, 'Mystische Sonnenfinsternisse,' *Himmel und Erde*, vii. (1895), 167.

LYNN, *Remarkable Eclipses* (London 1896).

LOCKYER, *Recent and Coming Eclipses* (London 1897).

KOBOLD, VALENTINER'S *Handwörterbuch der Astronomie*, i. (Breslau 1897).

BURRARD, *Eclipse 22d January 1898, with Charts of Eclipse Families* (Dehra Dun 1898).

For the literature of current eclipses, consult also the recent volumes of *Knowledge*, *The Observatory*, *Popular Astronomy*, *Journal British Astronomical Association*, *Publications Astronomical Society Pacific ;* and for technical reports, the *Proceedings* and *Transactions of the Royal Society. — D. P. T.*

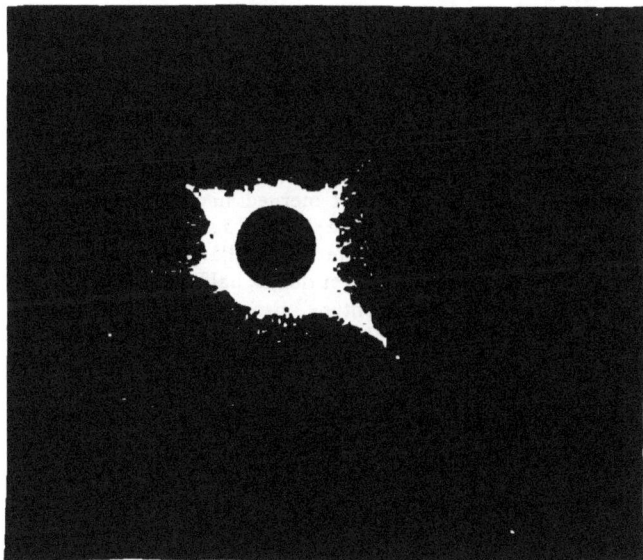

THE LONG CORONAL STREAMERS OF 22D JANUARY 1898

(From a photograph in India by Mr MAUNDER)

CHAPTER IX

THE SOLAR SYSTEM

THE Earth on which we live is one of a large number of bodies circulating in orbits round the Sun, and attracted to him by a force called gravitation, which, constant in itself, acts in such a way that its effect diminishes as the square of the dis-- tance increases. The revolving bodies themselves are endued with a similar attracting energy, but much smaller than that of the Sun in consequence of the much smaller amount of matter which they contain. The Sun, the central body of this great system, radiates light and heat to the whole with an intensity diminishing, like the force of gravity, in the same proportion as the square of the distance increases.[19]

[19] Independently of his light and heat, the Sun's supreme right to rule his family of planets is at once apparent from his superior size (see diagram on page 95), and from his vastly greater mass. Relative weights of common things readily give a notion sufficiently precise: let the ordinary bronze cent represent the weight of the Earth; Mercury and Mars, then, the smallest planets, would, if merged in one, equal an old-fashioned silver three-cent piece; Venus, a silver dime; Uranus, a gold double-eagle and a silver half-dollar (or, what is about the same thing in weight, a silver dollar, half dollar, and quarter dollar taken together); Neptune, two silver dollars; Saturn, eleven silver dollars; Jupiter, rather more than two pounds avoirdupois (37 silver dollars); while the Sun, outweighing nearly 750 times all the planets and their satellites taken together, would somewhat exceed the weight of the long ton. — *D. P. T.*

The bodies known to revolve round the Sun are divided into three classes, — planets, comets, and meteors. In their orbital revolutions, many of the

TYCHO BRAHÉ (1546-1601)

planets are attended by one or more companion bodies called moons or satellites. Also, some of the comets are known to consist of double or multiple portions. But the meteors mostly travel in vast shoals together, along orbits similar to those of comets, with which they seem to have a connection not yet fully understood.

KEPLER, by his laborious research upon the apparent motions of Mars (as obtained from the obser-

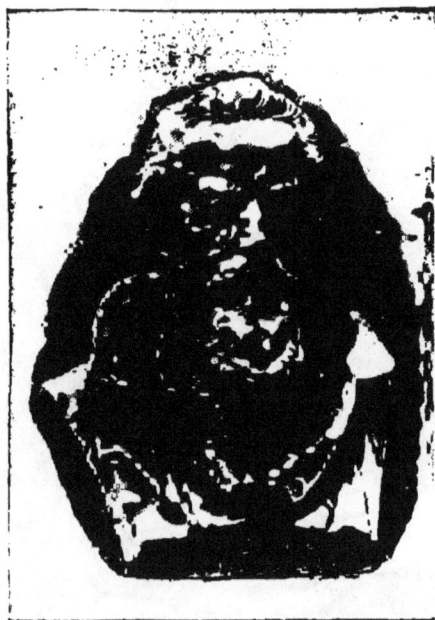

KEPLER (1571–1630)

vations of TYCHO BRAHÉ, and by extending to the other planets by analogy the conclusion derived therefrom) established three simple laws according to which these bodies move in the same general direction round the Sun. KEPLER's laws are : (1) The planetary orbits are ellipses, with the Sun in one focus ; (2) the planets move fastest when nearest the Sun, in such a way that the radius vector describes equal areas of the ellipse in equal times, ~~as the~~ figure shows ; (3) the squares of the periodic times of the planets' revolutions round the Sun are proportional to the cubes of their mean distances from him.

The first two of these laws were shown by NEWTON to be necessary consequences of the planets being attracted by the Sun with a force varying inversely as the square of the distance ; while the third proves

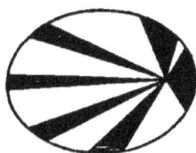

TO ILLUSTRATE
KEPLER'S 2d LAW
(*The six shaded areas are all equal*)

that the force of this attraction is the same upon each, the actual intensity being, as in the case of each

SIR ISAAC NEWTON 1643-1727)

separate planet, inversely proportional to the square of the distance from him.[20]

[21] The work of both KEPLER and NEWTON relates purely to laws of planetary motion, and is in no way whatever concerned

The planets vary very much in size, as is apparent from the opposite illustration, the largest (Jupiter) being more than 1,300 times greater than the Earth, while many of the small planets are so minute as to be visible only with the aid of powerful telescopes, and it is impossible to measure their real sizes. These small planets, however, form almost a class of themselves. Their mean distances from the Sun are all smaller than that of Jupiter, though greater than that of Mars, with a single exception; and their faintness arises not so much from their great distance as from their really small size.

All the planets revolve round the Sun in elliptical orbits of small eccentricity, not differing greatly from circles.[21] These planetary paths are inclined at va-

with physical causes underlying those laws. NEWTON himself says : ' The cause of gravity is what I do not pretend to know. . . . Gravity must be caused by an agent acting constantly according to certain laws, but whether this agent be material or immaterial, I have left to the consideration of my readers.' Consult Professor TAIT's *Properties of Matter*, pp. 131–138, CLERK MAXWELL's article on ' Attraction ' (*Encyclopædia Britannica*, 9th edition), and the admirable résumé of kinetic theories of gravitation by Mʳ W. B. TAYLOR in the *Smithsonian Report* for 1876. Research upon the mechanism of gravity belongs rather to the field of physics than astronomy, and the cause of gravitation is yet undiscovered. — *D. P. T.*

[21] In a subsequent chapter it will be shown how the law of gravitation, regnant through untold ages in the remote past, has brought about the evolution of the planetary system in its form at the present day. But a century and a half ago, it was a mere speculation what the future of such a system might be. Gravitation, that definite and powerful bond, mutual and interacting between the Sun and all his planets, maintaining each body in its individual orbit, and preserving continual order and harmony, — might it not relax its hold with the lapse of time, leaving all the planets, the Earth among them, to recede farther and farther into the cold of space, thus bringing all organic life to

rious angles to the ecliptic or

an end? Or might not the very potency of that same force, through orbital changes then known to be going on, eventually bring down all the planetary members of the system, with their retinues of satellites, upon the Sun in terrible cataclasm? Indeed, EULER, who made the first successful attempt to develop fully the significant results from the law of gravitation, had written a letter from Berlin in 1749 to WETSTEIN, stating his conclusion that the Earth's motion had been sensibly accelerated in the three centuries previous. He naturally accounted for this by the hypothesis of a subtile resisting medium, causing all the planets to travel round the Sun in slowly

Mercury

Venus

Earth and Moon

Mars and two satellites

Several hundred small planets

Jupiter and five satellites

Saturn and eight satellites

Uranus and four satellites

Neptune and one satellite

TO ILLUSTRATE THE RELATIVE SIZE OF SUN, PLANETS, AND SATELLITES

THE LENGTH OF THIS BLOCK REPRESENTS THE DIAMETER OF THE SUN ON AN EQUAL SCALE WITH THE PLANETS

plane of the Earth's orbit, and the courses of a few small planets are considerably more inclined to the

involving spirals, with the rapidity of their motion augmented at every revolution by increasing nearness to the Sun. Manifestly, then, the solar system could not last forever in its present state. The eminent French astronomers, LA GRANGE (whose portrait is opposite) and LA PLACE (page 246), attacked this great problem of stability of the planetary system; and by the application of mathematical methods and expedients, newly invented, solved it. The path in which each planet travels has certain geometrical characteristics; its form, its size, and its position relatively to the stars in space. Technically these are termed the *elements of the orbit*, and every orbit has six such elements. LA GRANGE and LA PLACE, proceeding on the hypothesis that all the planets were rigid bodies, proved that no interaction of gravity among them could ever alter the average size of their orbits; and that the values of the other elements could only oscillate harmlessly within certain narrow limits which their researches were competent to define. So long, then, as the inherent force of gravity remains constant, the solar system, in itself considered and independently of exterior influences, possesses all the elements necessary to absolute stability *in sæcula sæculorum*. The farther researches of POISSON and LE VERRIER in France, of SCHUBERT in Germany, and of STOCKWELL in America, have contributed greatly to our accurate knowledge of these important terms.

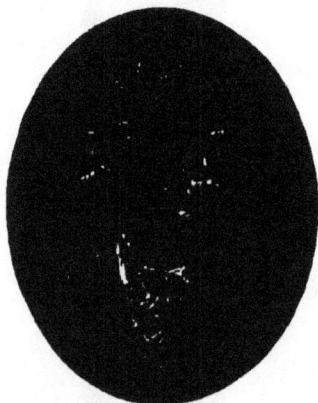

EULER (1707–1783)

The eccentricity of the Earth's orbit, then, could not go on increasing indefinitely in the future; so that our globe is forever insured against the possibility of retreating at one season so far from the central luminary that all the waters of the

ecliptic than the orbits of the large planets are. Also, the orbital eccentricities of the small planets present a wider range.

LA GRANGE (1736–1813)

Earth would freeze; nor at the opposite season could it ever approach so near the Sun that all forms of life would become extinct by reason of the excessive heat. It is the path of our own planet in which interest chiefly centres: its eccentricity, already given in Chapter VI as 0.01677, may vary between the limits 0.0047 and 0.0747, according to LE VERRIER. It is diminishing at the present time; that is, our yearly path round the Sun is now approximating more and more nearly to the circular form, and its nearest approach will take place about 24,000 years hence; after which the orbit will again grow more and more elliptical. The relatively minute fluctuations of an orbit are termed *secular variations*; and they may be crudely represented by holding a flexible and nearly circular hoop between the hands, now and then compressing it slightly, also wobbling it a little, at the same time slowly moving the arms one about the other.

LA PLACE'S mathematical investigations in his *Mécanique*

But, whether large or small, the planetary bodies
are all alike in partaking of the same easterly motion
in their orbits round the Sun. Comets (at least those
which are permanent members of the solar system)
also move about the Sun in elliptic orbits; but their
inclination and eccentricity are generally much greater
than those of the planetary orbits, and many which
travel in very elongated ellipses move in the reverse
direction from that of all the planets. Of the meteors,
those moving in regular orbits pursue paths like those
of many comets, swarms of them travelling in ellipses
nearly identical with known cometary orbits. Indeed,
it would seem not unlikely that many comets are composed of clusters of meteors loosely kept near together
by the feeble bond of attraction of the separate particles, similar discrete bodies being scattered along the
whole or a part of the orbit, but more thickly congregated in some portions than in others.

The large planets at present known are eight in
number: Mercury, Venus, the Earth, Mars, Jupiter,
Saturn, Uranus, and Neptune. The mean distances
of Mercury and Venus from the Sun, in proportion to
the Earth's, are 0.387 and 0.723, respectively; and
the mean distance of Mars from the Sun is 1.524, of

Céleste also led him to the discovery of the invariable plane of
the solar system, which passes through its centre of gravity,
and is determined by the dynamical principle that the sum of
the products of all the planetary masses by the projection of
the areas described by their radii vectores, in any given time, is
a maximum quantity. Relatively to the ecliptic, or ordinary
plane of reference, the invariable plane is defined as follows: —

    Longitude of its ascending node,   106°   14′   6″
    Its inclination to the ecliptic,   1   35   19,

according to STOCKWELL, one of the latest investigators who
have calculated its position. — *D. P. T.*

Jupiter 5.203, of Saturn 9.539, of Uranus 19.183, and
of Neptune 30.055 times that of the Earth.[22]

[22] Recalling that the distance of our globe from the Sun is
93,000,000 miles, these numbers readily give the approximate
distances in miles of all the large planets from the central
luminary. There has been devised no better illustration of
relative distances, magnitudes, and motions in the solar system
than the following, from Sir JOHN HERSCHEL'S *Outlines of
Astronomy:* 'Choose any well leveled field or bowling-green.
On it place a globe, two feet in diameter; this will represent
the Sun; Mercury will be represented by a grain of mustard
seed, on the circumference of a circle 164 feet in diameter for
its orbit; Venus a pea, on a circle of 284 feet in diameter; the
Earth also a pea, on a circle of 430 feet; Mars a rather large
pin's head, on a circle of 654 feet; the asteroids, grains of sand,
in orbits of from 1000 to 1200 feet; Jupiter a moderate-sized
orange, in a circle nearly half a mile across; Saturn a small
orange, on a circle of four-fifths of a mile; Uranus a full
sized cherry or small plum, upon the circumference of a circle
more than a mile and a half; and Neptune a good sized plum,
on a circle about two miles and a half in diameter. As to
getting correct notions on this subject by drawing circles on
paper, or, still worse, from the very childish toys called
orreries, it is out of the question. To imitate the motions of
the planets, in the above-mentioned
orbits, Mercury must describe its
own diameter in 41 seconds; Venus,
in 4^m 14^s; the Earth, in 7 minutes;
Mars, in 4^m 48^s; Jupiter, in 2^h 56^m;
Saturn, in 3^h 13^m; Uranus, in 2^h
16^m; and Neptune, in 3^h 30^m.'

Among the great astronomers
who, at the beginning of the present
century, turned their attention to
the important problem of determining a planet's motion from observation, GAUSS is pre-eminent,
and his finished work is embodied

GAUSS (1777-1855)

in his *Theoria Motus Corporum
Cælestium* (1809), translated by Admiral DAVIS (1857). The
large planets, with the exception of Uranus and Neptune, have

It has often been surmised that there is a planet (possibly several planets) nearer the Sun than Mer-

now been accurately observed so long a time, and the elements of their motion are so thoroughly well ascertained, that it is possible to predict their future positions with a precision which is satisfactory for most purposes. This is done through the instrumentality of *Tables* of their motion, which extend ordinarily a half century in advance of their date, and are capable of farther extension indefinitely. From such tables is prepared each year, by a tedious and complex arithmetical process, *The Nautical Almanac and Astronomical Ephemeris*, which gives the data required by navigators in conducting ships from port to port, and by astronomers in carrying on the operations of observatories. Some of these data are the accurate positions of all the planets among the fixed stars; and the degree of their precision depends upon (1) the perfection of the mathematical theory of their motion and (2) the accuracy with which the planets have been observed, chiefly during the past two centuries. It is at the great observatories maintained by the principal governments that such researches are systematically kept up: at Greenwich, founded 1675 (already shown on page 39); Berlin, founded in 1700; Paris, founded in 1668 (on page 141); Washington, founded in 1842 (page 101, an illustration of the new Observatory, completed in 1893), and elsewhere. The expense of maintenance of each of these establishments averages about $50,000 annually. Sir GEORGE AIRY, seventh Astronomer Royal at Greenwich (from 1835 to 1881), whose portrait is on page 102, was the most eminent of the modern astronomers who have devoted themselves to planetary observation. Since the time of his early predecessor, BRADLEY, in the middle of the 18th century, observations of the planets had been accumulating in unavailable form, and AIRY undertook and completed the prodigious task of calculating, or reducing them, as the technical expression is. This finished product of the observatory formed the basis of LE VERRIER's tables of all the principal planets of the solar system, completed in 1877, and published in the *Annales de l'Observatoire Impérial de Paris, Mémoires*. An entirely new system of tables has now for many years been in progress of construction under the immediate direction of Professor NEWCOMB, who gives in the Introduction to volume i, *Astronomical Papers of the American Ephemeris*

THE UNITED STATES NAVAL OBSERVATORY $\left\{\begin{array}{l}\text{Commodore ROBERT L. PHYTHIAN, U. S. Navy, } \textit{Superintendent}\\ \text{Professor WILLIAM HARKNESS, U. S. Navy, } \textit{Astronomical Director}\end{array}\right.$

(New Buildings erected at Georgetown Heights, District of Columbia, 1888–1893)

cury; but such objects must be very small, and none have yet been certainly discovered. The number of small planets now known is almost 450; many new ones are discovered every year, and there probably are more than 1000 in all. Whether there are planets more distant than Neptune it is impossible at present

SIR GEORGE AIRY (1801–1892)

(Washington 1882), a brief and lucid review of a century's progress in planetary tables. Also in his *Elements of the Four Inner Planets and the Fundamental Constants of Astronomy* (Washington 1895), he presents a general summary. — *D. P. T.*

to say ; but such bodies would not be visible without the aid of a powerful telescope.[23]

[23] Professor TODD in 1877 (from unexplained deviations of Uranus), and Professor FORBES in 1880 (from the aphelion distances of a family of comets thought to have been captured by planetary bodies far exterior to the present boundary of the solar system), derived two independent positions for the trans-neptunian planet which are in surprisingly good agreement; but, although the suspected region among the stars has been carefully searched, both optically and photographically, this planet is not yet visualized. — *D. P. T.*

AIRY, *Gravitation* (London 1834).

HANSEN, *Allgemeine Uebersicht des Sonnensystems*, in SCHU-MACHER'S ' Jahrbuch,' xv. (1837).

HIND, *The Solar System* (London and New York 1852).

MAEDLER, *Das Planetensystem der Sonne* (Leipzig 1854).

BREEN, *The Planetary Worlds* (London 1854).

KIRKWOOD and CHASE, Papers on planetary mechanics and harmonies in *Am. Jour. Science, Proc. Am. Phil. Society, Jour. Frank. Institute*, and *Proc. Am. Assoc. Adv. Science.*

LE VERRIER, *American Journal of Science*, xxxii. (1861), 222.

KLEIN, *Das Sonnensystem* (Brunswick 1871).

KAISER, Measures of all the planets with double-image microm-eter, *Annalen der Sternwarte in Leiden*, iii. (Hague 1872).

STOCKWELL, ' Secular Variations of Orbits of Principal Plan-ets,' *Smithsonian Contributions to Knowledge*, xviii. (1873).

ALEXANDER, ' Harmonies of the Solar System,' *Smithsonian Contributions to Knowledge*, xxi. (1875).

GLADSTONE, ' Chemical Constituents of the Solar System,' *Philosophical Magazine*, iv. (1877), 379.

LEDGER, *The Sun : Planets and Satellites* (London 1882).

SCHEINER, FROST, *Astronomical Spectroscopy* (Boston 1894).

BOYS, ' Newtonian Constant of Gravitation,' *Nature*, l. (1894), 330, 366, 400; *Phil. Trans.* clxxxvi. (A 1895), 1.

HILL, ' Progress of Mécanique Céleste,' *Observatory*, xix. (1896).

POINCARÉ, ' Stability of Solar System,' *Nature*, lviii. (1898), 183.

SEELIGER, ' Law of Gravitation,' *Pop. Astron.* v. (1898), 474, 544.

CHAPTER X

THE large planets of the solar system are classified as inferior planets (Mercury and Venus), and superior planets (Mars, Jupiter, Saturn, Uranus, and Neptune). As seen with a telescope, the inferior planets pass through all the phases of the Moon. The phases of Venus were first noticed by Galileo in 1610.

The orbits of Mercury and Venus lying wholly within that of the Earth, these planets can never come into apparent opposition to the Sun as seen from the Earth. But either body may come from time to time into line or conjunction with the Sun, called *inferior* conjunction when it is between Sun and Earth, and *superior* conjunction when beyond the Sun. If one of them when in inferior conjunction is near the node of its orbit, where it crosses the plane of the Earth's, the planet will be seen to pass like a round black spot across the Sun, and this phenomenon is called a

GASSENDI (1592–1655)

transit. The first time Mercury was ever seen on the Sun's disk was by GASSENDI at Paris, 7th November 1631. The transit of 7th November 1677 was well observed by HALLEY at Saint Helena. Dates of recent transits are 9th May 1891, and 10th November 1894.[24]

Mercury can be seen with the naked eye only when near greatest elongation from the Sun (which averages about 23° and never exceeds 28°), a little before sunrise or after sunset, as the case may be. He revolves round the Sun in 88 days at a mean distance of 36 millions of miles; but, as the eccentricity of his orbit is considerable, his actual distance from the Sun varies between 28½ and 43½ millions. Mercury is too

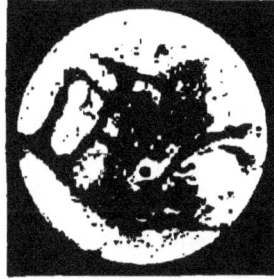

MERCURY (after SCHIAPARELLI)
Earth's diameter on the same scale =
3 inches

[24] As the Earth passes Mercury's descending node about 7th May, and the ascending node about 9th November, transits are possible near these dates only. Also this planet's path being very eccentric, and at its least distance from the Sun near the time of its passing the November node, there are about twice as many transits in November as in May, if long periods of time are considered. And while the greatest length of a May transit is 7ʰ 50ᵐ, Mercury's swifter motion near perihelion reduces the maximum duration of a November transit to 5ʰ 24ᵐ. About 13 transits of Mercury take place every century, the intervals being either 3½, 7, 9½, or 13 years. A period of 46 years (exact to about ¼ day) rarely fails of a return of any given transit. Professor NEWCOMB's researches on the motion of the Moon leading him to suspect that the Earth's rotation on its axis might be variable, he has investigated this question by means of the independent time-measure furnished by Mercury's motion round

near the Sun to allow us to see anything very distinctly on his surface. His diameter is about 3,000
miles. As shown on the preceding page, markings on
the surface of Mercury have been carefully observed
during several years by Professor SCHIAPARELLI of
Milan, indicating that the planet has a rotation on
its axis equal in duration to its revolution round the
Sun. Mercury's mass is about one tenth that of the
Earth, and his density is therefore much greater than
that of our planet.[25]

the Sun; but his critical discussion of 21 transits of the planet
observed from 1677 to 1881, afforded no conclusive evidence
of change in the length of the day. (*Astronomical Papers of
the American Ephemeris*, i. 465.) Following are the transits
during the half century, 1875–1925: —

Date	Eastern Standard Time of Mid-transit	Duration of Transit	Position of Transit-path on the Sun's Disk
1878 May 6	2^h 1^m P. M.	7^h 28^m	Considerably north of centre.
1881 Nov. 7	7 57 P. M.	5 17	Slightly south of centre.
1891 May 9	9 24 P. M.	4 47	Near south limb.
1894 Nov. 10	1 34 P. M.	5 13	Slightly north of centre.
1907 Nov. 14	7 7 A. M.	3 21	Near north limb.
1914 Nov. 7	7 5 A. M.	4 4	Near south limb.
1924 May 7	8 34 P. M.	7 47	Very nearly central

Transits of Mercury are chiefly useful in physical observations of the planet, its distance from the Earth being generally
too great to allow its throwing any light on the problem of the
Sun's distance, although the May transits, when the planet
comes within 50,000,000 miles of the Earth, might advantageously be photographed for this purpose. — *D. P. T.*

[25] Recalling Professor DARWIN's theory of tidal evolution
(*vide* Chapter XVI) it is important to observe that the near

Venus moves in an orbit more nearly circular than that of any other of the principal planets, and her distance from the Sun never varies much from 67½ millions of miles. She revolves round him in 225 days. The inclination of her orbit to the Earth's is greater than that of any other except Mercury's. The greatest elongation of Venus from the Sun amounts to about 45°, so that she may sometimes be seen for a very considerable part of the night, before sunrise or after sunset. In very ancient times she was called, when seen in the morning, Phosphorus, and when seen in the evening, Hesperus.

THE HEMISPHERES OF VENUS
(NIESTEN AND STUYVAERT)

The transit of Mercury in 1677 led HALLEY to suggest the observation of transits of Venus as the best means of obtaining the distance of the Sun. Venus had already been seen on the Sun by HORROX and

proximity of Mercury to the Sun, together with the small surface-gravity of the planet, render it *à priori* extremely probable that the periods of axial rotation and orbital revolution are identical. Also, observations by Mr DENNING in 1882, and in 1892 by Professor W. H. PICKERING at Arequipa, Peru, appear to confirm this view. Although Mercury seems to have no atmosphere, or one of extreme rarity, the markings upon its surface are exceedingly faint. Professor SCHIAPARELLI finds the axis of Mercury perpendicular to the path of its motion round the Sun. One side of the planet, then, must be constantly in full sunlight, the other remaining in total darkness ; but on account of the large libration, the Sun is forever hidden from only ⅔ of the planet's surface. — *D. P. T.*

CRABTREE in 1639; but no other transit would take place until 1761. The transit of that year, as well as the following transit in 1769, was extensively observed, and so have been those of 1874 and 1882. The next pair will occur in the years 2004 and 2012.[26]

1st December   5th December

VENUS NEAR INFERIOR CONJUNC-
TION IN 1890 (BARNARD)

VENUS, 20TH FEBRUARY
1891 (TROUVELOT)

The diameter of Venus is 7,700 miles, not quite equal to that of the Earth. Her mass, likewise smaller than that of our planet, indicates that her density also must be somewhat smaller. There is doubtless a very considerable atmosphere surrounding Venus, and loaded with clouds so dense as to render it very difficult to observe her surface distinctly, and to deter-

[26] As Venus revolves almost exactly 13 times round the Sun while the Earth is completing eight revolutions, the general law of recurrence of transits of Venus is quite simple. In centuries adjacent to the present, transits occur in pairs, eight years elapsing between the two transits of each pair, and the intervals between the midway points of the pairs being alternately 113½ and 129½ years. Usually, then, one pair will fall in each calendar century, and a June pair in one century will be followed by a December pair in the next, transits being possible only in the earlier days of these months, when the Earth is passing the planet's nodes. The 19th century pair having taken place only recently, no transit of Venus will occur in the 20th century, and the pair belonging to the 21st will happen

March 18   March 22   March 28

THE DISK OF MERCURY IN 1897 (LOWELL)

(*During 1896-97, Mr* PERCIVAL LOWELL *at Flagstaff, Arizona, and in Mexico, made a fine series of observations of Mercury's disk with his 24-inch Clark telescope. His first result was a confirmation of* SCHIAPAR-ELLI'S *conclusion, that the planet's axial and orbital periods are the same. Three of Mr* LOWELL'S *drawings are given above, and a combination of all of them enabled him to draw a complicated map of the planet's markings, a characteristic feature of which is a close cross-hatching, best explained by either cracks or corrugations as a result of the planet's cooling. His results were independently confirmed by Miss* LEONARD *and Mr* DREW. *Probably Mercury's small mass is incapable of retaining an atmosphere, absence of which is corroborated by the observations*)

October 23   October 25   October 25

THE DISK OF VENUS IN 1896 (LOWELL)

(*In August-October 1896, Mr* LOWELL, *assisted by Mr* DREW, *observed also the disk of Venus, and their work is confirmatory of* SCHIAPARELLI'S, *as to coincidence of axial and orbital periods. Three of Mr* LOWELL'S *drawings are here given, and his chart of the planet's surface subsequently constructed was independently confirmed by Mr* DOUGLASS *in 1898, using every precaution to prevent deception. The spoke-like markings always maintain the same relation to the terminator, and the planet's axis is perpendicular to its orbit plane. Notwithstanding an evident atmosphere of dazzling lustre, Mr* LOWELL *finds the markings invariably visible; that is, the atmosphere is cloudless. Probably the hemisphere of Venus always turned from the Sun is exceedingly cold, and the supposition of a polar hemisphere would explain the 'phosphorescence,' so-called, repeatedly observed on the perpetually unillumined regions of the disk. Precedent to incorporation with the body of established astronomical fact, these interesting observations of the inferior planets yet await full confirmation elsewhere. The Flagstaff observers insist, and reasonably, that an atmosphere of great tranquillity, not a telescope of great power, is the prime requisite for visualizing delicate planetary features*)

mine with certainty the time of her axial rotation. It was, indeed, thought that periodical changes had been seen indicating a rotation in about $23^h 21^m$; but such conclusion is now regarded by astronomers as very doubtful. As Professor NEWCOMB says, 'The circumstance that the deduced times of rotation in the cases both of Mercury and Venus differ so little from that of the Earth is somewhat

early in that century. The last pair of transits of Venus and the next to occur are as follows : —

Date — Eastern Standard Time	Duration of Transit	Limb of the Sun
1874 December 8, 11^h P. M.	$3^h 38^m$	North.
1882 December 6, noon.	5 37	South.
2004 June 8, 4 A. M.	6 2	South.
2012 June 5, 8 P. M.	6 20	North.

Of any December pair of transits (ascending node), the path of the earlier one always lies across the Sun's northern limb; and of the later one, across the southern limb. For any pair of June transits (descending node), the circumstances are reversed, the earlier one always crossing the Sun's south limb, and the later one the north limb. The shortest transit of Venus ever observed was that of 1874. Several centuries hence, when Venus passes centrally across the Sun, the maximum length of transit, equal to $7^h 58^m$, will be reached. As the inferior planets are always retrograding, or moving westerly, at the time of transit, it is important to notice that ingress always take place on the Sun's east limb, just the opposite of the solar eclipse, which always begins on the west side of the Sun.

Many observers in the 17th and 18th centuries thought they had discovered a satellite of Venus; but M. STROOBANT has satisfactorily disposed of nearly all these observations by other hypotheses. As none of the great telescopes of the present day reveal any such body, the existence of a satellite of Venus is extremely improbable. — *D. P. T.*

suspicious, because if the appearance were due to any optical illusion, or imperfection of the telescope, it might repeat itself several days in succession, and thus give rise to the belief that the time of rotation was nearly one day.' Professor SCHIAPARELLI'S recent observations would seem to show that Venus as well as Mercury rotates in the same time in which she revolves round the Sun ; but M. PERROTIN thinks that the rotation of Venus, though very slow, is probably not quite so slow as this.[27]

[27] Among the famous astronomers who observed Venus in the 17th and 18th centuries were CASSINI at Paris, BIANCHINI at Rome, SCHROETER at Lilienthal, and Sir WILIAM HERSCHEL in England. The mountains supposed to have been seen by SCHROETER were not visible to HERSCHEL, nor can they be seen at the present day. Near the middle of the 19th century, MAEDLER and DE VICO made careful studies of the planet, and many astronomers have in recent years turned their attention to Venus, but with slender avail. Among them Mr DENNING made several fine sketches in the spring of 1881 ; and MM. NIESTEN and STUYVAERT, observing at Brussels 1881–1890, saw a great deal of detail, as the illustration on page 107 shows, embodying all their drawings into a chart of both hemispheres of Venus. In the same year M. TROUVELOT published a series of observations made partly at Paris and partly at Cambridge, U. S., and extending through fifteen years. A pair of narrow white markings at the opposite edges of the disk were usually visible when properly illuminated, and they are thought to be snow caps at the planet's poles. Occasionally large gray spots could be seen covering equatorial regions. The terminator, sometimes seen as a wavy line, rather than straight or slightly elliptical, has by its roughness in certain parts assisted in affording an idea of elevations and depressions on Venus, and in estimating the time of rotation. While this is to be regarded as well determined, it is a little less than 24 hours, according to M. TROUVELOT. On page 108 is one of his sketches of the planet, near the time of greatest elongation. The little double drawing adjacent to it represents Venus as seen at times when she is very nearly between the Earth and Sun ; her dense at-

Mars is a much smaller planet than the Earth, his mean diameter being only 4,230 miles ; but the ellipticity of his figure is greater, the difference between his polar and equatorial diameters being about a hundredth part of the latter. The mass of Mars is about one ninth that of the Earth, and his density, less than that of Venus, is only about three fifths that of our globe. Mars revolves round the Sun in 687 days, at the mean distance of 141½ million miles.

MARS, 1ST AUGUST 1892 (FLAMMARION)

The disk of this planet is well seen with our telescopes at the favorable oppositions, and many maps have been drawn of the surface, which exhibits in many respects a striking analogy to that of the Earth. It differs, however, in this, that on Mars the proportion covered, or thought to be covered, by sea is considerably smaller than that which is apparently dry

mosphere appearing like a very slender sickle or crescent of silvery light, the horns of which Mᵣ BARNARD saw, 5th December, 1890, nearly meeting together. A similar observation had been made under more favorable circumstances by MAEDLER in 1849, and in December 1866, by LYMAN, both of whom saw the delicate atmospheric ring completely encircling the planet, and made measures of it which enabled them to calculate that the horizontal refraction in the atmosphere of Venus is 54'. As the corresponding quantity for the Earth is only 35', the greater density of the atmosphere surrounding Venus is readily inferred. — *D. P. T.*

land. The duration of the axial rotation of Mars is
.24^h 37^m 22^s.7, a constant now very accurately known.
The axis is inclined to the orbit at an angle of about
63°, somewhat less than in the case of the Earth; so
that the inclination of his equator to the plane of
orbital motion is greater by 3½° than it is on our
planet. Mars is surrounded by an atmosphere, probably of no great density.

MARS AND THE APPARENT ORBITS OF ITS SATELLITES

In August 1877 Professor HALL discovered that
this planet is attended by two satellites, to which he
afterward gave the names Phobos and Deimos.[28] The

[28] The illustration on the preceding page represents Mars
as drawn by M. FLAMMARION at Juvisy when the planet was
35,800,000 miles from the Earth, or very near its minimum distance. The minute satellites of Mars, so long in being discovered, but so easy to see with large telescopes once their existence
became certain, have been glimpsed with instruments of very
moderate capacity, one of the smallest being the 7¼ inch Clark
glass of the Amherst College Observatory, with which, near the
opposition of 1892, I saw both satellites without great difficulty.
Their apparent orbits are indicated in the above illustration.
At the favorable oppositions they are remarkably easy objects
in the Lick telescope; indeed, Professor KEELER in 1890 obtained a few successful observations of eclipses of Phobos by
the shadow of the planet. With the exception of the minute

inner of these (Phobos) is the brighter, and likely somewhat the larger of the two ; but neither of them probably exceeds ten miles in diameter. The period of the inner satellite round Mars is 7^h 39^m $15^s.1$; that of the outer, 30^h 17^m $54^s.9$. The distance of Phobos from the centre of Mars is only about 6,000 miles, so that its distance from the nearest point of the planet's surface must be less than 4,000 miles. Deimos is about two and a half times as far from Mars as Phobos, its distance from the planet's centre being approximately 15,000 miles.

𝛾′ Hundreds of small planets are now known to revolve round the Sun in elliptic paths lying between Mars and Jupiter. The smallest orbit is only a very little less than that of Mars, and the largest is not much smaller than the orbit of Jupiter. While Æthra (132) when at perihelion is actually nearer the Sun than Mars is at aphelion, no small planet yet discovered ever recedes so far from the sun as the perihelion distance of Jupiter. (433) Eros (p. 408) has the smallest mean distance

C. H. F. PETERS (1813–1890)

particles composing the dusky ring of Saturn, the inner satellite of Mars is the only secondary body in the solar system known to revolve round its primary in less time than the planet takes to turn round once on its axis, a seemingly strange relation which is easily accounted for by the secular action of tidal friction. So great interest attaches to Mars that I have thrown the fuller additions on this planet into a separate chapter, following the present one. — *D. P. T.*

S & T — 8

(1.457 in terms of the Earth's mean distance) ; while that of Thule (279) is the greatest, amounting to 4.247 on the same scale. The first four small planets were discovered in the early years of the 19th century (the first, Ceres, on the 1st January 1801), and these are probably the largest; but it is not likely that the diameter of any one of them exceeds 300 miles. After the discovery of the fourth (Vesta), in 1807, no more were found until 1845, when Astræa was detected.[29] Since then discovery has been nearly con-

---

[29] PETERS (portrait on the preceding page), the most successful American discoverer of small planets, calculated the diameters of 40 of these bodies found by him, the mean of which is 43 miles. Mr BARNARD in 1894 measured the largest with the Lick telescope as follows: Ceres, 485 miles ; Pallas, 304 miles ; Vesta, 243 miles. Mr ROSZEL estimates the united mass of the first 311 small planets at $\frac{1}{3200}$ that of the Earth. In a general way the group of these bodies ranges through $\frac{2}{8}$ of the broad interval between Mars and Jupiter, or about 280 million miles. Their orbits, by no means concentric, also exhibit wide divergence from a median plane. While the orbital inclinations of six small planets exceed 25°, the mean inclination to the ecliptic for the entire group is only 8°, not much more than that of Mercury. In general, orbits greatly inclined are also very eccentric. The periodic times of the small planets vary between 1.76 and 8.75 years. Regarding their cosmic origin, the explosion hypothesis of OLBERS was long ago abandoned, and it is now commonly considered that their discrete existence is sufficiently explained by the proximity of the excessive mass of Jupiter, whose perturbative influence in this region prevented the concentration of primordial planetary material into a single body. KIRKWOOD in 1866 directed attention to the fact that in those portions of the small planet zone where a simple relation of commensurability exists between the appropriate period of revolution and the periodic time of Jupiter, gaps are found, similar to those which separate the rings of Saturn. Minute intercomparison of their orbital elements reveals numerous instances of a near identity of paths about the Sun, or an apparent grouping, sometimes only in

tinuous; and of late years a large number have been found by photography. The number at present known is about 450.

pairs, but several times in families, one of which includes 11 members. Some of the outer members of the group suggest a transition between small planets and periodic comets; the path of Andromache (179), for example, being very like the orbit of TEMPEL'S comet. Minute and distant as these bodies are, the careful observations of Mr PARKHURST have established the fact of a phase (peculiar to each minor planet), which cannot be neglected in comparisons of brightness at different times. Studies of the small planet group have been made by MONCK, TISSERAND, NIESTEN, and others, among them KIRKWOOD, whose researches are embodied in his monograph entitled *The Asteroids or Minor Planets* (Philadelphia 1888). For complete data respecting the orbits of these bodies, consult the most recent issue 'of the *Berliner Astronomisches Jahrbuch*. The simple facts of interest connected with their discovery are given in a table at the end of this book.

The striking advantage of photography is shown in the rapidity and thoroughness with which the sky is now searched for new objects of this character; for example, M. CHARLOIS of Nice, the most successful of all discoverers, using a portrait lens of 6 inches diameter, and 32 inches focus, obtains every star to the 13th magnitude, on plates including more than 10° square on a side; that is, an area covered by 400 full moons arranged in a square, 20 on a side, with their disks just touching. Each plate is exposed 2½ hours, and its critical examination requires an equal amount of time; so that in five hours, a given region of the sky can be more thoroughly searched than in 90 hours by the old-fashioned method, at the eyepiece of a telescope. New discoveries of small planets are now double-lettered, provisionally, until observed at least five times; then a permanent number is assigned in the regular list. The first small planet found in 1894 is lettered *A Q*, and most of the recent discoveries are still without names. There is a growing recognition of the significance of the small planet group in the cosmogony; and it is hoped that an international agreement may soon be reached, looking toward the more even distribution of the vast labor of the calculations which the rapidly increasing discoveries of these bodies now entail upon the patience of Germany almost unaided. — *D. P. T.*

᠅ Exterior to the small planets are four bodies very much larger than the four planets interior to this group. Of these, the nearest to the Sun is the largest planet of all, Jupiter, often called the 'giant planet.' His bulk is 1,309 times as great as that of our globe, his mean diameter being 86,500 miles. But his mass being only 316 times that of the Earth (or $\frac{1}{1047}$ that of the Sun), his density is apparently less than a quarter that of our planet. He revolves round the Sun at a mean distance of $483\frac{1}{3}$ millions of miles in a period amounting to nearly twelve (11.86) years.[30]

[30] Jupiter, always during its visibility a magnificent planet among the stars, was first mentioned by PTOLEMY, who recorded its near approach to the star δ Cancri, 3d September, B. C. 240. About ten times as bright as α Lyræ and Capella, many modern observers, knowing just where to look, have found it visible in full daylight. Bright as Jupiter is, at times almost sufficient to cast a shadow, it would require 6,500 such stars to equal the lustre of the full Moon. Dr LOHSE, during a conjunction of Mars and Jupiter in 1883, found from his photographs of these planets that the intensity of the actinic or chemical rays reflected from Jupiter was 24 times stronger than that of Mars, and that the two hemispheres of Jupiter were by no means alike, the light from the southern hemisphere being twice as effective as that from the northern. Not until 1630, nearly a quarter-century after the invention of the telescope, were its conspicuous belts discovered by ZUCCHI, at Rome. The phase effect on Jupiter, though slight, is not difficult to detect when the planet is near quadrature and can be seen at a high altitude in twilight.

This giant planet, the most satisfactory of all celestial objects for scrutiny with a small telescope, has already been depicted on pages 30 and 95, as seen with instruments of moderate capacity. As the edge or limb of the planet is approached, not only the colors of the markings but their definiteness fades out completely. Area for area, Jupiter receives but $\frac{1}{27}$ the heat from the Sun that the earth does, and its own intensely heated condition must in large measure account for the excessive cloud. Professor NEWCOMB's value of the mass of Jupiter

The great ellipticity of Jupiter's figure attracts attention the moment the planet is seen with a moderately large telescope. This polar compression (or fraction representing the difference between polar and equatorial diameters divided by the latter) amounts to $\frac{1}{16}$, according to the best measures (that of the Earth is only $\frac{1}{293}$). The planet is surrounded by a thick atmosphere, in which are dense masses of cloud. The belts and spots on the apparent disk are interesting objects of study; and a large red spot, first noticed in 1878, has attracted very special attention during the last few years, undergoing remarkable changes in brightness, while retaining the same form and size. Observations of the spots have given the time of Jupiter's axial rotation with some approach to accuracy. Being about 9^h 56^m, it is less than that of any other planet whose rotation is known. The axis being nearly perpendicular to the plane of the orbit, the equator makes but a very small angle with the latter.[31]

adopted in his investigations of planetary motion, is $\frac{1}{1047.35}$ that of the Sun. Mr BARNARD'S recent measures with the Lick telescope give for the equatorial diameter of the planet 90,190, and for the polar, 84,570 miles. These correspond to apparent diameters $38''.52$ and $36''.11$, respectively, for a distance of Jupiter from the Earth equal to 5.20 times that of the Earth from the Sun. — *D. P. T.*

[31] Jupiter's rotation on his axis, first determined by CASSINI about 230 years ago, presents many points of resemblance to that of the Sun. The equatorial regions revolve most rapidly, and bright spots on that part of the planet have quite uniformly given a rotation in about 9^h $50\frac{1}{3}^m$. Receding from the equator, either north or south, the atmospheric markings revolve in longer periods, but the exact relation to Jovian latitude is not yet established. Following are a few of the better recent determinations for regions other than equatorial : —

As soon as telescopes were directed to Jupiter, it was seen that he had four attendant moons or satel-

			h	m	s
1880.	SCHMIDT	. . . . .	9	55	34.4
1885.	HOUGH	. . . . .	9	55	37.4
1886.	MARTH	. . . . .	9	55	40.6
1887.	WILLIAMS	. . . . .	9	55	36.5
1898.	DENNING	. . . . .	9	55	37.7

The last depends upon 29 years' observation of the great red spot; and should this object eventually prove to be actually permanent part of the body of the planet, Mr DENNING'S value must be very near the truth. So rapid is this rotation, and so great the size of Jupiter, that a point on its equator travels nearly eight miles a second.

These differing periods of rotation have suggested a theory, now accepted by many astronomers, that entire zones of the Jovian atmosphere, as well as individual spots in it, may differ in rotation period, the dense envelope perhaps drifting in sections quite independent of each other. The great white equatorial belt is at times nearly free from markings of any kind, and it usually exceeds all other features in brilliance, the markings upon it being often exceedingly faint, as shown in the accompanying illustration. Small spots of an intenser white, and about 2500 to 3500 miles across, are sometimes seen upon it, and they are presumably at the highest

JUPITER, 20th JULY 1889 (KEELER)

AS SEEN WITH THE 36-INCH LICK TELESCOPE

(*Earth on the same scale, the size of a pin-head*)

level of all the cloud formations in Jupiter's atmosphere. These bright spots going through a complete rotation in 5½ minutes less than that of the great red spot, must pass it at a speed exceeding 250 miles per hour; and in six or seven weeks they make a complete revolution round the planet, relatively to that well-known marking. This vast whirling cloud mass, banding the planet's equator, is about 20,000 miles in breadth, and has been termed the 'hurricane girdle.' Also, occasional small white spots appear on the light strip south of the southern

lites. Their discovery is generally attributed to
GALILEO, who noticed them first on 7th January 1610,

equatorial belt, which is usually more active than the northern
one, its detailed configuration often changing rapidly. The
principal belts are variable in color and intensity of tint, their
hue being sometimes brownish, coppery, or purple. Usually
the darkest and most permanent regions of these great belts
are their edges adjacent to the planet's equator. Lesser changes
in form and intensity are frequent, but long observation has established the essential permanence of their general outlines.

The belts on Jupiter's tropics are not uniform in color in the
two hemispheres, the southern one having a strong reddish tint,
and the northern one shading into blue and gray. They seem
to be regions relatively little disturbed, and have no connection
with the rapidly whirling equatorial belt, except through the
oblique streamers, often seen trailing over them, as Professor
KEELER'S drawing shows. The extensive polar regions adjacent to both poles of the planet beyond these median belts are
sometimes seen banded nearly to the pole, and the prevailing
tint of the northern region is bluish.

It is not to be inferred that the very great differences of illumination in the apparent disk of Jupiter indicate anything
more than fleeting cloud-configurations. But while this atmospheric envelope may be said to resemble that encircling our
own globe, it is by no means identical with it ; in fact its metallic
vapors render it more nearly like a solar atmosphere than a
terrestrial one. The intense heat of the great globe, its activity
and its rapid rotation, all tend to complicate the atmospheric
movements and configurations, about which there is much remaining unexplained, and which must continue so until the
variable markings have been more critically and persistently
studied. Members of the British Astronomical Association
have banded together for detailed research upon this planet's
surface, having made as many as 200 drawings in a single
year ; and Mr STANLEY WILLIAMS'S painstaking investigation
of an excellent series of photographs of Jupiter is effecting
substantial advances.

Careful studies of Jupiter's spectrum by Dr HUGGINS in
England, and Dr VOGEL in Germany, have revealed a dark
band in the red, which may indicate some substance in the
planet's atmosphere not present in our own. The fluctuations

and gave in his *Sidereus Nuncius* an account of his
subsequent observations, proving them to be satel-

of long period in Jupiter's atmosphere are matched by detail
variations of intensity in the lines of its spectrum. Whether or
not Jupiter may be slightly self-luminous, is, according to D‍ʳ
SCHEINER, a question upon which the spectroscope furnishes
but little evidence.

The great red spot, excellently shown in Professor KEELER'S
drawing on p. 118, was probably first recorded in recent times
by GLEDHILL and MAYER in 1869. Ten years later many
observers had noticed it, and there are indications that CASSINI
at Paris observed it in 1685. The gigantic size of Jupiter may
be imagined from the magnitude of this object, the superficial
area of the elliptical spot alone exceeding the surface area
of the whole Earth. About the middle of 1883 it nearly disappeared; was so faint the year after as to be very difficult to
observe, and in 1885 a white cloud apparently covered its central portions, making it appear like a flattened ring. In this
year and the one following, the intensity of the spot was slightly
recovered; but onward from that time it grew gradually less
and less easy to see until 1891, when there was again a partial
return to its former visibility. So that fluctuations of this
familiar object are now well recognized. To Professor WILSON
in October 1891, and Mʳ WILLIAMS at the end of 1892, it appeared quite pale in the middle regions, as if a bright elliptic
ring. Also it seemed to be very diffuse, as if covered in part
with a veil of dense mist. If this object is periodic, it must now
be near another minimum, as Mʳ BARNARD reports it as very
faint in the Lick telescope in 1894–95. Often this spot has been
observed to exercise a strikingly repellent effect upon cloud-
markings near it. The great red spot, and its variations of
form and visibility, have for years been studied, not only by
many foreign astronomers, but by Professor HOUGH with the
18½ inch Chicago telescope, whose measures place this strange
marking at a considerable south latitude, and make its length
about 30,000 miles, and its width 8000 miles. Many astronomers think that this persistent object is simply a vast fissure in
the outer atmospheric envelope of Jupiter, through which at
times are seen the dense red vapors of interior strata, if not the
actual surface of the planet; but its slow drift in longitude, and
its slight shift in latitude are unwelcome obstacles to this

lites. SIMON MARIUS (or rather MAYR) claimed to have discovered them a few weeks before GALILEO ; but this was contested, and at any rate he did not publish his observations until later. MAYR's proposed names for the satellites are not much used, lest perhaps such acceptance should seem to imply recognition of his claim to the first discovery. GALILEO named them jointly the Medicean stars, in honor of the Grand Duke of Tuscany, and individually after members of his family ; but these designations have been dropped by general consent, as all were discovered at the same time, and were called the First, Second, Third, and Fourth Satellite respectively, reckoning according to their distance from Jupiter outward, and this nomenclature has been found quite sufficient.

THE SATELLITES OF JUPITER

Number of Satellite	Name	Distance from Centre of Jupiter	Synodic Period or Mean Interval between Eclipses				Diameter	Mass in Terms of Jupiter
		Miles	d	h	m	s	Miles	
I	Io	261,000	1	18	28	35.945	2,500	0.0000169
II	Europa	415,000	3	13	17	53.735	2,100	0.0000232
III	Ganymede	664,000	7	3	59	35.854	3.550	0.0000884
IV	Callisto	1,167,000	16	18	5	6.928	2,960	0.0000425

A fifth satellite, probably not exceeding 100 miles in diameter, was discovered by Mr BARNARD at the

theory. PROCTOR maintained that the real globe of Jupiter lies far within the globe we see and measure, and that the atmospheric envelope is perhaps 10,000 miles in depth. Professor HOUGH thinks that all the observed phenomena can be best accounted for by assuming that Jupiter is still in a gaseous condition. — *D. P. T.*

Lick Observatory, 9th September 1892.   Its distance
from the centre of Jupiter is only about 112,000 miles,
and it revolves around him in 11^h 57^m 22.^{s}7.[32]

THE LICK OBSERVATORY ON MOUNT HAMILTON, CALIFORNIA
(*Elevation of the summit plateau,* 4200 *feet above sea level*)
(*Professor* JAMES E. KEELER, *Director*)

Next beyond Jupiter, and next in size, is Saturn, the
equatorial diameter of which is about 73,000 miles,

[32] The possession of one of the practical handbooks named
on page 396 will add greatly to the interest of the young observer; and the *Nautical Almanac* will be requisite in identify-
ing and following the ever changing configurations of the four
bright satellites, as they undergo their frequent eclipses by
dropping into the planet's shadow, and pass alternately behind
the planet and in front of it, their little shadows dotting the
brilliant disk like a transit of Mercury or Venus in miniature.
The opposite drawing represents such a phenomenon; to an eye

so that its bulk is rather more than half that of the
giant planet, and would contain the Earth's about

placed anywhere within the black spot, a total eclipse of the
Sun takes place. It is no unusual thing for two such solar
eclipses to happen upon Jupiter at the
same time (though of course not at the
same place), caused by two different
moons. During a Jovian year a specta-
tor on Jupiter who could transport himself anywhere at will on that planet
might observe 8,800 eclipses of Sun and
moons.

JUPITER, 18TH FEBRU-
ARY 1874 (KNOBEL)

From LA PLACE'S relation between
the mean motions and longitudes of the
three inner satellites, it is impossible
that all of them can ever be eclipsed at the same time, but
it sometimes happens that while two of these satellites are
eclipsed the fourth is also undergoing eclipse; and the remaining satellite, being of necessity on that side of the planet opposite its shadow, may be projected upon the disk of Jupiter,
where it will usually be invisible. Also this interesting phenomenon may happen from numerous other conditions of
configuration. One of the longest recorded instances when
this great planet was seemingly devoid of its wonted retinue of
satellites was that observed by the writer at Amherst College
Observatory, 21st March 1874, when no satellites could be
seen for nearly two hours.

Markings upon the satellites of Jupiter were repeatedly seen
by the earlier observers, DAWES, LASSELL, and SECCHI, the first
of whom gives drawings of what he saw, in the *Monthly Notices
Royal Astronomical Society*, xx. (1860), p. 246; and in 1897 these
bodies were intently scrutinized with the fine 24-inch Clark
refractor in Arizona, belonging to Mr LOWELL. From observations of their surfaces Mr DOUGLASS reports the disks of satellites I and III striped with series of narrow bands, which he
has charted, and which have enabled him to ascertain the true
period of rotation of these satellites on their axes. He finds
the period of I to be $12^h 24^m.0$, and confirms Professor W. H.
PICKERING's earlier observations (p. 125), assigning an equality between axial and revolution periods to both III and IV.
Also the little disks have been carefully observed with the

760 times.   But its density is not much more than
half that of Jupiter, and the mass of the latter is equal

great Lick telescope, two drawings made with which are here
shown.   Mʳ BARNARD has made interesting studies of the first

JUPITER'S THIRD SATELLITE

As drawn independently by CAMPBELL
(*left*) and SCHAEBERLE (*right*), 23d
October 1891

satellite which, having an
equatorial region relatively
bright, is sometimes seen very
much elongate, when pro-
jected on a dark belt of Jupiter; while on a bright belt,
the satellite's central region
coalescing with it, the two
relatively dark polar regions
of the satellite may give the
deceptive appearance of a
double body.   Professor W. H. PICKERING at Arequipa, Peru,
in 1892, and at the Lowell Observatory in Arizona, 1894, has

EYE-END OF THE LICK TELESCOPE
(*showing the 6-inch finder, and mechanical accessories for handling the
telescope*)

to three or four times that of Saturn. Indeed, the mass of Jupiter is more than double that of all the

seen surface detail upon all the satellites except the second. His observations indicate that the equator of the first satellite is appreciably elliptical, not circular, and that it revolves on its axis in a retrograde direction. Satellite I exhibits many striking peculiarities of form not easy to account for. II is more difficult to observe, and its axial period is not known. The corresponding periods of III and IV are the same in duration as their periods of revolution round Jupiter, therefore resembling the relation of our Moon to the Earth. A white spot was several times seen near the north pole of IV.

M͏ʳ BARNARD'S new fifth satellite was discovered with the telescope here shown, the chief instrument of the great observatory founded by JAMES LICK in 1875 (p. 122). It has not yet been seen with any telescope of less than 18½ inches aperture; and the eclipses of this minute body, whose successful observation would have so great significance in exact astronomy, appear to be beyond the optical reach of all telescopes at present in existence. M. TISSERAND, applying the principles

THE LICK TELESCOPE

with which Jupiter's 5th satellite was discovered

(*Lens 3 ft. in diameter. Tube 57 ft. long*)

other planets put together, while his volume is about one and a half times as great as the aggregate of all the rest. The compression of the figure of Saturn is the greatest of all the planets, and amounts to as much as one tenth. Saturn revolves round the Sun in 29½ years, at a mean distance from him equal to 886 millions of miles.

Belts are seen on Saturn similar to those on Jupiter, but they are much fainter on account of the greater distance from both Sun and Earth. Nevertheless, by the aid of white spots in the equatorial belts in 1894, Mr WILLIAMS found the axial period of that region to be 10^h 12^m 35^{s}8. The axis of Saturn makes an angle of about 62° with the plane of its orbit, so that the planet's equator is inclined about 28° to that plane.[33]

---

of celestial mechanics to the motion of this satellite, finds that its short period, combined with the great equatorial protuberance of Jupiter, produces a swift motion of the major axis of the satellite's orbit, so rapid, indeed, that the orbit makes a complete revolution in about five months. — *D. P. T.*

[33] Saturn, always shining as a dull reddish star of about the first magnitude, is the farthest planet from the sun of those which have been known from the remotest antiquity. Its light as a whole is subject to some variation, owing to our position relatively to the Saturnian rings: when the Earth is near their plane, Saturn appears to be 2½ times less bright than when our globe is at the greatest elevation of 26° above or below the plane of the rings; therefore, its brightness will slowly increase until 1899, and then as slowly diminish till 1907. The mass of Saturn is $\frac{1}{3500}$ that of the Sun, as determined by Professor A. HALL, jun^r, from observations with the Yale heliometer in 1885-7. So slight is the density of the planet that it would float in water; probably, as surmised in the case of Jupiter, the real Saturn is very much smaller than the ball we measure, and surrounded by a gaseous envelope many hundred miles in depth. The ball of Saturn has sometimes been observed to assume an abnormal figure, very much flattened at

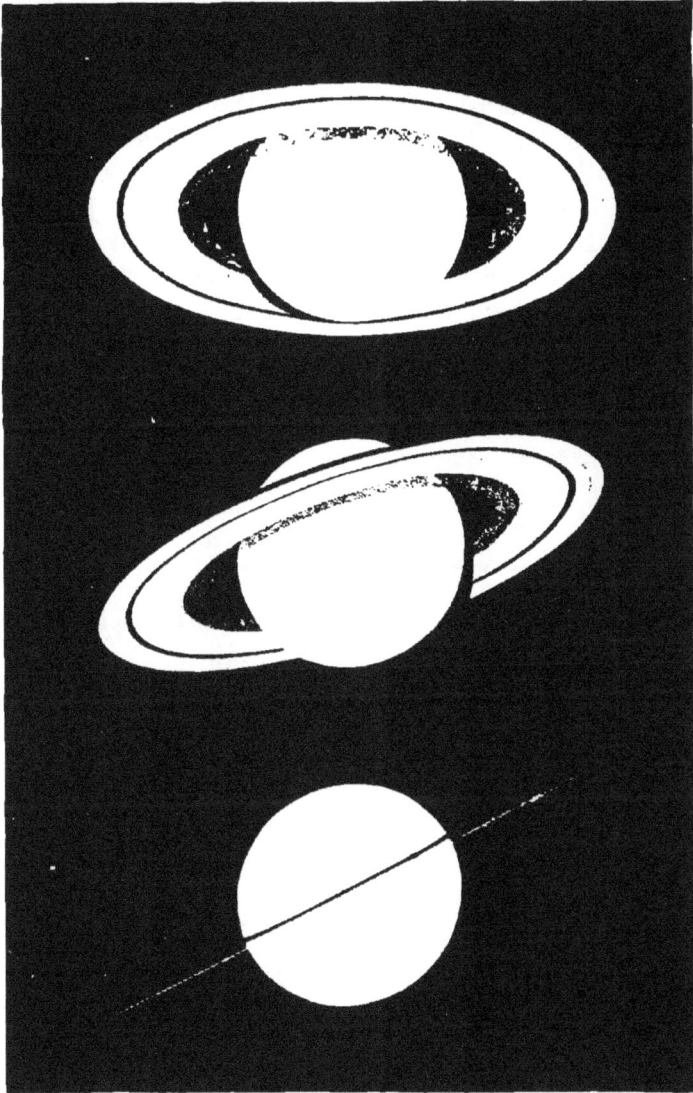

SATURN, WITH ITS RINGS IN VARIOUS PHASES

(Scale, about 70,000 miles to the inch. Neglecting unimportant technical details, these are the appearances of the planet, as viewed with an inverting telescope; in 1891 and 1907 (lower figure), in 1895 and 1902 (middle), and in 1899 (upper figure). From 1892 to 1907 the northern face of the ring is turned toward the Earth; and the southern face is similarly turned from 1907 to 1921)

There are now known to be eight satellites revolving round Saturn. Owing to the confusion which arose from the order of discovery being different from that of distance from the planet, it became necessary to give them names, and those proposed by Sir JOHN HERSCHEL have been generally accepted. Of their actual sizes, only very rough estimates are possible : Titan, the largest, is probably about 3,500 miles in diameter, Japetus 2,000, and Rhea 1,500 ; while each

the poles, and very much bulged out at the four regions about midway between poles and equator, giving it roughly the shape of a square with rounded corners. This phenomenon, first observed by Sir WILLIAM HERSCHEL in 1805, and several times verified by other astronomers, is known as the 'square-shouldered aspect of Saturn ; ' but no satisfactory explanation of so strange an anomaly has yet been advanced.

Careful studies of this planet were made by the BONDS at Harvard College Observatory, 1848-51, and by LASSELL with his twenty-foot reflector at Malta in 1852-53; also by DAWES in 1855-56, M. TROUVELOT in 1872-75, Mʳ RANYARD in 1883, and M. TERBY of Louvain in 1887. In more recent years, several astronomers at the Lick Observatory, among them Professor KEELER and Mʳ BARNARD, have employed the 36-inch telescope in observing this unique planet to the greatest advantage, and the fine illustrations on page 127 embody substantially everything that can be certainly seen under the best conditions. Excellent photographs of Saturn showing clearly the belts on the planet and its system of rings have been taken by Professor PICKERING, the brothers HENRY, and others. The absorption at the edge of the ball, and the differing brightness of the rings are well brought out; but, as in the case of Mars and Jupiter, the photographs hitherto taken do not show the minute details secured by close optical scrutiny of these bodies directly with the telescope. No phase of Saturn's ball is perceptible, even in the largest instruments. The spectrum of the ball is difficult to delineate ; but according to HUGGINS, SECCHI, VOGEL, and JANSSEN, the first of whom observed it in 1864 and photographed it in 1889, its lines are practically identical with those of Jupiter. — *D. P. T.*

THE SATELLITES OF SATURN

Name of Satellite	Name of Discoverer	Date of Discovery	Distance from Centre of Saturn	Sidereal Period of Revolution
			Miles	d h m s
Mimas	W. Herschel	17 Sept. 1789	117,000	0 22 37 5.7
Enceladus	W. Herschel	28 Aug. 1789	157,000	1 8 53 6.9
Tethys	J. D. Cassini	21 Mar. 1684	186,000	1 21 18 25.6
Dione	J. D. Cassini	21 Mar. 1684	238,000	2 17 41 9.3
Rhea	J. D. Cassini	23 Dec. 1672	332,000	4 12 25 11.6
Titan	C. Huygens	25 Mar. 1655	771,000	15 22 41 23.2
Hyperion	W. C. Bond	16 Sept. 1848	934,000	21 6 39 27.0
Japetus	J. D. Cassini	25 Oct. 1671	2,225,000	79 7 54 17.1

of the four interior satellites, besides Hyperion, is probably less than 1,000 miles in diameter.[34]

[34] The visibility of the Saturnian satellites corresponds to the order of their discovery, a very small telescope sufficing to show Titan, also Japetus in the western portion of its orbit, where it is 4½ times as bright as on the other side of the planet. Mimas and Hyperion, the satellites last discovered, are always difficult objects, even in very large telescopes. The orbits of the five inner satellites are sensibly circular. At the great Russian observatory shown on the following page, Dr HER-MANN STRUVE, son of the late Director of the Observatory, Dr OTTO STRUVE, has for many years been prosecuting a thorough investigation of the Saturnian system, both mathematically and by means of observations with the 30-inch telescope (page 133). He has brought to light a sensible acceleration in the motion of Mimas, the innermost satellite, during the last few years. Also its orbit is inclined 1° 26′ to the equator of Saturn; and as its nodes have a motion of about 1° each day, the orbit returns to its previous position at the end of every year. Dr STRUVE's researches farther have verified from observation the remarkable connection existing between the orbits and motions of Mimas and Tethys, and of Enceladus and Dione, previously developed from theory by Professor NEWCOMB. Also the latter has investigated the motion of Hyperion, the seventh satellite, the perisaturnium

THE IMPERIAL OBSERVATORY OF RUSSIA — Dʳ O. BACKLUND, *Director*

(*Founded by the Emperor Nicholas the First in 1838 at Pulkowa, near St Petersburg*)

Saturn is unique among the planets in the possession of a ring, or rather system of concentric rings,
surrounding it within the orbits of all its satellites.
When GALILEO first saw Saturn with a telescope, he
was astonished to see what he thought a small body
adhering to it on each side, which afterward disappeared, recalling to his mind the old myth about
Saturn devouring his own children. HUYGENS was
the first to explain the real nature of the appendage
and its varying appearance according to the position of Saturn with regard to Sun and Earth. This
he did in 1656, at the end of a pamphlet announc-
ing his discovery of the satellite Titan the year
before, which contains a Latin anagram, afterward

of whose orbit has an extraordinary regression in a period of
about 18 years. The perisaturnium is that point in the path
of a satellite of Saturn which is nearest the planet; and as
these points ordinarily partake of an advance motion, Hyperion
has afforded a novelty in celestial mechanics, to which the
ordinary mathematical methods were found entirely inapplica-
ble. The cause of this anomaly was traced to the perturbative
action of Hyperion's massive inner neighbor, Titan, three
times the motion of which nearly equals four times that of
Hyperion. The mass of Titan is $\frac{1}{4700}$ that of Saturn. The
orbit of Japetus, the outermost of all the satellites, is inclined
about 10° to the plane of Saturn's ring system. There is much
uncertainty about the size of all these bodies. PICKERING'S photometric determinations in 1877–78 making them less than half
the diameters above given. Hyperion is the smallest satellite,
and is probably not over 500 miles in diameter. Mimas comes
next in size, its diameter being about ¼ greater than that of
Hyperion. Eclipses of some of the satellites, and transits,
particularly of Titan, Rhea, Tethys, and Dione, and of their
shadows across the ball of the planet, have occasionally been
observed, near the times (about 15 years apart) when the Earth
is passing the planes of their orbits; but these phenomena
require telescopes of exceptional power. — D. P. T.

interpreted to mean 'Annulo cingitur tenui plano nusquam cohærente, ad eclipticam inclinato.' (It is surrounded by a thin, plane ring, nowhere adhering to it, and inclined to the ecliptic.) The explanation is given in his *Systema Saturnium*, published in 1659, in which he refers to the observations of WILLIAM BALL in England as confirming his own. In 1676 CASSINI in France perceived that there was a division in the ring. The inner ring of these two is perceptibly brighter than the outer, and it appears that CAMPANI at Rome, so early as 1664, had noticed the greater brightness of the interior portion of the ring, though he failed to recognize an actual division between the two parts thus distinguished. Other smaller divisions have been noticed since, particularly one in the outer ring by ENCKE in 1837. But the most remarkable recent discovery in this system of rings is that technically called the 'dusky ring,' inside the two bright rings. This was first seen in 1850 by the BONDS of Cambridge, U. S., though there are several indications that a shading of the inner bright ring toward the planet had been noticed long before. GALLE had in fact called especial attention to it under this description in 1838. LASSELL compared the dusky ring to 'something like a crape veil covering a part of the sky within the inner ring'; and its transparency was afterward noticed.[35]   The

---

[35] According to Dr OTTO STRUVE's notation, adopted by astronomers generally, the rings are called *A*, *B*, and *C*, the last being the dusky ring, and *A* the outermost one. The luminous rings of Saturn differ greatly in brightness, the outer one being much fainter; and having at times a very narrow, and probably non-permanent, division about midway in it. The inner bright ring, area for area, is several times brighter than the outer one; and the innermost, or dusky ring appears to be

D. CASSINI

(JEAN DOMINIQUE CASSINI (1625-1712), *whose chief claim to distinction rests on his discoveries in the Saturnian system, was earlier known as a student at Genoa, and a professor at Bologna, where his first astronomical work was a discussion of the comet of* 1652. *Close observation upon Jupiter and Mars enabled him to ascertain their periods of rotation in* 1665. *Astronomical refraction, the distance of the Sun, libration of the Moon, the zodiacal light, and measurement of a meridian arc were among the fertile subjects of his investigation.* CASSINI'S *son and great-grandson were also directors of the Paris Observatory down to* 1845, *and his grandson likewise was engaged in the pursuit of science, though never director.* CASSINI *also constructed tables of the motion of Jupiter's satellites. Although a lifelong observer, he availed himself of but few of the instrumental advances set on foot in his day*)

Dʳ OTTO VON STRUVE AT THE EYE-PIECE OF THE 30-INCH TELESCOPE
AT PULKOWA

(Object glass by ALVAN CLARK & SONS ; *mounting by the* REPSOLDS)

Saturnian ring system, which has in so many ways been an enigma to astronomers, is now regarded as consisting of an innumerable multitude of very small satellites, so arranged as to present at a distance the appearance of a thin flat ring, with several narrow divisions in its breadth. Doubtless the peculiar

immediately joined to it, the interior bright ring generally seeming to be darker on its inner edge, as if shading into the dusky ring. KIRKWOOD has explained the existence of divisions in the rings as due to perturbations in their substance caused by satellite attraction exerted in narrow zones where there would be a relation of simple commensurability with the periods of certain satellites, chiefly Mimas; these divisions being analogous to the well-known gaps in the group of small planets caused by the perturbative action of Jupiter.

The partial transparency of the dusky ring, so well shown in the illustrations on page 127, where the outline of the ball of Saturn is readily seen through this 'crape veil,' was excellently demonstrated by observations made by Mr BARNARD with the 12-inch telescope of the Lick Observatory, 2d November 1889, during an eclipse of Japetus in the shadow of the Saturnian system. As the satellite passed through the shadow of the dusky ring, its light grew fainter and fainter on approaching the shadow of the bright ring, showing that the particles of this semi-transparent veil are more thickly strewn as the bright ring is approached. That the latter is fully as opaque to the Sun's rays as the globe of Saturn itself was convincingly shown by the complete disappearance of Japetus immediately on entering the shadow of the bright ring. Also the blackness of its shadow on the ball emphasizes the opacity of the substance composing this ring. In the same year Professor KEELER examined the spectrum of the Saturnian rings with the 36-inch telescope of the Lick Observatory, but found no trace whatever of any such bands as are strongly marked in the spectra of Jupiter and the ball of Saturn; so that there can be no atmosphere about the ring. Also in 1895, by a neat adaptation of DOPPLER's principle to an interpretation of the distorted spectra of the planet and its ring-system, he demonstrated anew the meteoric constitution of the rings; for he proved that their inner edges revolve more swiftly round the planet than their outer edges do. — *D. P. T.*

appearance of the dusky ring is due to the tiny satellites of which it is composed being much more scattered than those forming the bright rings. The outer
diameter of the exterior ring amounts to 173,000
miles, and the breadth of this ring is somewhat more
than 10,000 miles. The outer diameter of the interior
bright ring is 145,000 miles, and its breadth is 16,500
miles. The distance of Mimas, the innermost satellite, from the exterior ring is 31,000 miles, and an
interval of nearly 10,000 miles divides the dusky
ring from the ball of the planet. The rings and
satellites all revolve around Saturn in planes making
only small angles with the planet's equator.[86]

[86] Saturn's ring is sometimes totally invisible, except in very
large telescopes; and then, even with these, it can be seen only
as depicted in the lower figure on page 127. This most interesting phenomenon takes place when the Earth reaches the plane
of the ring (for this object is so thin that it does not fill an
appreciable angle when seen edge on); when the plane of
the ring is directed toward the Sun (for then its edge only is
illuminated, and the reflected light is very feeble); and when
Sun and Earth are on opposite sides of the ring (for its unillumi-
nated side is then toward us). As the plane of the ring always
remains parallel to itself, the epochs of disappearance are
nearly 15 years apart, this interval being about one half the
periodic time of Saturn. The two disappearances of the rings
in 1861–62, many weeks in duration, were excellently observed.
The subsequent recurrences, in 1878 and 1891–92, were short,
and could not be well seen because Saturn was too near conjunction with the Sun. But about the middle of 1907 will
occur the next disappearance, with a repetition in a general
way of the phenomena of 1861–62, and with Saturn very favorably placed for observation.

Our present knowledge of the constitution of the Saturnian
rings has been attained through the researches of LA PLACE,
who showed that the rings could not be solid, because even the
attractions of the external satellites would be sufficient to rup-
ture them; of the late Professor BENJAMIN PEIRCE of Harvard
College, whose investigation of the dynamics of the rings led him

Beyond the orbit of Saturn are two large planets, Uranus and Neptune, the existence of which was not

BENJAMIN PEIRCE (1809–1880)

to the view that they must be fluid; and of CLERK MAXWELL whose researches in 1857 showed the inconclusiveness of both previously mentioned hypotheses, and established on a firm basis the present theory of a vast shoal of clustering meteors, all revolving round Saturn in orbits of their own, as if actual satellites. Particles on the inner edge of the dusky ring, then, must revolve round the planet in 5^h 50^m; while the periodic time of those making up the outer edge of the outer ring is 12^h 5^m. Probably the rings are slightly eccentric about the ball. According to H. STRUVE, the entire mass of the ring system cannot exceed $\frac{1}{314}$ that of the planet; BESSEL had, in 1831, estimated it as $\frac{1}{118}$, and M. TISSERAND in 1877, as $\frac{1}{525}$.

The meteoric constitution of the rings is farther demonstrated from what is technically termed 'ROCHE'S limit,' equal (for each planet) to 2.44 its radius; and Professor DARWIN has pointed out that no large satellite can circulate round a planet inside this limit, because the known action of tide-producing forces would tear it into small fragments. As the Saturnian ring system lies wholly inside this critical limit, even the periphery of the outer ring lying about 2,500 miles within it, the conclusion is justifiable that the rings are composed of particles too small to be disrupted by planetary tides — of 'dust, rocks, and fragments.' Secular changes in the ring system are doubtless taking place; and the observations of the middle of the 20th century may, on comparison with those of the past, suffice to disclose such slowly progressive variations: perhaps a ninth satellite is in process of formation. — *D. P. T.*

known to the ancients. Uranus was discovered by Sir WILLIAM HERSCHEL while observing at Bath, 13th March 1781, and at first supposed by him to be a slowly moving comet. Its planetary character was established as soon as the nature of its orbit became approximately known; and the principal credit of pointing this out is due to LEXELL, whose name is more generally known by connection with the comet of 1770, the remarkable story of which is told in a later chapter. HERSCHEL proposed to name the new planet Georgium Sidus, or the Georgian Star, in honor of his royal patron, GEORGE the Third, while others preferred to call it after the discoverer; but when the name Uranus was suggested, it soon obtained general acceptance. This planet takes 84 of our years to revolve round the Sun, from which it is distant about 1,782 millions of miles. It was seen several times before HERSCHEL'S discovery, but always supposed to be a fixed star; the first of these observations was by FLAMSTEED in 1690. Its mean diameter is 32,000 miles.

Some recent observations made by Professor SCHIAPARELLI at Milan and by Professor YOUNG at Princeton tend to confirm results obtained by MÄDLER at Dorpat in 1842–43, that the polar compression of Uranus is about one twelfth, or greater than that of any other planet except Saturn. Markings have been noticed on its surface which indicate that Uranus has, like Jupiter and Saturn, a rapid rotation, performed in a period of about ten hours. This rotation, however, takes place in a direction nearly perpendicular to the plane of the planet's orbit. The density of Uranus is a little less than that of Jupiter, and nearly one fourth that of the Earth; while its mass is rather less than the twentieth part of that of the 'giant planet.'

Early in 1787, Sir WILLIAM HERSCHEL discovered two satellites revolving round Uranus; he afterward thought he had discovered four more, but the existence of these has never been confirmed. Small fixed stars near the planet were probably mistaken

THE SATELLITES OF URANUS

Name of Satellite	Date of Discovery	Distance from Centre of Uranus		Sidereal Period of Revolution			
		In Miles	In Arc	d	h	m	s
			"				
Ariel	14 Sept. 1847	120,000	13.8	2	12	29	21.1
Umbriel	8 Oct. 1847	167,000	19.2	4	3	27	37.2
Titania	11 Jan. 1787	273,000	31.5	8	16	56	29.5
Oberon	11 Jan. 1787	365,000	42.1	13	11	7	6.4

for satellites. But two other satellites, additional to HERSCHEL'S first two, were discovered by LASSELL toward the end of 1851; and on removing his telescope temporarily to Malta the following year in order to profit by its transparent sky, he succeeded in determining their periods with considerable accuracy.

URANUS IN 1884
(*The Brothers* HENRY)
(*Earth on same scale, the size of a small letter 'o' in the foot-note opposite*)

These two, which are interior to HERSCHEL'S two, were named Ariel and Umbriel; while the others received the designations Titania and Oberon. It is not possible to measure accurately the sizes of these distant and comparatively small objects, but Ariel and Umbriel are estimated to be about 500 miles, and Titania and Oberon about 1,000 miles in diameter. Their motions are nearly perpendicular to the plane of the planet's orbit, and are performed in the reverse direction to that of all known planets, and of the satellites of all the planets interior to Uranus.[37]

[37] Uranus on clear, moonless nights is bright enough to be seen without a telescope; but one must know from the *Ephemeris* just where to look, as it is not far from the limit of visibility with the naked eye. Its stellar magnitude is $5\frac{1}{2}$, and does not vary greatly from opposition to conjunction. This exceeding faintness, relatively to Saturn, is partly due to the remoteness of Uranus, which is the only planet whose distance is more than double that of its next interior neighbor. The inclination of the orbit of Uranus to that of the Earth is the least of all the planets, being only $\frac{3}{4}°$. Professor YOUNG, who observed this distant member of the solar system in 1883, with the 23-inch Princeton telescope, saw faint belts upon its disk, and found its ellipticity to be $\frac{1}{14}$. Professor SCHIAPARELLI the year before had found the degree of oblateness expressed by $\frac{1}{11}$; and both series of observations indicate that the planet's equator coincides with the orbits of the satellites. M. PERROTIN, formerly of the Observatory at Nice, and his assistant, the late M. THOLLON, discovered on Uranus in 1884 a white spot, whose reappearances indicated a rotation of the planet in about ten hours, a result needing confirmation. Five years later, with the 30-inch refractor at Nice, the belts seemed only a few degrees inclined to the plane of the satellite orbits, and the oblateness was not less than $\frac{1}{20}$.

The paths of the satellites of Uranus are sensibly perfect circles; and in particular the orbits of Oberon and Titania Professor NEWCOMB (*Washington Observations* for 1873) found to be more nearly circular than in the case of any of the large planets of our system. Also these orbits have no discernible mutual inclination. The motions of the satellites lead to a mass of Uranus equal to $\frac{1}{22800}$ that of the Sun. Professor PICKERING's determinations of the diameters of the outer satellites from photometric measures in 1878 gave, for Titania 590 miles, and for Oberon 540 miles. Indications are that the inner satellites, Ariel and Umbriel, have a diameter about half as great; and probably the combined mass of the satellites does

We now come to Neptune, the most distant known planet of all. When Bouvard was forming his Tables of planetary movements, about 40 years after the discovery of Uranus, he found it impossible to reconcile the observations of that planet since its discovery with the earlier observations made at various times when it was supposed to be a fixed star. Therefore, in forming his Tables, he rejected the latter altogether, and made use of the former only; but stated in doing so that he left it to future time to determine whether the difficulty arose from inaccuracy in the older observations, or whether it depended on some extraneous and unperceived influence which might have acted on the

not exceed $\frac{1}{15000}$ that of the planet. At present the apparent orbits of the satellites of Uranus are ellipses of small eccentricity, and large telescopes will now show these bodies at every point of their paths about the planet. In 1903 the Earth will be near the pole of these orbits, so that they will appear almost perfectly circular. The outer, or Herschelian satellites have been seen under perfect atmospheric conditions with a 6-inch telescope; but the inner ones, Ariel and Umbriel, require for their certain observation the largest of telescopes and the keenest of eyes. Variability in the light of Ariel is suspected.

The spectrum of Uranus has been critically surveyed by Dr Huggins with the assistance of photography, and by Dr Vogel, Professor Keeler, and Mr Taylor. Professor Keeler's optical observations of the spectrum of Uranus with the 36-inch Lick telescope give ten broad diffused bands, from C to F, indicating strong absorption by a dense atmosphere differing greatly from ours; and the presence of these bands accounts for the invariable sea-green tint of the planet. The substance producing the absorption is not yet identified, and the darkest band, about midway between C and D, is also shown in exactly the same position in the spectra of Jupiter and Saturn. So great is the light-gathering power of the Lick glass that even the outer satellites of Uranus, faint as they are, gave spectra which Professor Keeler could just recognize as continuous. — *D. P. T.*

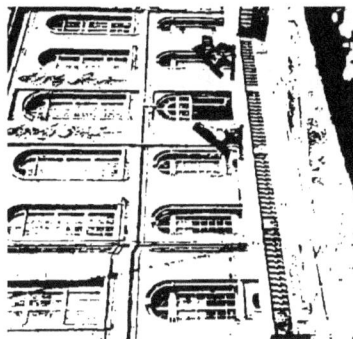

planet. The question as to the nature of the cause was soon settled by the rapidly increasing deviation of Uranus from the course marked out for it by the Tables based upon observations between 1781 and 1821; nor could it reasonably be doubted that the cause was the perturbing attraction of a planet still farther from the Sun, and never (so far as was known) hitherto observed.

LE VERRIER (1811–1877)

But the problem of calculating the place of an unknown planet by simple knowledge of its influence upon another, staggered the few mathematical astronomers who were capable of attacking it, most of whom were already deeply engaged in labors too essential and exacting to bear an interruption of the length which its solution would seemingly demand. In 1843–45, however, it was taken up by two young mathematicians, one in England the other in France, whose names are now famous wherever astronomy is studied. The former, JOHN COUCH ADAMS, for a long time Professor of Astronomy at Cambridge, and

ADAMS (1819–1892)

Director of the Observatory there, died 21st January 1892 ; the other, URBAIN JEAN JOSEPH LE VERRIER, was for many years Director of the National Observatory at Paris, where he died 23rd September 1877. We cannot enter here into the history of their recondite investigations ; the approximate place of the unknown planet was carefully determined by both independently, and CHALLIS at Cambridge began searching for it by its motion, 29th July 1846. During several weeks he mapped down all the stars visible in a considerable tract of the heavens around the place indicated by ADAMS, with the intention of comparing afterward the places of all these stars and ascertaining which of them had moved; and he thus obtained several positions of the planet. But before he had completed his charts in this way, news arrived that GALLE at Berlin had discovered the planet, 23rd September 1846, on looking for it in the place pointed out by LE VERRIER, who had suggested that the planet might be distinguished by its disk among the fixed stars in the neighborhood of the calculated place. Its appearance at once showed that it was the object of search; also it was wanting in a map (then recently made by BREMIKER) of the stars in that part of the heavens, and which had just been received at the Berlin Observatory. The next evening, 24th September, the alteration in its place put its planetary character beyond a doubt. Subsequently the name ' Neptune ' (one of the names suggested for Uranus when a designation for that planet was under discussion) was by common consent adopted.

Neptune occupies $164\frac{3}{4}$ of our years in revolving round the Sun, at a distance of 2,792 millions of miles, about 30 times that of the Earth. The eccentricity of its orbit is the smallest of those of all the principal planets, with the single exception of Venus. The diameter of Neptune is about 34,800 miles, so that its volume or bulk is 85 times that of the Earth, but only about $\frac{1}{15}$ that of Jupiter. Its density is nearly the same as that of Uranus, and its mass is to that of the Sun in about the proportion of 1 to 19,380. Being at so great a distance, it has not yet been found possible to perceive any spots or markings on the surface ; so that its time of axial rotation is unknown to us. Like the other large planets, it probably rotates much more rapidly than the Earth.

Only one satellite of Neptune is known. This was discovered by LASSELL, 10th October 1846, very soon after the discovery of the planet itself. Being visible at so great a distance, it is probably the largest satellite in the solar system. Like the satellites of Uranus, it moves in a reverse direction to that of the planetary motions, and its orbit is very much inclined to those of the planets (about 55° to that of Neptune itself). Its distance from the centre of Neptune is 225,000 miles, and its time of revolution is 5ᵈ 21ʰ 2ᵐ 38ˢ.9. If this planet has other satellites, as is possible, they must be much smaller, since none have hitherto been detected with the powerful telescopes which have been brought into use in recent years.[88]

[88] So remote is Neptune that its disk, although more than 4½ times the actual diameter of the Earth, shrinks to an angle no greater than that which a nickel 5-cent piece would fill, if held up a mile distant. According to Professor PICKERING, its stellar magnitude is 7.63, so that it can never be visible to the unaided eye.

Portraits of both the theoretical discoverers of Neptune have been given on pages 142 and 143. For the fullest account of their great discovery it is necessary to consult the original papers; but Dʳ GOULD has given a complete résumé in his *Report on the History of the Discovery of Neptune* (Washington 1850). As in the case of Uranus, so with Neptune, it was soon found that the new planet had been accidentally observed as a star (by LALANDE in 1795), so that observations of Neptune now extend through more than a century, or an arc of about 225° round the Sun as a centre. Among American astronomers who performed a zealous and important part in the researches on Neptune's orbit, immediately after its discovery, must be mentioned PEIRCE and WALKER; and in 1866 were published Professor NEWCOMB'S *Tables of Neptune*, which provided a good representation of the planet's motion for about 15 years. In 1877, LE VERRIER'S *Tables* appeared, founded upon many years more observations than the preceding; but already, in less than 25 years, the planet is beginning to deviate widely from their prediction, and the theory of Neptune's motion is at the present time undergoing a second revision by Professor NEWCOMB. Neptune's path round the Sun, like that of Jupiter, is but slightly removed from the invariable plane of the solar system, this plane in fact lying about midway between the orbits of Neptune and Jupiter.

Probably the sensibly circular path of Neptune's satellite is inclined by a large angle to the planet's equator; and the motion of its orbit plane, as established by observation, has been accounted for by TISSERAND, who supposes an oblateness of Neptune equal to about $\frac{1}{85}$, the equatorial protuberance causing the orbit plane to partake of a regressive motion round the pole of the planet in a period exceeding 500 years. At the present time the path which the satellite seems to describe about its primary is so open that this minute object is at no time overpowered by the planet's rays; any good telescope above 12 inches of aperture will show it. The diameter of Neptune's satellite, as determined by Professor PICKERING from photometric measures in 1878, is 2260 miles. Critical search for an additional satellite, both optically

F. F. TISSERAND (1845–1896)

and by means of photography, has shown that there is no such body, unless its dimensions are very minute. Neptune's spectrum closely resembles that of Uranus, and probably the planet is surrounded by a dense atmosphere. — *D. P. T.*

The extensive popular literature of the planets can only be indicated here by tabulated reference to the pages of POOLE'S *Indexes* to periodical literature : —

Subject of Reference	POOLE'S *Index to Periodical Literature*				FLETCHER'S *Index to General Literature* (Boston 1893)	Annual Literary *Index for*		
	Vol. 1 1802–81	1st Supplement 1882–86	2d Supplement 1887–91	3d Supplement 1892–96		1897	1898	1899
Planets	p. 1014	p. 343	p. 333	p. 445	p. 223	p. 100	p. 104	
Mercury	824	283	274	. . .	. . .	82	. . .	
Venus	1366–7	459	448	602	299	135	140	
Mars	804	277–8	269	357	184	81	. . .	
Asteroids	69	25	24	33	223	. . .	104	
Jupiter	699	239	233	306	157	69	71	
Saturn	1152	386	375–6	503	253	113	116	
Uranus	1359	456	446	. . .	297	. . .	. . .	
Neptune	906	309	. . .	. . .	223	. . .	. . .	

A thorough study of the planets will be greatly facilitated by reference to the following important papers, for the most part original sources. These lists need, however, to be supplemented by reference to current numbers of *The Astrophysical Journal*, 1895–1898 : —

INTRA-MERCURIAN PLANETS

LE VERRIER and LESCARBAULT, *Comptes Rendus*, xlix. and l. (1859–60). Also LE VERRIER, *Ibid.*, 1876–78.

WOLF, *Handbuch der Mathematik*, ii. (Zürich 1872), 326.

WATSON, *Am. Jour. Science and Arts*, cxvi. (1878), 230, 310.

SWIFT, *American Journal of Science and Arts*, cxvi. (1878), 313.

PETERS, *Astronomische Nachrichten*, xciv. (1879), 321, 337.

LEDGER, *Lecture on Intramercurian Planets* (Cambridge 1879).

TISSERAND, *Annuaire du Bureau des Longitudes* for 1882, p. 729.

NEWCOMB, *Astron. Papers of Am. Ephemeris*, i. (1882), 474.

HOUZEAU and LANCASTER, *Bibliographie Générale de l'Astronomie*, ii. (Bruxelles 1882), 1090, contains nearly 150 titles.

BAUSCHINGER, *Bewegung des Planeten Merkur* (Munich 1884).

CHAMBERS, ' Vulcan,' *Astronomy*, i. (Oxford 1889), p. 75.

CORRIGAN, *Popular Astron.* iv. (1897), 414; v. (1897).

MERCURY

SCHROETER, *Hermographische Fragmente* (Göttingen 1815–16).
LE VERRIER, 'Tables of its Motion,' *Annales de l'Observatoire de Paris, Mémoires* v. (1859).
VOGEL, *Bothkamp Beobachtungen*, ii. (1873), 133.
HARKNESS and others, 'Transit 1878,' *Wash. Obs.* 1876, App. ii.
HOLDEN, *Index of . . . Transits of Mercury* (Cambridge 1878).
TODD, 'Satellite,' *Proc. A. A. A. S.* xxviii. (1879), 74.
NIESTEN, 'Transits 1600–2000,' *Annuaire de l'Obs. Roy. Bruxelles*, 1881, p. 192.
SCHIAPARELLI, *Atti dell' Accad. dei Lincei*, v. (1889), ii.; *Pop. Sci. Monthly*, xxxvii. (1890), 64; *Astron. Nachr.* cxxiii. (1890), 241.
CLERKE, 'Rotation,' *Jour. Brit. Astron. Assoc.* i. (1890), 20.
HARZER, 'Motion perihelion,' *Astron. Nachr.* cxxvii. (1891), 81.
AMBRONN, 'Diameter,' *Astron. Nachr.* cxxvii. (1891), 157.
TROUVELOT, *Les Planètes Vénus et Mercure* (Paris 1892).
MÜLLER, 'Magnitudes,' *Publ. Obs. Potsdam*, viii. (1893).
NEWCOMB, *Elements of the Four Inner Planets and the Fundamental Constants of Astronomy* (Washington 1895); 'Small Planets between Mercury and Venus,' p. 116.
LOWELL, *The Atlantic Monthly*, lxxix. (1897), 493.
LOWELL, *Popular Astronomy*, iv. (1897), 360.
LOWELL, 'New Observations,' *Mem. Am. Acad.* xii. (1898), 433.
VILLIGER, *Ann. k. Sternwarte*, iii. 301 (Munich 1898).

VENUS

SCHROETER, *Aphroditographische Fragmente* (Helmstedt 1796).
LE VERRIER, 'Tables of its Motion,' *Annales de l'Observatoire de Paris, Mémoires*, vi. (1861).
LYMAN, C. S., *Am. Jour. Sci.* xliii. (1867), 129; ix. (1875), 47.
KAISER, 'Diameters of Planets,' *Leyden Observations*, iii. (1872).
SAFARIK, *Report British Assoc. Adv. Sci.*, 1873, p. 404.
VOGEL, *Bothkamp Beobachtungen*, ii. (1873), 118.
PROCTOR, *The Universe and the Coming Transits* (London 1874).
FORBES, *Transits of Venus* (London 1874).
SCHORR, *Der Venusmond* (Brunswick 1875).
HARTWIG, *Publ. Astron. Gesellschaft*, xv. (Leipzig 1879).
PERROTIN, *Comptes Rendus*, cxi. (1890), 587.
TERBY, *Bulletins Acad. Royale de Belgique*, xx. (Brussels 1890).
SCHIAPARELLI, *Rend. del R. Ist. Lombardo*, xxiii. (1890), ii.
CLERKE, A. M., *Jour. Brit. Astron. Assocn.* i. (1890), 20.

NEWCOMB, ' Transits,' *Ast. Papers Am. Eph.* ii. (1890), 259.
AUWERS, ' Diameter,' *Astron. Nachr.* cxxviii. (1891), 361.
LÖSCHARDT, *Sitz. k. Akad. Wiss.*, c. (Vienna 1891), 537.
NIESTEN, *Rotation de la Planète Vénus* (Brussels 1891).
TROUVELOT, *Nature*, xlvi. (1892), 468.
CLERKE, E. M., *The Planet Venus* (London 1893).
FLAMMARION, *Comptes Rendus*, cxix. (1894), 670.
BRENNER, *Astronomische Nachrichten*, cxxxviii. (1895), 197.
SCHIAPARELLI, *Astronomische Nachrichten*, cxxxviii. (1895), 249.
VOGEL, ' Planetary Spectra,' *Astrophys. Jour.* i. (1895), 196, 273
LOWELL, *Popular Astronomy*, iv. (1896), 281.
ANTONIADI, ' Rotation,' *Jour. Brit. Astr. Assoc.* viii. (1897), 43.
LOWELL, *The Atlantic Monthly*, lxxix. (1897), 327.
FLAMMARION, *Knowledge*, xx. (1897), 234, 258.
CHANDLER, ' Rotation,' *Popular Astronomy*, iv. (1897), 393.
STONEY, ' Atmospheres,' *Astrophysical Journal*, vii. (1898), 25.
DOUGLASS, ' Markings,' *M. N. Roy. Astron. Soc.* lviii. (1898), 382.

MARS

For the literature of this planet see page 178.

THE SMALL PLANETS

GOULD, *The American Journal of Science*, vi. (1848), 28.
D'ARREST, *Ueber das System der kleinen Planeten* (Leipzig 1851).
NEWCOMB, *Mem. Am. Acad. Arts and Sciences*, viii. (1861), 123.
STONE, E. J., *M. N. Royal Astron. Society*, xxvii. (1867), 302.
CALLANDREAU, *Revue Scientifique*, xviii. (1880), 829.
NIESTEN, *Annuaire de l'Observatoire Royal de Bruxelles* for 1881.
SVEDSTRUP, *Astronomische Nachrichten*, cxv. (1886), 49.
PARKHURST, *Annals Harvard College Observatory*, xviii. (1888).
LIAIS and CRULS, *Annales de l'Obs. de Rio de Janeiro*, iv. (1889).
SCHMIDT, R., *Publ. Astron. Society Pacific*, ii. (1890), 238.
TISSERAND, *Popular Science Monthly*, xxxix. (1891), 195.
KIRKWOOD, *Proc. Am. Phil. Society*, xxx. (1892), 269.
BACKLUND, *Mém. Acad. Sci. St Pétersbourg*, xxxviii. (1892).
TODD, *Astronomy and Astrophysics*, xii. (1893), 313.
CALLANDREAU, *Comptes Rendus*, cxviii. (1894), 751.
ROSZEL, *J. H. Univ. Circ.* xiii. (1894), 67 : xiv. (1895), 23.
CHARLOIS, *Bulletin Astron.* xi. (1894) ; xiii. (1896).
ZENGER, *Bulletin Société Astronomique de France* (1895), 243.
MASCART, ' Small Planets,' *Bulletin Astron.* xv. (1898), 235.
CROMMELIN, ' Planet D Q,' *Knowledge*, xxi. (1898), 250.

JUPITER

SOMERVILLE, *Mechanism of the Heavens* (London 1831).

DE DAMOISEAU, *Tables Écliptiques Satellites* (Paris 1836).

ENGELMANN, *Helligkeitsverhältnisse der Jupiterstrabanten* (Leipzig 1871).

LOHSE, *Bothkamp Beobachtungen*, i. (1872); ii. (1873). Also *Publ. Astrophys. Observ. Potsdam*, i. (1879); iii. (1882).

MARY SOMERVILLE (1780–1872)

GLASENAPP, *Сравненіе Наблюденій Затмьній Спутниковъ Юпитера (С.-Петербургъ 1874).*

DOWNING, Abstract of foregoing, *Observatory*, xii. (1889), 173.

ROSSE, *Monthly Notices Royal Astronomical Society*, xxxiv. (1874).

BREDICHIN, *Ann. de l'Observatoire de Moscou*, ii.–vi. (1875–80).

TODD, *Continuation of* DE DAMOISEAU'S *Tables Écliptiques to* 1900 (Washington 1876).

LE VERRIER, 'Tables of its Motion,' *Annales de l'Observatoire de Paris, Mémoires*, xii. (1876).

DENNING, *Science for All*, part xxx. (London 1880), 169.
SOUILLART, *Memoirs Royal Astronomical Society*, xlv. (1880). 1.
TROUVELOT, *Proc. Am. Acad. Arts and Sciences*, viii. (1881), 299.
KEMPF, ' Mass,' *Publ. Astrophys. Observ. Potsdam*, iii. (1882).
DENNING, ' Summary of Markings and Rotation-periods,' *Nature*, xxxii. (1885), 31; *Observatory*, ix. (1886), 188; xi. (1888), 88, 406; xiv. (1891), 329.
WILLIAMS, *The Observatory*, ix. (1886), 231.
LAMEY, *Comptes Rendus*, civ. (1887), 279, 613.
SOUILLART, *Mém. Académie des Sciences*, xxx. (Paris 1887).
BACKLUND, ' Satellites,' *Bulletin Astronomique*, iv. (1887), 321.
BARNARD, *Publ. Astron. Society Pacific*, i. (1889), 89.
LANDERER, *Estudios geometricos sobre el sistema de los satélites*, (Barcelona 1889). *Comptes Rendus*, cxiv. (1892), 899.
WILLIAMS, *Zenographic Fragments* (1889).
WAUGH, *Jour. Brit. Astron. Association*, i. (1890).
KEELER, *Publications Astron. Society Pacific*, ii. (1890), 286.
HILL, ' A New Theory of Jupiter and Saturn,' *Astron. Papers Am. Ephemeris*, iv. (Washington 1890).
SCHAEBERLE, CAMPBELL, *Pub. Ast. Soc. Pacific*, iii. (1891), 359.
CLERKE, E. M., *Jupiter and his System* (London 1892).
SCHUR, *Astronomische Nachrichten*, cxxix. (1892), 9.
TISSERAND, ' Fifth Satellite,' *Comptes Rendus*, cxvii. (1893), 1024.
PICKERING, W. H., *Ast. and Astro-Phys.* xii.–xiii. (1893–94).
WILLIAMS, FREEMAN, and others, *Mem. Brit. Astron. Assoc.* i. (1893), 73; ii. (1894), 129.
HOUGH, ' Constitution of Jupiter,' *Astronomy and Astro-Physics*, xiii. (1894), 89; *Popular Astronomy*, ii. (1894), 145.
BARNARD, *Astron. and Astrophysics*, xiii. (1894).
MAUNDER, GREEN, *Knowledge*, xix. (1896), 4, 5.
POTTIER, ' Satellites,' *Bulletin Astronomique*, xiii. (1896), 67, 107.
FLAMMARION, *Bulletin Soc. Astron. France*, July 1896.
BÉLOPOLSKY, ' Rotation,' *Astron. Nach.* cxxxix. (1896), 209.
DOUGLASS, ' Satellite III,' *Pop. Astron.* v. (1897). 308.
BRENNER, ' Observations 1895–96,' *Vienna Acad. Sci.* 1897.
DENNING, ' Red Spot,' *Nature*, lviii. (1898), 331.
WILLIAMS, ' Rotation,' *M. N. Roy. Ast. Soc.* lviii. (1898), 10.
DENNING, *M. N. Roy. Ast. Soc.* lviii. (1898), 480, 488.

SATURN

BOND, G. P., ' Rings of Saturn,' *Am. Jour. of Sci.* lxii. (1851), 97.
DAWES, *American Journal of Science*, lxxi. (1856), 158.

WATSON, ' Ring.' *Mo. Not. Roy. Astr. Soc.* xvi. (1856), 152.
MAXWELL, *Stability of Saturn's Rings* (Cambridge 1859).
PROCTOR, *Saturn and its System* (London 1865).
PEIRCE, ' Saturnian System,' *Mem. Nat. Acad. Sciences,* i. (1866).
HIRN, *Anneaux de Saturne* (Colmar 1872).
BESSEL, *Abhandlungen,* i. (Leipzig 1875), pp. 110, 150, 319.
TROUVELOT, *Proceedings Am. Acad.* iii. (1876), 174.
LE VERRIER, ' Tables of its Motion,' *Annales de l'Observatoire de Paris, Mémoires,* xii. (1876).
PICKERING, E. C., *Annals Harvard Observatory,* xi. (1879), 269.
HALL, ' Six inner Satellites,' *Washington Observations* 1883.

TROUVELOT, *Bulletin Astron.* i. (1884), 527 ; ii. (1885), 15.
BAILLAUD, *Bulletin Astronomique,* i. (1884), 161 ; ii. (1885), 118.
MEYER, *Le Système de Saturne* (Geneva 1884).
POINCARÉ, *Bulletin Astronomique,* ii. (1885).
SEELIGER, *Mém. de l'Acad. Sciences de Bavière* (Munich 1887).
GORE, J. E., *Planetary and Stellar Studies* (London 1888).
ELGER, *Month. Notices Royal Astron. Society,* xlviii. (1888), 362.

WATSON (1838–1880)

PERROTIN, ' Rings,' *Comptes Rendus,* cvi. (1888), 1716.
STRUVE, H., *Observations de Poulkova, Supplément,* i. (1888); ii. (1892). Also *M. N. Roy. Astron. Society,* liv. (1894), 452.
TISSERAND, *Annales de l'Observatoire de Toulouse,* ii. (1889) ; *Bulletin Astronomique,* vi. (1889), 383, 417.
HALL, A., jr., *Trans. Obs. Yale Univ.* i. (1889), 107.
DARWIN, ' The Rings,' *Harper's Magazine,* lxxix. (1889), 66.
ANDING, *Astronomische Nachrichten,* cxxi. (1889), 1.
OUDEMANS, *M. N. R. A. S.* xlix. (1889), 54 ; *Astron. Nach.* 3074.
STROOBANT, *Bulletins de l'Académie Royale Belgique,* xix. (1890).
TROUVELOT, *Bulletin Astronomique,* vii. (1890), 147, 185.
NEWCOMB, *Astron. Papers Am. Ephemeris,* iii. (1891), 345.
STRUVE, H., *Bulletin Acad. Sci. St Pétersbourg,* ii. (1891).
EICHELBERGER, *The Astronomical Journal,* xi. (1892), 145.
GREEN, FREEMAN, *Mem. Brit. Astron. Assoc.* ii. (1893), 1.

WILLIAMS, *Month. Not. Roy. Astron. Society*, liv. (1894), 297.
OLBERS, ' Ring,' *Sein Leben und Seine Werke* (Berlin 1894).
PRITCHETT, *Ueber die Verfinsterungen der Saturntrabanten* (Munich 1895).
BUCHHOLZ, ' Eclipse of Japetus,' *Astron. Nach.* cxxxvii. (1895).
KEELER, 'Ring Spectra,' *Astrophys. Jour.* i. (1895), 416; ii. 63.

OLBERS (1758–1840)

WITT, *Himmel und Erde*, ix. (1896), 75, 121.
BARNARD, *Pop. Astron.* v. (1897), 285; *M. N. R. A. S.* lvi. (1896), 14; lviii. (1898), 217.

URANUS

NEWCOMB, 'Orbit of Uranus,' *Smithsonian Contributions to Knowledge*, xviii. (1872).
NEWCOMB, 'The Uranian and Neptunian Systems,' *Washington Observations* for 1873, Appendix I.
LE VERRIER, 'Tables of its Motion,' *Annaies de l'Observatoire de Paris, Mémoires*, xiv. (1877).
SCHIAPARELLI, *Astronomische Nachrichten*, cvi. (1883), 81.

YOUNG, *Astronomische Nachrichten*, cvii. (1883), 9.

PERROTIN, *Comptes Rendus*, xcviii. (1884), 718, 967.

HENRY, *Comptes Rendus*, xcviii. (1884), 1419.

GREGORY, 'Uranus,' *Nature*, xl. (1889), 235.

LOCKYER, 'Spectrum,' *Astron. Nach.* cxxi. (1889), 369.

HUGGINS, W. and M. L., *Proceedings Royal Society*, xlvi. (1889), 231; also *M. N. Royal Astronomical Society*, xlix. (1889), 404.

KEELER, *Astronomische Nachrichten*, cxxii. (1889), 401.

TAYLOR, 'Spectrum,' *M. N. Royal Astron. Soc.* xlix. (1889), 405.

BARNARD, *The Astronomical Journal*, xvi. (1896).

NEPTUNE

LOOMIS, 'History of its Discovery,' *Am. Jour. Sci.* lv. (1848), 187. Also *Recent Progress of Astronomy* (New York 1850).

GOULD, *History of the Discovery* (Washington 1850).

HERSCHEL, J. F. W., 'Perturbation of Uranus by Neptune,' *Outlines of Astronomy* (London 1865), p. 533.

NEWCOMB, 'Tables of its Motion,' *Smithsonian Contributions to Knowledge*, xv. (Washington 1867).

LE VERRIER, 'Tables of its Motion,' *Annales de l'Observatoire de Paris, Mémoires*, xiv. (1877).

PEIRCE, *Ideality in the Physical Sciences* (Boston 1881), App. B.

TISSERAND, 'Satellite,' *Comptes Rendus*, cxviii. (1894), 1372.

STRUVE, H., 'Satellite,' *Mém. de l'Acad. Sciences*, xlii. (St. Petersburg 1894).

LIAIS, 'Historia Descubrimiento,' *Anuario del Observatorio Astronómico Nacional de Tacubaya para el Año de* 1895, p. 247.

ADAMS, *Scientific Papers* (Cambridge, Eng. 1896), i.

NEWTON'S BIRTHPLACE AT WOOLSTHORPE, LINCOLNSHIRE

(On one end of the house are shown the sundials NEWTON made when a boy)

CHAPTER XI

THE RUDDY PLANET

MARS, the earliest observation of which, B. C. 356, is its occultation by the Moon recorded by ARISTOTLE, was first scrutinized with the telescope by GALILEO, who discovered the planet's phases in 1610. The earliest sketch of its surface was made by FONTANA, in 1636, though little or no detail was made out until 1659, when HUYGENS observed its markings clearly enough to show him that the period of rotation of Mars was about the same as that of the Earth. Seven years later CASSINI discovered the well-known polar caps, and made the first determination of the axial period, $24^h 40^m$, now known to be only $\frac{1}{800}$ part in error. Among other prominent observers of the 17th century were RICCIOLI, HOOKE. and HEVELIUS; and in the _. '- BIANCHINI, SCHROETER, and Sir WILLIAM HERSCHEL, who, beginning in 1777, first detected, in 1783, the fluctuating dimensions of the polar caps with the Martian seasons, determined the inclination of the planet's axis to its orbit, and measured the oblateness of Mars.

ARAGO (1786-1853)

Early in the 19th century came FLAUGERGUES, HARDING, and ARAGO who observed the planet for 36 years; and in 1830–40, BEER and MAEDLER, who drew the first map of Mars, with the markings then known (among them *Lacus Phœnicis*) carefully set down in Arean longitude and latitude. Since their day Martian investigations have rapidly increased in fulness and importance, the chief observers down to the very favorable opposition of 1862 being Sir JOHN HERSCHEL, KAISER, SECCHI who studied the colors of different regions, Lord ROSSE, LASSELL, and LOCKYER whose fine sketches inaugurated the modern standard of excellence.   From that time onward, there have been critical observations by DAWES, KNOBEL, TROUVELOT, GREEN, TERBY, NIESTEN, LOHSE, and others, culminating in the classic labors of SCHIAPARELLI, begun at the memorable oppositions of the planet in 1877 and 1879.   So abundant and careful have been the drawings of this planet in the past that many excellent maps of its entire surface have been made, by KAISER (1864), PROCTOR (1867), GREEN (1877), DREYER (1879), and in particular by SCHIAPARELLI (1888), supplanting all others, from his elaborate sketches with the 18-inch Merz refractor of the Brera Observatory at Milan.   Dr WISLICENUS of Strassburg has done excellent service by applying the micrometer to the more conspicuous markings, thereby establishing their accurate positions; and it is no exaggeration to say that the areography of Mars is now better known than the geography of immense tracts of our own planet.

Mars, although the nearest to the Earth of all the planets, Venus alone excepted, is an object by no means easy to observe.   To realize the extent of the difficulty, even under the best imaginable conditions of instrument and atmosphere, let any one with no previous knowledge of the Moon attempt to settle precise markings, colors, and the nature of objects on our satellite by simple scrutiny with an ordinary opera-glass. Indeed, the Moon would, for many technical reasons, prove the less perplexing, although the two cases would be quite parallel in so far as mere geometry is concerned.

SKETCHES AND VARIATIONS

Personal differences affect all sketches of Mars unduly, and the delicate and changing local colors add much uncertainty. Still, the leading features of the planet's surface are well made out, and their stability leaves no room to doubt their reality as permanent planetary crust.   Here, unfortunately, the great advance in photographic application to astronomical requirements has not yet helped much, although excellent plates have been obtained by Professor PICKERING, and at the Lick Observatory as well: the texture of the sensitized film is too coarse, and the image of the planet is too small and faint.   The border of its disk is brighter than the interior, as the Lick photographs show;

but this brightness is far from uniform, and the variation is probably caused by the surface features of the planet. Also variations of color in the markings of Mars depend upon the diurnal rotation of the planet and the angle of vision ; and changes of apparent brightness in certain regions are as well established as in the case of the Moon. *Argyre I*, for example, first observed by DAWES. in 1852, has often been seen by SCHIAPARELLI to be reddish yellow when near the central meridian, while again, after the planet's rotation has carried it round to the limb, it has appeared so brilliant as to be mistaken at first for a polar cap. This skilful observer has also noticed similar changes, progressive in character, covering extensive continental regions. A small white spot, *Nix Atlantica*, first observed by him in 1877, then almost square, and more brilliant than any other part of the planet, became at subsequent oppositions nearly round, fluctuating in appearance and brilliancy, and in 1888 had entirely disappeared.

KAISER (1808-1872)

Mindful of these obstacles in our own day, it is easy to see why the early telescopists failed to discern very much : all had not only that ever-present foe of the astronomer, the atmosphere, to battle against, but the imperfections of their instruments were a farther and most serious handicap. Formerly observations of a near approach of Mars to the Earth would be made only from localities where observatories were previously established ; but the exceeding present interest in our neighboring planet led to the building and maintenance of an observatory with a large and perfect telescope and a corps of astronomers at a mountain elevation in a clear and steady atmosphere, with the especial intent of a critical survey of the ruddy planet, during its recent and very favorable presentation in 1894–95. Frequent allusion will be made in this chapter to the excellent work of Mr PERCIVAL LOWELL (at the observatory bearing his name, and temporarily located at Flagstaff, Arizona, elevation 7250 feet above sea level), and of his coadjutors, Professor W. H. PICKERING and Mr DOUGLASS.

GENERAL TOPOGRAPHY

The most casual observer of Mars would not fail to notice the striking difference between the brightness of the two hemispheres—the northern chiefly bright, and the southern pronouncedly dark. In fact, the light and dark regions are each approximately a hemisphere in area. Allowing that the cosmic conditions of Mars are still suitable for the existence of water and an atmosphere, though of no great density, it is an easy surmise, and perhaps the true one, that the lighter regions of the planet's face disclosed by the telescope are land and the darker water. The southern hemisphere of Mars, then, is principally water, and the northern continental land, precisely as in the case of the Earth (page 24); but while, on our globe, the proportion of land and water is about four to eleven, the surface of Mars shows water and land in very nearly equal amounts, with a slight preponderance of the latter.

Mars, in its general topography, presents no analogy with the present scattering of land and water on the Earth. Fully one third its surface is the great *Mare Australe*, strewn with many islands, and apparently intersecting the continents by numerous divisions, narrowing northward and suggesting gulfs. There seems little reason to doubt that the northern regions, with their predominant orange tint, in some places a dark red and in others fading to yellow and white, are really continental; and Professor SCHIAPARELLI thinks that the Arean continents appear red and yellow because they are so in fact, atmospheric action having little or no influence upon their hue. Different from the Earth, the lands of Mars are mostly massed in a single vast continent, spotted with the great lakes *Mare Acidalium* and *Lacus Niliacus*, the former of which is found to be variable in area. Other extensive regions, for example, the islands of *Mare Australe* and *Mare Erythraeum*, sometimes appearing yellow like continents, and again dark brown, are thought by SCHIAPARELLI and NIESTEN to be vast marshes, the varying depth of water causing the diversity of color.

The seas of Mars, generally brown mixed with gray, also vary in intensity from light gray to deep black; it seems probable that only the darkest of dark regions may actually be deep water. The true aquatic character of the suspected seas would best be established by watching for the reflection of the solar image from them; and M. FLAMMARION has calculated that under favorable circumstances the Sun, thus reflected, ought to

shine as a star of the third magnitude ; but no such phenomenon has yet been observed. Professor PICKERING'S latest observations incline him to the opinion that the permanent water
area upon Mars, if it exists at all, is extremely limited.

AXIS AND POLAR CAPS

The axis of Mars pierces the northern heavens about midway between the two bright stars α Cephei and α Cygni
(Deneb), and no precessional motion of the pole is yet made
out. The waxing and waning of the polar caps, being largest
near the end of the Martian winter and smallest near the end of
summer, are phenomena which have long been well established.
A brilliant white, they are generally thought to be composed
of snow and ice ; and their existence is a most convincing
argument for the reality of a Martian atmosphere, capable of
transporting and diffusing vapor.

The northern cap is centred on the north pole of the planet
almost exactly; and its gradual melting, observed by SCHIAPA-
RELLI in 1882, 1884, and 1886, appeared to create a temporary
zonal ocean, from which radiated many darkish streaks, con-
necting southward with numerous round patches, called 'lakes'
by him, and 'oases' by Mr LOWELL. Our knowledge of Mars
has mainly been derived from a study of its surface at the
very favorable oppositions, about 15 years apart, when the
planet was near its perihelion, and consequently near its least
distance of 34 million miles. Always at those times the
inclination of the planet's axis brings the south pole into
view, while the northern is turned from us. At the intermediate oppositions, when the planet's north pole is best
visible, Mars is not far from aphelion, and least favorably situate, in point of distance ; there is, however, a compensation in
that the great telescopes of the world, located for the most part
in northern latitudes, are better placed for observations of the
planet. But its apparent diameter being so much reduced, and
the difficulties of studying its surface so greatly augmented, relatively few observers have made a study of so unpromising an
object as Mars on these occasions, when its distance usually
exceeds 60 million miles ; and our knowledge of the northern
hemisphere is proportionally incomplete.

Mr KNOBEL, among others, undertook to obliterate this inequality during the opposition of 1884, and three of his sketches
during that period are reproduced here. No irregularity in

the polar spot was detected. The next favorable oppositions for observing its north polar regions will occur in 1899, 1901, and 1903, when, with the planet's pole tilted toward the Earth, it is especially desirable that the phenomena of the melting of the northern polar cap be closely watched by experienced observers with the most powerful telescopes.

Curiously, while the north polar snows cap the exact pole quite concentrically, and as far as the 85th parallel of latitude all around, the south circumpolar snows do not centre about the pole, but round a point now nearly 200 miles removed from

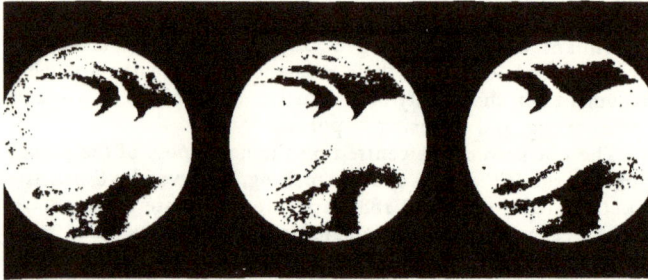

11th February   11th February   10th February

MARS AS DRAWN BY KNOBEL IN 1884

(*North polar cap near its greatest visibility*)

the planet's true pole, toward *Mare Erythraeum*. The well-established variation in the distance of the centre of the south cap from the true pole is unaccounted for; it was as much as 8° in the elder HERSCHEL'S time (1783), and even 20° as measured by LINSSER in 1862, but only 3° at the opposition in 1892, according to Professor COMSTOCK, whose observations very clearly bring out not only the rapid shrinking of the snow-cap — its diameter diminishing about fifteen miles daily — but a progressive shifting of its centre, which would seem to show an unsymmetrical shrinking. At the beginning of the Martian summer the polar cap reached down to latitude 70°, and its diameter was 1200 miles. In three months it had shrunk to one seventh this diameter. Probably the north cap was all the while increasing, but this pole was turned away from the Earth, and consequently invisible.

All the changing conditions of 1892 were most scrupulously observed by Professor PICKERING, then located at the Harvard Observatory in the Andes of Peru (frontispiece), where, with a 13-inch telescope he had much better opportunity than any other astronomer for observing the planet at its nearest approach to the Earth. Mars was faithfully followed on every night but one, from 13th July to 9th September, and the apparent alterations in the polar cap, even from night to night, were very marked. As the snow began to decrease, a long dark line made its appearance near the middle of the cap, and gradually grew until it cut the cap in two. This white polar area (and probably also the northern one), with the progress of its summer season becomes notched on its edge, dark interior spots and fissures form, isolated patches separate from the principal mass, and later seem to dissolve and disappear. It is much as if one were located at the distance of Mars and viewing with a telescope the effect of the advancing summer season upon the ice and snow of circumpolar regions of a nearly cloudless Earth.

Evidently the fluctuations of the polar caps are the key to the physiographic situation on Mars, and at the opposition of 1894, as well as that of 1892, the southern one has been critically scrutinized. The south pole reached its maximum dip of 24° toward the Earth on 22d June 1894. The rate of its progressive diminution is by no means constant, and its area, sometimes increasing and again diminishing, is at other times intersected by one or more dark narrow markings; naturally interpreted as heavy falls of snow in the one case, and in the other as the melting of snow in the polar valleys. These fluctuations were subjected to exact measurement by M. FLAMMARION and others; and the average rate of diminution was not very different from that found by Professor COMSTOCK in 1892. One 'great rift' of the polar cap, very conspicuous, Mᵣ LOWELL compared to the appearance of 'a huge cart-track coming down to one over the snow.' Its breadth he estimates at 220 miles, and length 1,200. Surrounding this polar cap is a narrow, dark line, nearly uniform in breadth, which is, Mᵣ LOWELL surmises, 'clearly water at the edge of the melting snow — a polar sea, in short.' Concerning the dark line dividing the polar cap, which made its appearance in the same place at the corresponding Martian season in 1894, he farther says: 'It started apparently not far from the centre of the snow-cap, and increased in width and length till it opened into the polar sea in longitude

160°. Later its opposite end came out in longitude 330°. Meanwhile it was spreading even faster in the middle. While it was thus eating into the snow, some brilliant points, shining like stars, were observed in the snow between it and the outer edge of the cap on several successive mornings. They undoubtedly were the far-off glisten of snow-slopes. The subsequent behavior of the spots from which they came bore witness to this. For the rift grew till it formed a large lake in the midst of the snow not far from the geographical pole. Other rifts also appeared and ate into the cap, but the spots that had shone out so brilliantly still remained as snow islands, though daily diminished in size. At last the smaller of the two chief portions into which the cap had been split dwindled entirely away, and the other became a tiny patch eccentrically placed.'

In consequence of the one-sided situation of the southern cap, the planet's true pole is always uncovered at the height of the summer season ; and until the opposition of 1894, the minimum size of this cap was that observed by SCHIAPARELLI in 1879, when its diameter shrank to less than 150 miles. But the observers at Flagstaff were treated to a genuine surprise, when, in October 1894, Mʳ DOUGLASS witnessed the complete disappearance of this polar cap, undoubtedly the first phenomenon of this character ever recorded; it was verified also by Professor WILSON at Northfield, M. BIGOURDAN of Paris, and others. The region formerly occupied by it became in no respect different from the east and west limbs of the planet.

THE TERMINATOR

As Mars is the nearest of the outer planets, its phase at times is very marked, being about like that of the Moon when two days before or after full. (See the illustration on page 172). Its terminator, the line marking the limit of visibility on the incomplete side of the disk, may therefore be well observed; and the phenomena of this region should tell us much about the general character of the planet's surface.

Bright projections on the terminator of Mars were first observed by Mʳ KNOBEL, who, on several occasions in 1873, described and delineated a white spot in this position, 'glistening as brightly as the polar ice.' In 1884, however, with Mars at a corresponding phase, he could detect no bright spot in the same areographic location. Similar markings were discerned by M. TERBY in 1890. Also, in August 1894, MM. PERROTIN

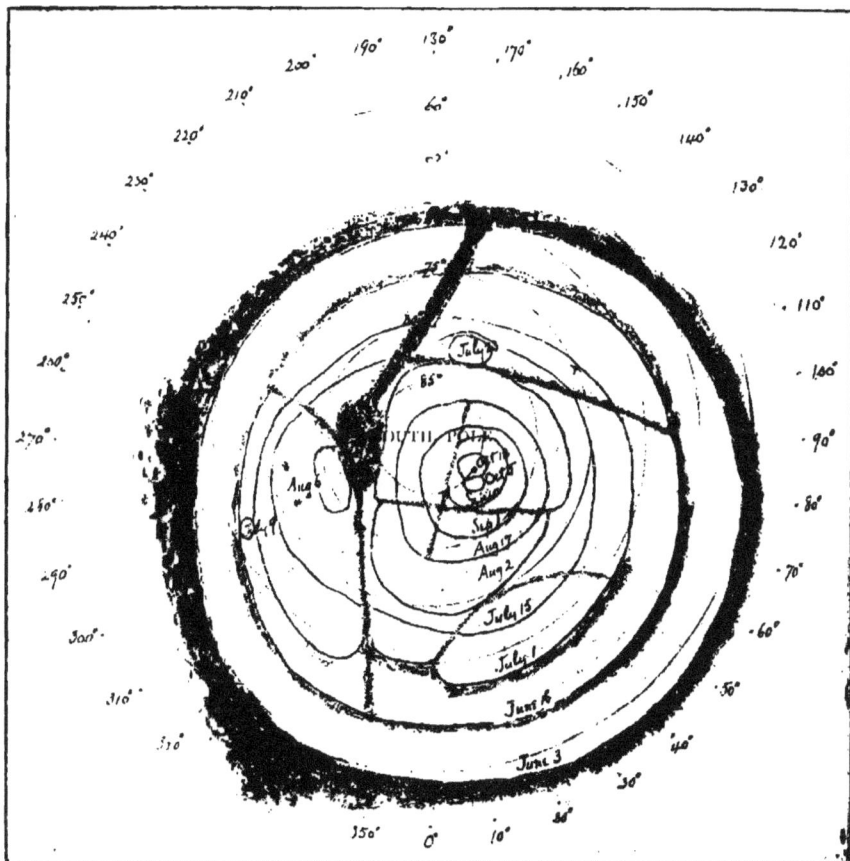

DWINDLING OF THE SOUTH POLAR CAP IN 1894 (LOWELL)

(The planet's pole of rotation is at the middle point of the square, and the eccentric position of the cap is shown, as well as the gradually diminishing breadth of the dark belt that fringed the polar cap at all times)

and JAVELLE, and Mr WILLIAMS saw brilliant projections beyond the terminator, which, in case they are constant in position, are probably due to Martian mountains, or clouds, or a combination of both. Certain gray regions of large area, as watched by Professor PICKERING in passing the terminator, were notched to different depths, indicating hills and valleys, and not the surface of an ocean. At the recent oppositions, irregularities of the Martian terminator have been carefully looked for, particularly at the Lowell Observatory, where, on 30th June 1894, Mr DOUGLASS detected local flattenings in this line of demarcation, 20° to 40° in extent. Mr LOWELL finds them almost invariably upon that part of the terminator where the darker of the dark regions was then passing out of sight. These chord-like flattenings are still unaccounted for. Also there were projections and small notches, similar to those on the lunar terminator, though much less pronounced, and indicating mountains of perhaps 3,000 to 4,000 feet elevation. About two months later, Professor PICKERING sketched a series of unusually marked elevations and depressions upon the terminator (page 172). Eventually data of this character will suffice for a map of approximate contours, but great care is necessary in distinguishing those projections which the observations of 1896–97 indicate as perhaps due to cloud.

Evidently the most persistent scrutiny of this ever-changing line of sunrise and sunset can alone enlighten us as to configurations of the Martian surface, which without doubt is relatively flat in comparison with the present rugged contour of Earth and Moon. Only when the crowding enigmas of the Moon have dissolved shall we be ready to pronounce upon Martian appearances with certainty. Particularly must the marked changes of lunar objects with the direction of their illumination be critically interrogated as an auxiliary key to the situation on Mars. The striking systems of lunar 'streaks' vanish utterly as our satellite recedes from the full: and this simple fact shows the danger of attaching too great significance to apparent phenomena of the Martian surface which may turn out to be mere phantom changes of illumination.

THE LAKES, OR OASES

It is the vast orange-tint area of the northern hemisphere, probably little diversified with hills and valleys, which the keen eye of Professor PICKERING first dotted with numerous darkish regions of relatively small extent, and which are connected together by the complex system of *canali*, intersecting the northern Martian continents, as if the lines of a huge system of triangulation. If these dark spots are lakes, they are not permanent reservoirs, like terrestrial ones, as sometimes they wholly disappear. Also their apparent doubling, or gemination from time to time is not unknown. Mr Lowell adduces excellent reasons for regarding them as oases in the vast Arean deserts, and their visibility as due to the growth of vegetation with the advance of spring. On pp. 176–77 are twelve charts of Mars, obtained by photographing a globe on which all the known features of the planet had been drawn by Mr Lowell, through

MARS AS SEEN WITH THE 30-INCH TELESCOPE
AT NICE, 2D JULY 1888 (PERROTIN)

whose courtesy they are here presented. They give comprehensively the important results of his recent studies of the planet, and show the striking system of the spots admirably.

Lacus Phoenicis, one of the southernmost, on very fine nights with medium power, appears as a small spot, almost black and nearly round, resembling a shadow of a Jovian satellite near the middle of its transit. A little farther south is one of the best known regions of the planet, also a dark, oval, isolated spot, about 530 miles in its longest diameter, and formerly called very appropriately the 'Oculus,' or eye of Mars. First observed by MAEDLER in 1830, charted as *Terby Sea* on GREEN's map, it is now known as *Solis Lacus*, after SCHIAPARELLI. As early as June 1890 this keen observer saw it divided into two distinct parts, and on 2d September 1894, the Lick telescope enabled Mr SCHAEBERLE to divide it into three regions, all separate

and very dark. Mʳ Douglass has still farther subdivided it
(page 166). Schiaparelli has often seen a large part of the
planet's disk, particularly near the polar regions, sprinkled with
minute white spots, for example, in January 1882, the region between *Ganges* and *Iris;* and these phenomena, so important in
studying the physical constitution of the planet, are often within
the reach of moderate telescopes. But the smallest of the true
'oases,' or dark spots, which are about 60 miles in diameter,
are even more difficult to see than their intersecting 'canals,'
and both oases and canals have been observed to fluctuate together in visibility, as if they were parts of one intimately connected system.

THE MARTIAN 'CANALS'

Although faint markings of this character were first indicated on the map of Beer and Maedler (1840), and Dawes
detected eight or ten of them in 1864, they are generally regarded
as the original discovery of Schiaparelli during the remarkable opposition of 1877, when he made the first extended triangulation of the Martian surface, and mapped most of the
canali known at the present time. Many observers fail utterly
to detect these delicate markings ; others, practised in the observation of them (Mʳ Williams, for example, who verified
nearly all of them in 1894), would call the plainer ones conspicuous, and the average of them not difficult. But at least
they require a steady atmosphere, and a perfect telescope with
a trained eye behind it. Not even then are they always visible.
Their appearance and degree of visibility are variable, often in

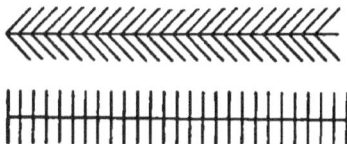

PROBABLE DETAIL OF 'CANALS,' IF ARTIFICIAL

periods of only a few days' duration. Seemingly capricious
even the general law of this variation is not yet fully understood.
Generally only a few of the *canali* are visible at once. Owing to
the changing physical aspects, as to Mars's season and orbital
position with reference to the Earth, under which they may be

presented, given markings of this type may for a long time remain invisible; but nearness of the planet is not essential to their
visibility, for they were seen at the remote opposition of 1896–7.

If it is granted that the *canali* have an artificial origin, and
were constructed for the very reasonable purpose of irrigation,
one would naturally expect their minute structure to be somewhat as shown schematically on the previous page; straight and
relatively narrow channels through the middle, intersected, at
intervals more or less regular, by a multitude of short cross-
channels, throughout their entire length. Indeed, says SCHIA-
PARELLI, they very often look like gray bands deepest in intensity at the middle and shaded at the edges. Such a theory
would obviate the necessity of supposing that the *canali* are
really as wide as they look; because at our great distance, the

THE 'SOLIS LACUS' REGION OF MARS
(*As drawn by* DOUGLASS, *9th October* 1894. *Scale,* 1 *inch* = 1200 *miles*)

apparent width of a 'canal' would simply be equal to the mean
breadth of the system. Optically it is a question of resolva-
bility which ought not to be beyond the power of the largest
existing telescopes with a perfect atmosphere.

Straight in their course and rarely deviating from a great
circle, but very different in length (from 300 to 4,000 miles),
they join one another at various angles like spokes in the
hub of a wheel; and intersect the great northern continent

with a marvellous network of fine darkish stripes. Mr LOWELL describes their color as a bluish green, identical with that of the seas with which they connect. While each *canale* maintains its own width throughout its entire length, the breadth of these markings is by no means uniform ; perhaps 20 miles for the narrowest, and ten times that amount for the broadest, of which the *Nilosyrtis* is probably the easiest of all to see. But although minor changes are suspected, and beginning to be made out, the *canali* form a permanent configuration of the Martian surface ; and every *canale*, as is apparent from Mr LOWELL'S charts, page 176, connects at its ends with another marking of like character, or with darker ones, possibly seas.

Preferably, however, they converge toward the small spots named lakes, or oases ; for example, seven *canali* converge in *Lacus Phoenicis*, eight in *Trivium Charontis*, six in *Lunae Lacus*, and six in *Ismenius Lacus*. About 50 of these dark spots, or lakes, are now known, including the recent additions of PICKERING and LOWELL. Often the canals open out into an expanding trumpet-shaped region — *e. g.*, mouth of *Nilosyrtis* named *Syrtis Major*, an area which exhibits seasonal changes, as if due to vegetation. SCHIAPARELLI'S critical studies of these markings, and their changes of width and colors at the oppositions of Mars in 1882, 1884, and 1886, when the planet's northern pole was turned toward us, have led him to the simple and natural interpretation that the *canali* form a veritable hydrographic system, for distributing the liquid mass of the melting snows. Whether one views this marvellous and intricate system as a whole, or in some detached portion of high detail, as in the opposite sketch of the *Solis Lacus* region by Mr DOUGLASS, it is difficult to escape the conviction that the *canali* have, at least in part, been designed and executed with a definite end in view.

A dozen of the *canali* became visible to Mr LOWELL at Flagstaff in 1894, ten weeks preceding the summer solstice of Mars's southern hemisphere ; and there were abundant verifications by many other astronomers in widely separate localities and with instruments

MARS, 14th OCTOBER 1894
(BRENNER)

(*The elongated chain-link region is the Mare Cimmerium*)

of varying capacity, — Professor WILSON, Northfield, Minnesota, and the Mount Hamilton observers, 30 *canali* by Herr BRENNER in Istria, 40 by M. ANTONIADI (who, at M. FLAMMARION'S observatory at Juvisy, made numerous fine sketches of Mars), and 50 by Mr WILLIAMS of Brighton, who saw nearly all those marked on SCHIAPARELLI'S map, and a few new ones in addition. But this record was far outstripped at the Lowell Observatory, where some of the canals appear on more than one hundred separate drawings, and 183 canals were seen in all. Mr LOWELL has made out a progressive change in the tint of the *canali*, which he regards as seasonal, and probably due to actual water, and vegetation fed by it. Naturally, extensive irrigation and agricultural operations on a large scale would seem the most likely explanation of the *canali*, especially when we reflect that upon Mars, doubtless a world farther advanced in its life history than our own, in the first place, erosion may have worn the continents down to a minimum elevation, making artificial water-ways easy to construct ; also with its vanishing atmosphere and absence of rains, the necessity of water for prolonging the support of animal and vegetable life could only be met by conducting water from one part of the planet to another in channels artificial or partly so.

In our present knowledge of the *canali*, this seems a plausible explanation ; but many years of patient investigation are yet required, especially with larger telescopes on perfect mountain sites and by the most painstaking observers. Particularly are the best conditions for telescopic research paramount, because the *canali* manifest themselves chiefly in the northern hemisphere of Mars. To ascertain the true significance of the *canali*, however, does not necessarily seem to be forever beyond the power of man. The argument of the 'canals' is fully presented by Mr LOWELL in his interesting volume on *Mars* (Boston 1895) ; and since its publication he has employed his 24-inch Clark telescope in the study of Mars during its opposition in 1896–97. The north polar cap was first seen in August 1896, practically central over the planet's north pole, and covering a stretch of 50° in latitude, equivalent to 1800 miles.

Seasonal changes are well illustrated in the colored plate (see Frontispiece), showing Hesperia central at three epochs months apart : in early spring the polar cap has just begun to shrink ; in early summer canals are beginning to develop from south to north ; in late summer they are fully developed and the polar cap has vanished.

[30-inch Equatorial] [Observers' houses and Library] [Small Equatorial]

THE NICE OBSERVATORY, FOUNDED BY M. BISCHOFFSHEIM OF PARIS — M. PERROTIN, *Director*

DOUBLING OF THE 'CANALS'

Most striking of all the phenomena of Mars is the periodic doubling of the *canali*, also discovered by SCHIAPARELLI, who first detected in 1879 the gemination of *Nilus*. Two years later this occurrence presented itself extensively, and he established it as a characteristic feature of the Arean continents. At about the Martian equinoxes, just before and after the apparent inundation of the northern continent, occurs this surprising phenomenon — never shown by all the *canali* simultaneously ; but under the proper seasonal conditions, the gemination makes its appearance irregularly and in isolated instances. A few of the *canali* even have never been seen doubled, of which the *Nilosyrtis* is one. The gemination takes place rapidly, in a few hours or days at the most ; and the original and previously single marking becomes paralleled with precision by its duplicate, the distances between the faint parallel lines being as small as 25 miles in some cases, and ranging up to 350 miles in others. Reference is again made to Mr LOWELL's excellent charts, pp. 176–77, on which *Ganges* and *Nectar* (chart No. 3) are typical instances of greatest and least distance between twin canals. Ordinarily the two bands appear equal, as well as parallel, and of the same color and intensity ; but the gemination of course is not to be seen at all, except at suitable seasons, and even then it is visible only with difficulty.

This strange transformation from single markings into double ones, was many years doubted by astronomers generally ; but having now been verified on many occasions, and independently, by M. TERBY, Mr WILLIAMS, the Lick astronomers, and many other observers, it has been most reliably confirmed by M. PERROTIN, formerly director of the great observatory on Mont Gros, depicted on page 169. One of his excellent sketches has already been given on page 164. The geminations are found to vary at different recurrences, so that they cannot be fixed formations on Mars, as the *canali* themselves are. After some months of visibility, the twin marking fades out, and is seen no more till the favorable epoch again recurs. If SCHIAPARELLI's theory is the true one, the duplication can occur only between the spring and autumn equinox of the northern hemisphere. A recent opportunity, thus, was in 1890, and a later one occurred in January and February 1895 ;

but already Mr LOWELL, from his lofty station in Arizona, had on 19th November 1894 seen *Phison* and *Euphrates* double, and Mr WILLIAMS in the same season had recorded the gemination of *Ganges, Eunostos, Cerberus* and others. Again the gemination duly put in an appearance in 1897, and was depicted in many drawings by Mr LOWELL. The width of each of the double canals was 25 to 40 miles, and their distance apart 135 to 170 miles, from centre to centre. The oases were about 200 miles in diameter, and nearly all of them were seen to lie exactly between the twin canals. Among others who saw the double canals at this opposition was M. CERULLI, one of whose drawings is here given, showing the

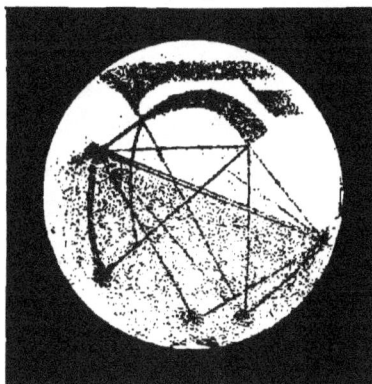

MARS, 17TH JANUARY 1897 (CERULLI)
(*The long double canal is Eumenides-Orcus*)

Eumenides-Orcus plainly double. It is somewhat curious that those observers who have seen the gemination best are most puzzled as to the cause of it. In fact, no plausible explanation of this wonderful phenomenon has yet been advanced, although many crude guesses have been hazarded. M. ANTONIADI has sought to account for it by maladjustment of focus. That it cannot be due to optical illusion or double refraction seems quite certain; rather the twin canal appears to have in each case a real existence.

The sketch on page 172 shows a portion of Mars not commonly drawn, which with ordinary seeing affords scarcely any detail, except mere outlines. At the same time a number of the more interesting and difficult features are shown. Descrip-

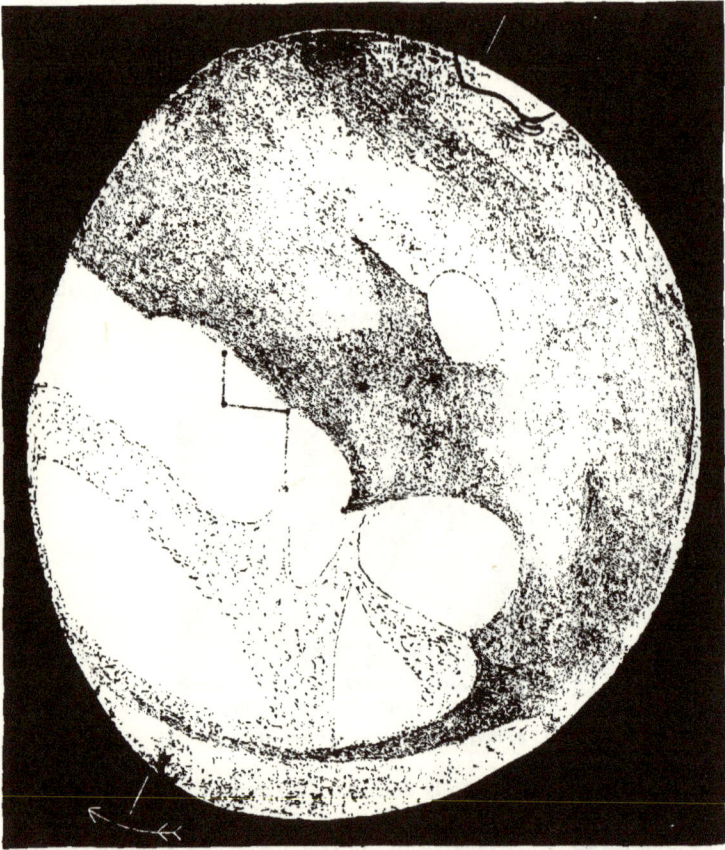

MARS AS SEEN AT THE LOWELL OBSERVATORY, ARIZONA

16th August 1894, *under fine atmospheric conditions (by Professor* W. H. PICKERING, *with the* 18-*inch Brashear telescope, magnifying power* 420)

The scale is $\frac{1}{73,800,000}$; *that is,* 1 *millimetre* = 73.5 *kilometres* = 45½ *miles. The linear value of* 1 *second (of arc) is* 6 *millimetres* = ¼ *inch (nearly) The two left hand 'lakes,' connected by a nearly vertical 'canal'* 280 *miles long, are* 1 *second apart. The phase of Mars is near its greatest value. The arrow below indicates the direction of motion in consequence of the planet's rotation. Above and below the disk are two hair lines marking the position of the planet's axis, which passes through the upper or south polar cap, whose displacement from the true pole is well shown. Into the seemingly minute area of one of the nearly central dark spots, or 'lakes,' could be crowded* 30 *cities the size of New York.*

tive of this drawing, Professor PICKERING says, '*Syrtis Minor* has nearly reached the middle of the disk, and *Syrtis Major* has just come around the limb. . . . The southern [upper] snow cap is bounded by a uniform black line, of the width shown in the sketch, and presumably due to water. Near it is a small isolated patch of snow, probably located upon a mountain range. . . . On the terminator is noticed a projection which is unusually high, and has unusually steep ends. Near the centre of the disk are seen six lakes and three canals. They are all difficult objects. The diameter of the lakes was about o.''1 [36 miles], and the breadth of the canals o.''05 [18 miles]. These canals are rather blacker than they should have been, and the southern lake is rather too conspicuous. In the northern hemisphere are seen the hazy, radiating surfaces which probably precede the formation of the canals. South of the northern cloud the dark regions were brownish, north of it, gray; while the *Syrtis Major* region was greenish gray. The northern hemisphere and the limb were yellow, the radiating bands very faint gray, and the extreme north bounded by the canal and the fine line, light green.'

Still, great as is the amplification in this drawing, it will help to present in a forcible manner the difficulties with which astronomers must contend, if it is remembered that the area of the full Moon's apparent disk is 4000-fold that of the little disk of Mars at his nearest.

MARTIAN ATMOSPHERE

Likelihood of an atmosphere encircling Mars is inferred from temporary obscurations of well-known and permanent markings of the planet's surface, frequently observed and satisfactorily attributed to clouds. An important occurrence of this character was recorded by Mr DOUGLASS, 24th September 1894, when the western half of *Elysium* appeared as if veiled for a time by a cloud, subsequently dissipated. Also Mr WILLIAMS saw *Mare Cimmerium* itself occluded, with considerable variations in the extent of the cloud envelope from day to day. In October 1894, the same careful observer saw what he considered to be cloud, or mist, affecting the visibility of the extensive land region north of *Mare Cimmerium ;* and he thinks these atmospheric occlusions are common and extensive. Mars, however, seems never to be covered, as the Earth generally is, with vast cloud areas obliterating its continents and oceanic features, but his skies appear to be nearly perpetually clear, in every climate and every zone. Now and then a few whitish spots, changing their form and position, though rarely extending over a very wide area, frequent the islands of *Mare Australe*, and *Elysium* and *Tempe* of the northern continents. Their variation during the Martian day is well known, and SCHIAPARELLI thinks them a thin veil of fog rather than a true nimbus of the type yielding terrestrial rains; or possibly a temporary condensation of vapor as dew or hoar frost.

The spectroscope, in the hands of SECCHI, HUGGINS, RUTHERFURD, VOGEL, and MAUNDER, has given certain evidence of absorption lines in the spectrum of Mars not due to our own atmosphere; and the inference has always been irresistible that the ruddy planet is surrounded by a gaseous envelope. Dr HUGGINS in 1867 detected lines in the spectrum due to the presence of watery vapor; also in the same year, embracing an opportunity when Mars and the Moon were at the same altitude, he found the spectra of the two bodies practically identical, while, on applying photography in 1879, no lines or modifications peculiar to the planet's spectrum made their appearance. Professor CAMPBELL, in July 1894, when Mars and the Moon were again near together, repeated the observation of Dr HUGGINS by making direct and critical comparison of their spectra under improved conditions of drier atmosphere and more powerful apparatus, with the identical result that the two spectra were alike in every particular. This may be interpreted as meaning, of course, that Mars has no more atmosphere than the Moon itself, long known to be devoid of such an envelope.

But the obstacles to such interpretation are by no means slight. For example, as the planet turns on its axis, carrying the spots from centre to edge of the disk, they gradually melt from view, just as they would if seen through a greater thickness of atmosphere. The temporary obscurations of certain parts of the disk, supposably by clouds, will have to be otherwise accounted for. The amount of water, too, must be inappreciable, let alone the difficulties of explaining that periodic and well-established expansion and shrinking of the polar caps, if there is no atmosphere to act as a medium in the formation and deposition of snow. Professor CAMPBELL does not consider the spectroscopic observations as proving that Mars has no atmosphere whatever, similar to our own, but simply that they set a superior limit to its extent. In the latter part of 1894, both Dr HUGGINS and Dr VOGEL repeated their examination of the spectrum of Mars, resulting in essential verification of their earlier researches; and Professor KEELER, in the winter of 1896–97, obtained photographic tests at Allegheny which, as far as they go, agree with the visual results of Professor CAMPBELL. But judgment upon the spectroscopic evidence of a Martian atmosphere must be suspended until the planet is again very favorably placed for observation in 1907. Meanwhile, photometric observations by Dr MÜLLER show that the behavior of Mars, as to its phases, distinctly indicates an atmosphere comparable in density with that of the Earth. Also Mr LOWELL, in discussing a fine series of observations of the diameter of Mars, made at his observatory by Mr DOUGLASS in 1894, finds an unmistakable twilight arc of 13°, — independent evidence of an atmosphere. The dimensions of Mars are 4184 miles for the polar diameter, and 4206 for the equatorial, the polar compression of the planet being $\frac{1}{200}$. Dr SCHUR, however, in 1896–7 finds it as great as $\frac{1}{47}$ with the heliometer.

In considering the possibility of an atmospheric envelope surrounding our neighbor planet, a few fundamental facts ought to be kept in clear and constant view. Mars is a planet intermediate in size between Moon and Earth: twice the diameter of the Moon equals the diameter of Mars; and twice the diameter of Mars approximately equals the diameter of the Earth. As to masses the Moon is $\frac{1}{81}$, and Mars about $\frac{1}{9}$, the mass of our globe. We have here an abundant atmosphere: the Moon has none; so it would seem safe to infer that Mars may have an atmosphere of slight density, — not dense enough for the spectroscope to detect, but dense enough to account for the observed phenomena of the Martian disk, otherwise hard to explain. Indeed, MATTIEU WILLIAMS, from considerations of the relative size and mass of the Earth and Mars, calculated that the ruddy planet is entitled to $\frac{1}{20}$ the atmosphere that our globe has, and that the mercurial barometer would stand at about 5$\frac{1}{4}$ inches at the sea level on Mars.

The climate of Mars, then, would seem likely to resemble that of a clear season on a very high terrestrial mountain, a climate of extremes, with great changes of temperature from day to night. But safe speculation regarding the possibility of organic life upon the ruddy planet really hinges upon the selective absorption of the Martian atmosphere, and whether it aids the planet, as our atmosphere does, in storing heat by preventing its radiation. And until it has become known whether the Earth's surface has yet reached, or has passed, that maximum of secular temperature, toward which it must tend so long as it continues to receive more heat from the Sun than it radiates, obviously there is little use in speculating whether some other planet may or may not have passed that critical epoch. The inequality of Martian seasons is such that in the northern hemisphere, the cold season lasts 381 days, and the hot only 306. The polar caps attain their maxima three or four months after the winter solstice, and their minima about the same time after the summer solstice; and this lagging affords a strong argument for a Martian atmosphere with heat-storing properties similar to the Earth's.

It should be observed that the existence of fluid water upon Mars would imply that the temperature of Arean climates is comparable with our own. The relations of the planet's axis to its orbit are similar to ours; but the supply of direct solar heat is about one-half as great per square mile as that of the Earth. Owing to the eccentricity of Mars's path round the Sun, the surface of the planet receives at perihelion about one-half more of solar heat than at aphelion; so that the southern summers must be much hotter and the winters colder than those of the northern hemisphere. The length of summer, twice that of the terrestrial season, may amply suffice to melt all the ice and snow, an unusual condition of things which actually took place in October 1894.

But the aspect of the entire situation is somewhat modified by a recent paper (1898) by Dr JOHNSTONE STONEY of Dublin, whose investigations of the phenomena of planetary atmospheres began about 35 years ago. His method is based on the kinetic theory of gas; according to which, if free molecules travel outward, as they often must, with velocities exceeding the limit that a planet's gravity can control, they will effect a permanent escape, forever after travelling in orbits of their own. So Dr STONEY succeeds in accounting for the practically entire absence of atmosphere from the Moon, and of free hydrogen and helium from the Earth's encircling medium. Applying the same theory to Mars, he is led to the significant inference that water cannot in any of its forms remain upon that planet. Without water, there can of course be no vegetation such as we know; and in its absence much free oxygen is unlikely. Under these circumstances, analogy to the Earth suggests to him that the atmosphere of Mars consists mainly of nitrogen, argon and carbon dioxide.

It is part of the province of physical science not only to

0° — Fastigium Aryn

30° — Margaritifer Sinus

60° — Ganges

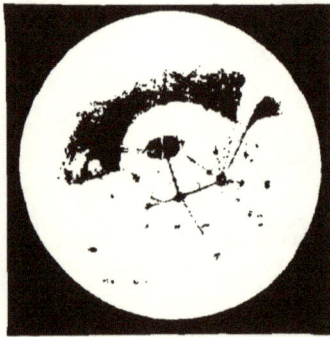

90° — Solis Lacus

120° — Nodus Gordii

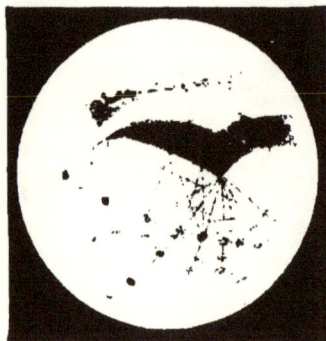

150° — Mare Sirenum

THE PLANET MARS IN

(At every 30° of Martian longitude, and
earthward. Below each presentation is
together with the name of the most prominent

180° — Atlantis

210° — Trivium Charontis

240° — Hesperia

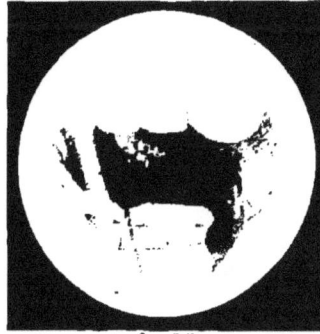

270° — Libya

300° — Syrtis Major

330° — Phison and Euphrates

NOVEMBER 1894 (LOWELL)

with the planet's south (upper) pole tilted 23°
given the longitude of the central meridian,
marking upon it)

observe and record, but to explain, natural phenomena. How then shall we explain the observed and regular fluctuations of polar caps, and the seasonal development of canals and oases? As the hypothesis of water affords the readiest and most natural explanation, we are driven to the important conclusion that the secular dissipation of water on Mars, while constantly progressing through countless ages, is not yet complete, although his original store of water already exhibits marked signs of approaching exhaustion.

From the motions of Earth and Mars given in previous chapters, it is easily found that the ruddy planet goes 17 times round the Sun in 32 years, or, more accurately, 25 times in 47 years. These relations then indicate when Mars may be best observed in the future, as the following table shows, in addition to those epochs of the past half century when this planet was exceptionally well placed for observation : —

FAVORABLE TIMES FOR OBSERVING MARS

Years	Mars nearest the Sun on	Least Distance of Mars from the Earth		Angular Diameter	Declination of Mars	Average Meridian Altitude in the U.S.
		On	In miles			
1845	August 30	August 18	34,610,000	30″	20°S.	28°
1860	Sept. 16	July 22	36,270,000	29	27 S.	21
1862	August 4	Sept. 29	37,660,000	28	2 N.	50
1877	August 21	Sept. 2	34,950,000	30	12 S.	36
1892	Sept. 7	August 6	35,020,000	29	24 S.	24
1894	July 26	Oct. 13	40,000,000	26	9 N.	57
1907	Sept. 24	July 12	39,000,000	27	20 S.	28
1909	August 13	Sept. 20	36,000,000	29	4 S.	44

All delicate questions concerning the planet's surface, its southern hemisphere in particular, must now be tabled for many years. Meanwhile, the northern regions are being faithfully interrogated as they come more and more within reach ; and the intervening years will afford opportunity for thoroughly discussing all the earlier drawings, and, in the light of large and freshly garnered harvests, welding them together into a homogeneous system, capable of consistent and reasonable interpretation.

In additon to the references for this planet already given on page 147, consult first of all FLAMMARION's comprehensive work entitled *La Planète Mars et ses conditions d'habitabilité : synthèse générale de toutes les observations* (Paris 1892). Also the following, comprising the bulk of the more important Martian literature, both popular and technical, chiefly since 1885 : —

WILLIAMS, 'Meteorology,' *Fuel of the Sun* (London 1870).

GREEN, At Madeira in 1877, *Mem. R. A. S.* xliv. (1879), 123.

HARTWIG, ' Diameter,' *Publ. Astron. Gesell.* xv. (Leipzig 1879).

YOUNG, ' Diameter,' *Am. Jour. Sci.* cxix. (1880), 206.

SCHRÖTER, *Areographische Beiträge* (Leiden 1881).

MAUNDER, *The Sunday Magazine*, xi. (1882), 30, 102, 170.

TROUVELOT, *Comptes Rendus*, xcviii. (1884), 788 ; *L'Astronomie* iii. (1884), 321 ; *Observatory*, vii. (1884), 369.

BAKHUYZEN, ' Rotation,' *Ann. Sternwarte Leiden*, vii. (1885).

GREEN, ' Northern hemisphere,' *M. N. R. A. S.* xlvi. (1886), 445.

PERROTIN, ' Les Canaux,' *Bulletin Astron.* iii. (1886), 324.

STANISLAS MEUNIER, *Pop. Sci. Monthly*, xxxi. (1887), 532.

FIZEAU, JANSSEN, PERROTIN, TERBY, *Comp. Rend.* cvi. (1888).

FLAMMARION, PERROTIN, *Comptes Rendus*, cvii. (1888).

MAUNDER, ' Canals,' *The Observatory*, xi. (1888), 345.

NIESTEN, *Bulletins Acad. Roy. Belgique*, xvi. (1888), Nr 7.

FLAMMARION, ' Variations,' *Bull. Soc. Astron. de France* (1888).

PERROTIN, ' Canals,' *L'Astronomie*, vii. (1888), 366.

TERBY, *Ciel et Terre*, ix. (1888), 271, 289 ; xii. (1891) ; *Mémoires couronnés Acad. Belgique* (1888); *Bull. Acad. Belgique*, xx. (1890) ; *l'Astronomie*, ix. (1890).

WILSON, ' Canals,' *Sidereal Messenger*, viii. (1889), 13.

GERIGNY, ' Les Marées,' *L'Astronomie*, viii. (1889), 381.

YOUNG, *The Presbyterian Review*, x. (1889), 400.

HALL, HOLDEN, *The Astronomical Journal*, viii. (1889), 79, 97.

SCHIAPARELLI, *L'Astronomie*, viii. (1889), 19, 42, 89, 124 ; *Himmel und Erde*, i. (1889), 1, 85, 147.

MEISEL, *Astronomische Nachrichten*, cxxi. (1889), 371.

SCHEINER, *ib.* cxxii. (1889), 251.

WISLICENUS, *ib.* cxx. (1889), 241 ; cxxvii. (1891), 161.

PICKERING, ' Photographs,' *Sid. Mess.* ix. (1890), 254, 369.

RITCHIE, *ib.* ix. (1890), 450.

WILLIAMS, A. S., ' Canals and markings,' *Jour. Brit. Astron. Association*, i. (1890), 82.

FLAMMARION, *L'Astronomie*, x. (1891), xi. (1892).

LOHSE, *Publ. Astrophys. Observatorium Potsdam*, Nr 28 (1891).

LOHSE, MAUNDER, NIESTEN, *Jour. Brit. Astron. Association,* ii. (1892), 417, 423, 507.
TISSERAND, 'Diameter,' *Bulletin Astronomique,* ix. (1892), 417.
FLAMMARION, PERROTIN, MEUNIER, *Comp. Rend.* cxv. (1892).
LOCKYER, J. N., *Nature,* xlvi. (1892), 443.
BARNARD, COMSTOCK, PICKERING, TERBY, WILSON, YOUNG, *Astronomy and Astro-Physics,* xi. (1892).
FLAMMARION, LOCKYER, PICKERING, *Nature,* xlvii. (1892).
BALL, *In Starry Realms* (London 1892), chapter xii.
BALL, *The Fortnightly Review,* lii. (1892), 288 ; *In the High Heavens* (London 1893), chapter vi.
PEAL, 'Distribution of land and water,' *Jour. Brit. Astron. Association,* iii. (1893), 223.
COMSTOCK, 'South Polar Cap,' *Astron. Jour.* xiii. (1893), 41.
KEELER, *Memoirs Royal Astronomical Society,* li. (1893), 45.
ABETTI, 'South Polar Spot,' *Astron. Nachr.* cxxxiii. (1893), 25.
MAUNDER, *Jour. Brit. Astron. Association,* iv. (1894), 395.
DOUGLASS, LOWELL, PICKERING, SCHAEBERLE, SCHIAPARELLI, and others, *Astronomy and Astro-Physics,* xiii. (1894).
FLAMMARION, *L'Astronomie,* xiii. (1894), 447.
LOWELL, *Popular Astronomy,* ii. (1894).
MEYER, *Die Physische Beschaffenheit* (Berlin 1894).
CAMPBELL, 'Atmosphere,' *Publ. Astron. Society Pacific,* vi. (1894), 228, 273.
ANTONIADI, LOWELL, *Nature,* li. (1894), 40, 64, 87, 259.
HUGGINS, KNOBEL, WILLIAMS, *The Observatory,* xvii. (1894).
LOCKYER, W. J., *Nature,* l. (1894), 476, 499.
YOUNG, 'Diameter,' *The Astronomical Journal,* xiv. (1895), 165.
BRENNER, Sketches 1894, *Astron. Nachr.* cxxxvii. (1895), 49.
ELLERY, *The Astrophysical Journal,* i. (1895), 47.
LOWELL, *Mars* (Boston 1895).
SCHIAPARELLI, *Osservazioni astronomiche e fisiche sull' asse di rotazione et sulla topografia del pianeta Marte* (Rome 1897). Also 4 previous memoirs by same author.
PERROTIN, 'Zones,' *The Observatory,* xx. (1897), 132.
BRENNER, *Abhandl. Kön. Akad. Wiss. Berlin,* 1897.
LOHSE, *Publ. Astrophys. Observatorium Potsdam,* xi. (1898).
STONEY, 'Atmosphere,' *Astrophys. Jour.* vii. (1898), 44.
CERULLI, *Marte nel* 1896–97 (Collurania 1898).
LOWELL, *Annals Lowell Observatory,* i. (Boston 1898).

Bibliographic lists in the 'Smithsonian Report' each year comprise many additional titles. For papers since 1895 consult the bibliographies in *The Astrophysical Journal.*

THE PLURALITY OF WORLDS, ETC.

The absorbing popular interest of this subject, and the serious attention accorded it by many eminent writers, render a brief list of the better literature worthy of the space here given it.

HUYGENS, *Cosmotheoros* (*Opera*, ii. p. 641) (1755).

FONTENELLE, *Entretiens sur la Pluralité des Mondes* (Paris 1852) ; with notes by PROCTOR, *Knowledge*, vi. (1884); vii. (1885).

WHEWELL, *The Plurality of Worlds* (London 1853).

BREWSTER, *More Worlds than One* (London 1855).

SMITH, H. J. S., *Oxford Essays* (1855), 105.

FLAMMARION, *La Pluralité des Mondes habités* (Paris 1863).

PROCTOR, *Other Worlds than Ours* (London 1870).

SEARLE, G. M., *The Catholic World*, xxxvii. (1883), 49 ; lv. (1892), 860.

MILLER, W., *The Heavenly Bodies* (London 1883).

PROCTOR, ' Life in Other Worlds,' *Knowledge*, vii. (1885).

BURR, ' Are heavens inhabited ? ' *Presby. Rev.* vi. (1885), 257.

PORTER, *Our Celestial Home* (New York 1888).

SEARLE, G. M., *Pub. Astron. Society Pacific*, ii. (1890), 165.

FLAMMARION, *L'Astronomie*, x. (1891).

GUILLEMIN, ' Communication with the planets,' *Popular Science Monthly*, xl. (1892), 361.

BALL, ' Life in other worlds,' *The Fortnightly Review*, lvi. (1894), 718 ; *McClure's Magazine*, v. (1895), 147.

FLAMMARION, *No. Am. Rev.* clxii. (1896), 546.

JANSSEN, *Popular Science Monthly*, l. (1896), 812.

SERVISS, *ib.* lii. (1897), 171.

For the periodical literature, much of it worthless, consult under *Worlds*, the indexes of POOLE and FLETCHER, often cited ; also, MATSON'S *References for Literary Workers* (Chicago 1892) contains, at pp. 410–412, abundant references to papers on the plurality of worlds, an interminable subject.

CHAPTER XII

COMETS

W HEN KEPLER had shown that the planetary orbits are ellipses, and NEWTON had proved that this is a necessary consequence of their being attracted toward the Sun with a force varying inversely as the square of the distance from him, it was natural to surmise that comets also might be moving round the Sun, in ellipses so much more eccentric that these bodies would be visible only in those parts of their elongated orbits which are nearest to the Sun and Earth. As the law of equable description of areas would of course hold good in their case, causing them while in these nearer parts of their courses to move much more rapidly than in any other part, the comets would obviously be visible during only a small portion of their whole orbital revolution.

NEWTON applied these principles to the splendid comet of 1680 (visible a few years before the publication of the first edition of the *Principia* in 1687), and found that it was moving in an ellipse of so eccentric a character as to approach very nearly in form to a parabola. It was thought by himself and by HALLEY that it might be identical with great comets recorded as having been seen in B. C. 44, A. D. 531, and A. D. 1106, and that the period was about 575 years in length. Subsequent investigations have not confirmed this particular conjecture, and it is probable that the actual period of that remarkable comet (which made such an exceptionally close approach to the Sun) amounts to, not hundreds, but thousands of years; so that any previous appearances most likely took place before historic dates. But in regard to a fine comet which appeared two years later, in 1682, HALLEY was able, after calculating the elements of its orbit, to show that they were almost precisely the same as those of comets observed in 1531 and 1607, so far at least as could be decided by the observations accessible to him. Hence in this case there was ground for concluding that all these appearances were one and the same comet; that its period was about 75 or 76 years in duration, and that it would probably return in 1758 or 1759. It did so return, being first seen by PALITZSCH, near Dresden, on Christmas Day, 1758; and it has ever since been known by the name of the illustrious astronomer who had so confidently predicted its return. It passed its perihelion at that appearance, 12th March 1759, and at the next return, 16th November 1835. According to PONTÉCOULANT, it will be due at perihelion again in 1910. The distance of HALLEY's comet from the Sun varies between 0.58 and 35.3, in terms of the Earth's mean distance; that of Venus, expressed in the same way, being 0.72, and of Neptune, 30.05.

HALLEY'S COMET IN 1835 (STRUVE)

A very large proportion of the known comets travel in parabolas, or in ellipses of such great eccentricity as to be undistinguishable from parabolas during the time of visibility.[39]   Following are a few particulars about the most remarkable comets which are moving in elliptical orbits of moderate eccentricity.

[39] The trustworthy records of cometary apparitions through out all past time are about 1000 in number, and the increase is very rapid at the present day, because many skilful observers with suitable telescopes are continually searching for comets. Nearly one half of the entire number (rather more than 400) have been subjected to mathematical calculation, and the paths of their motion determined.   It is only by the agreement among the elements of these paths that the identity of two or more comets can be proved; for the physical appearance and characteristics of a given comet are generally very different at the different returns.   When a comet is first discovered, if remote from the Sun, its appearance will usually resemble that of the PONS-BROOKS comet (page 188, *left*) ; and a large proportion of telescopic comets never develop beyond this simple stage.   But, if the comet is a great one, on approaching nearer a nucleus will begin to aggregate in that part of the hazy mass on the farther side from the Sun ; from this centre the wonderful phenomena of the *coma*, or head, soon manifest themselves, unfolding in envelopes or sheaths, as in COGGIA'S comet of 1874 (page 205), or more often opening toward the Sun, like the

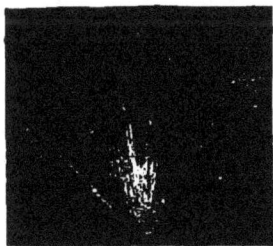

THE GREAT COMET OF 1861
(WILLIAMS)

HEAD OF THE GREAT COMET OF 1861
HIGHLY MAGNIFIED (SECCHI)
(*The tail is directed downward*)

separate parts of a fan, the handle of which is at the nucleus of the comet. This structure is well shown in the adjacent

Of the periodic comets, HALLEY'S, by far the most interesting, has been traced with high probability for

illustration of the head of the great comet of 1861. (A general view of the entire object, as it appeared in the northern heavens, is given above in miniature. This remarkable body, discovered 13th May 1861 by Mʳ TEBBUTT of New South Wales, had a tail which appeared to stretch one third the way round the heavens. The Earth and Moon passed through the tail of this body, 30th June 1861, with no apparent effect save a peculiar sky glare. According to Sir JOHN HERSCHEL, this comet far exceeded in brilliancy all other comets that he had ever seen, even those of 1811 and 1858. It recedes from the Sun to a distance nearly twice that of Neptune, in an elliptical orbit, with a period of 410 years.) On the side of the nucleus opposite the fan, and almost invariably directed away from the Sun, begins the development of the tail of the comet, usually the most striking characteristic of a great comet. Cometary tails sometimes appear almost straight, and others have varying degrees of curvature. A typical tail is that of BROOKS's comet (1886 V), shown in the accompanying illustration. Other comets (for example, DONATI's, the sixth comet of 1858, shown on page 203) have tails both straight and curved. When a comet has passed its perihelion, the tail will, in nearly all cases, be found to have swung round and changed its apparent position with respect to the comet's motion; so that it will still remain true that the tail is directed from the Sun, and the comet will move away from the Sun with its head following the tail. The early astronomers described the tail in this position as the 'beard.' Generally the median line of the tail of a large comet is dark; though this sometimes changes to a brighter streak, and vice versa, as in the case of COGGIA's comet of 1874.

BROOKS'S COMET (1886 V)

(*7th May* 1886)

nearly 1900 years. DION CASSIUS tells us of a comet seen in B. C. 12, about the time of the death of the

The tails of comets appear to be made up from the material particles first projected toward the Sun from the nucleus in a manner incompletely understood; afterward curving over and forming the cup-shaped sheaths of the coma, and then repelled by the Sun to form the trains of various types. It seems likely that these filmy, evanescent objects exist as hollow cones in space. BESSEL in Germany, NORTON in America, and in particular BREDICHIN in Russia, have contributed most to establish a theory of comets' tails which accounts for nearly all the facts of observation. BREDICHIN'S theory is that the straight tails, like those of the comets of 1858 and 1861, are probably composed of hydrogen, the Sun's action of repulsion upon which would exceed twelve times the attraction of gravitation; the moderately curved tails (the usual type) are hydrocarbons in gaseous forms, with an action of repulsion slightly in excess of gravity along the inner edge of the tail, and increasing to $2\frac{1}{2}$ times that amount on the outer edge; and a few comets exhibit still a third type of tail, — short, sharply curved, probably due to the vapor of heavier substances (iron, chlorine, and sodium), upon which the repulsive effect is relatively weak, varying between $\frac{1}{10}$ and $\frac{1}{3}$ that of the Sun's gravitative action. In rare instances a comet has been known to exhibit tails of all three of these types. Probably the tail of a comet is formed by an expenditure of the substances composing the nucleus, in such a manner that the particles once leaving the nucleus are never returned to it. The larger comets, then, must grow smaller and smaller at every return to perihelion. Undoubtedly this is the true explanation of the faint and tailless condition of most of the short period comets, whose encounters with solar disrupting influences have been frequent.

The visibility of comets chiefly depends upon two conditions, — their intrinsic lustre, and their position in space relatively to Sun and Earth. While some are seen for a few days only, and the average visibility is about three months in duration, on rare occasions a comet can be observed throughout an entire year, and the great comet of 1811 was visible for 17 months. In nearly all cases, the curve traversed during visibility is only a small part of the entire path, so that a degree of uncertainty exists in cometary orbits which does not obtain in determining the paths of planets. — *D. P. T.*

great Roman general Agrippa, and the Chinese annals also mention a comet seen at a date corresponding to this. Dr Hind has shown that this was probably the first appearance of Halley's comet. It is also probably referred to by Josephus as seen in A. D. 66, when the Jewish rebellion broke out which led to the destruction of Jerusalem four years afterward by the Roman army under Titus. The Chinese records also speak of a comet seen at that time, as well as of another in A. D. 141, which was probably the same. In A. D. 218 it appears to have been noticed both in China and Europe, being followed (according to Dion Cassius) by the death of Opilius Macrinus, the Emperor, soon after his defeat at Irumæ, near Antioch, by Elagabalus. In the years 295 and 373 we are also able to recognize with much probability Halley's comet in accounts furnished by Chinese annalists; and it can hardly be doubted that this comet was seen in A. D. 451, not only in China, but also in Europe (about the time of the great defeat of the Huns under Attila near Châlons-sur-Marne by Ætius and Theodoric). In A. D. 530 or 531, during the reign of Justinian, 'a very large and fearful comet' was seen, which (as already mentioned) Newton and Halley thought to be an appearance of the comet of 1680. The fuller observations of the Chinese have become accessible to us since their time, and show that it was more likely the comet of 1682.[40]

[40] 'Ugly monsters' that comets always were to the ancient world, the mediæval church perpetuated the misconception with such vigor that even yet these gauzy, harmless visitors from the interstellar spaces have a certain 'wizard hold upon our imagination.' This entertaining phase of the subject is invitingly treated in President Andrew D. White's scholarly 'History of the Doctrine of Comets' (*Papers of the American Historical*

Like some previous appearances, the returns of
A. D. 608 and 684 were, so far as is known, recorded
only in the Chinese annals; but in A. D. 760 a remarkable comet was chronicled, not only by their
annalists but by a Byzantine historian in the reign
of CONSTANTINE COPRONYMUS, and Dʳ HIND thinks it
little short of a certainty that this too was a return
of HALLEY'S comet. Also, appearances of it were
probably seen both in Europe and China in A. D. 837
and 912; likewise in A. D. 989, as related in the
Chinese records only. The next return was in the
year of the Norman Conquest. On the Bayeux
tapestry there is a representation of some people gazing at a comet which appeared soon after Easter
A. D. 1066, while England was threatened with two
invasions at once, — both taking place in succession,
but with very different terminations. This comet also
was doubtless HALLEY'S, and it was seen again in 1145,
1223, 1301, 1378, and 1456. In the latter year it
seems to have been very conspicuous, making a great
sensation in Europe, where the Turks, who had taken
Constantinople only three years previously, were advancing into Hungary, when they were defeated and
repulsed at Belgrade by the famous HUNYADI. At the
succeeding return in 1531, the comet was apparently
less brilliant; at any rate our knowledge of its path on
that occasion depends entirely upon the observations
of PETER APIAN, or BIENEWITZ. In 1607 this comet

Association, vol. ii.); while for a very extensive collection of
the literature of comets, both historical and scientific, reference
may be had to the *Catalogue of the Crawford Library of the
Royal Observatory, Edinburgh* (1890), pp. 89-142. The titles
of many other important papers and treatises on comets are
given at the end of this chapter. — *D. P. T.*

was observed by the illustrious KEPLER ; and it was by
a comparison of elements principally deduced from
his observations in that year, and APIAN'S in 1531, that
HALLEY identified these comets with the one observed by himself in 1682.

Two other comets have periods a little shorter than
HALLEY'S, but their history cannot be traced farther
back than the apparitions preceding the last. One
of them, found at Marseilles, 20th July 1812, by PONS,
the most successful of all discoverers of comets, became visible for a few days to the naked eye, and

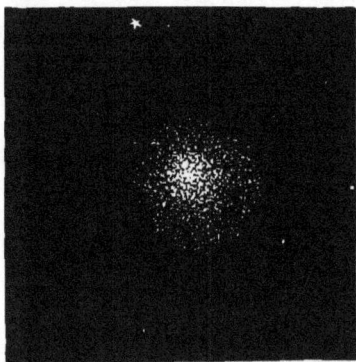

PONS-BROOKS COMET, 26th OCTO-
BER 1883

(*A typical telescopic comet ; nucleus ill
defined, coma undeveloped, tailless*)

(PONS-BROOKS COMET)
(*Drawn by Dr SWIFT
29th December 1883*)

passed its perihelion on 15th September, the very day
on which the conflagration of Moscow broke out during the French occupation of that city under NAPO-
LEON. ENCKE'S calculations made its period 70½
years. Rediscovered by Mr BROOKS at Phelps, New
York, 1st September 1883, renewed observations
showed that the period was slightly longer, and that

it would arrive at perihelion in the January following, — which it did on the 25th of that month. The next return will be due in 1955. The other comet was discovered by OLBERS at Bremen, 6th March 1815, passed its perihelion 26th April, and was calculated to have a period of about 72 years. It was rediscovered by Mr BROOKS at Phelps, 25th August 1887, and passed its perihelion 8th October; so that its period is not quite 72½ years, and it may be expected again early in 1960.

All other comets of this class have much shorter periods, and with one exception, they attain a distance from the Sun which never much exceeds that of Jupiter. ENCKE's comet is the most interesting. First discovered by MÉ-CHAIN at Paris in 1786, and again independently by CAROLINE HERSCHEL at Slough in 1795, and by THULIS at Marseilles in 1805, it was each time supposed to be a different body. PONS, the first to see it in November 1818, also imagined that it was a new comet; but at that return ENCKE (who always called it PONS's comet) showed that it was moving in an orbit with a period of only about 3⅓ years, and that it was identical with those just mentioned. He predicted another return in the spring of 1822, which was observed in that year at Paramatta, New South

CAROLINE HERSCHEL
(1750–1848)

Wales; and it has been seen at every subsequent
return, passing perihelion on the last occasion in May
1898.   Its next return is due in the summer of 1901.
Of all the periodic comets, ENCKE'S approaches
nearest the Sun.   When in perihelion, it is very near
the orbit of Mercury; and in 1835 it made a rather
close approach to the planet itself, affording the
means of determining the value of Mercury's mass.
Even in aphelion, ENCKE'S comet does not recede so
far from the Sun as several of the small planets.[41]

[41] This small, though very important comet, the seventh
discovered by Miss HERSCHEL, one of the most indefatigable
of the early searchers for comets, was formerly just visible
without the telescope, but at late returns it has been fainter, so
that it cannot now be seen without optical aid.   At no two
returns has its physical appearance been observed to be the
same.   As drawn by W. STRUVE in 1828, it was almost struc-
tureless, being very like the typical telescopic comet shown on
page 188.   In 1848 it had two
tails, one about a degree long,
pointing away from, and another much smaller directed
toward, the Sun.   In 1871
ENCKE'S comet exhibited
many freaks, — the nucleus at
one time having the shape of
a star-fish, seen obliquely at
one side of the globular mass
of the comet, and later a single straight filament or tail
shot out a long distance from
the coma.   At subsequent returns no physical phenomena
of especial interest have been
developed; it has been grow-

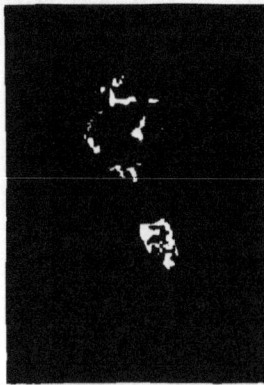

ENCKE (1791–1865)   ing irregularly fainter, and at
its last return in 1894–95, it
was first recorded, 31st October of the former year, on a photographic plate, by Dr MAX WOLF of Heidelberg, and during

Another very remarkable periodic comet is that
known as BIELA's, the periodicity of which was detected at its return in 1826, when it was discovered

the next few days it was near the limit of visibility with the
30-inch telescope at Nice.

At the return of this comet in 1838, ENCKE, the eminent German astronomer, then director of the Observatory at Berlin,
was enabled to establish a remarkable progressive diminution
in its period, amounting to about $2\frac{1}{2}$ hours at each return to the
Sun, and this circumstance led to the immediate renewal of
the old hypothesis of a resisting medium filling all space; not
dense enough to interfere with the motions of massive bodies,
like the planets, but of such nature as might be conceived to
retard the motion of tenuous bodies like comets. Subsequent
observations have fully confirmed this acceleration from a period of $1212^d.79$ in 1789 to $1210^d.44$ in 1858, as ascertained by
ENCKE in the latter year; but there is no way of finding out
whether a resistance to the comet's motion takes place everywhere throughout its orbit, or only when near the Sun. Probably the latter is true. The resisting medium, however, is not
accepted by astronomers generally, because it is felt that its
action upon other comets should be appreciable; but no such
effect upon any other comet has ever been detected. (For the
literature of the resisting medium, consult HOUZEAU'S *Bibliographie Générale*, vol. ii. col. 680–686.) ENCKE continued his
computations upon this comet down to his death in 1865; and
since that time his labors have been continued by VON ASTEN
and BACKLUND of Pulkowa. According to their researches,
the progressive shortening of the comet's period, although still
going on, was suddenly, about 1868, reduced to only half its
former rate; so that the theory of a resisting medium must
either be modified, or abandoned altogether. Professor YOUNG
has suggested a regularly recurring encounter with a cloud of
meteoric matter; and others have suggested a possible change
in the physical character of the comet. Especially marked in
the case of ENCKE'S comet has been that apparent contraction
of bulk (as also exhibited in several other comets) on approaching the Sun, and the subsequent expansion when receding from perihelion; when in this region ENCKE'S comet is
only $\frac{1}{20}$ its visible diameter when it first comes into view. The
general explanation proposed by Sir JOHN HERSCHEL is that

by BIELA at Josephstadt in Bohemia. First found,
however, in 1772 by MONTAIGNE at Limoges, it was
also independently discovered (being supposed, as in
1826, to be a new comet) by PONS in 1805. The
period was determined in 1826 to be about 6¾ years,
and the comet was accordingly observed again in 1832,
in 1845-46, and in 1852. At the second of these
appearances it was seen to have separated into two portions, the relative
brightness of which fluctuated considerably, — the investigations of HUB-
BARD, of Washington, afterward showing that this disintegration probably
occurred in the autumn of 1844. Both
portions returned, but at a somewhat
greater distance from each other, in
1852. But since then the comet has
not been seen at all, — at any rate as

BIELA'S DOUBLE COMET

(19th February 1846)

a comet, — though it is supposed that
a shower of meteors sometimes seen
about the end of November, when the Earth passes
near the comet's orbit, may form part of its dispersed
material. (See page 219.)

PONS discovered another comet at Marseilles, 12th
June 1819, and ENCKE's investigations showed that it
was moving in a short ellipse with a period of about
5½ years. It was not, however, seen again until
1858, when it was rediscovered as a new comet by

near the Sun a large part of the cometary substance is rendered
invisible by evaporation, just as a cloud or fog might be.

At the recent near approach of ENCKE's comet to the planet
Mercury in 1891, Dr BACKLUND determined anew the mass of
that planet, making it $\frac{1}{9,700,000}$ that of the Sun, a value probably too small. — *D. P. T.*

WINNECKE at Bonn, who, after determining its orbit, noticed its identity with the discovery made by PONS nearly forty years previously. Hence it is generally called WINNECKE's comet. It was observed again in 1869 and 1875, but not in 1863 and 1880, when its positions were very unfavorable. It was observed at the return of 1886, and again in 1892, passing its perihelion in June. The period having increased in length, the next return came early in 1898.

A faint comet found by M. FAYE at Paris in November 1843, with a period of about 7½ years, has been observed at every subsequent return, the last in 1895, when it was rediscovered by M. JAVELLE of Nice, 26th September, about six months before perihelion passage. It is next due in 1903.

BRORSEN's comet, discovered at Kiel in 1846 (period about 5½ years), and not seen in 1851 and 1863, was observed in 1857, 1868, 1873, and 1879. As it was not seen in 1884 and 1890, some catastrophe has perhaps overtaken it. Mr DENNING's comet of 1894 may be a portion of it.

In 1851 D'ARREST at Leipzig discovered a small comet. It was found to be moving in an ellipse with a period of about 6½ years, and was observed again (but only in the southern hemisphere at the Cape of Good Hope) in the winter of 1857-58. In 1864 it was invisible, being unfavorably placed; but it was observed at the returns of 1870 and 1877, though not in 1884. It was, however, seen again in 1890, and last in the summer of 1897. The next return is due in the autumn of 1903.

Another comet of more than double this period, seen at several returns to perihelion, though not consecutive, is known as TUTTLE's, from its discovery by

S & T — 13

Mr H. P. TUTTLE at Cambridge, Massachusetts, in January 1858, when its periodicity was determined. Previously discovered by MÉCHAIN at Paris in 1790, its period is about 13¾ years, so that it had returned four times without having been noticed. It was, however, observed in 1871, and again in 1885, when it passed perihelion 11th September. Another return is due in the summer of 1899. TUTTLE's comet is near the Earth's orbit when in perihelion, and wanders beyond the orbit of Saturn before reaching aphelion.

Two comets of short period were discovered at Marseilles and Milan by TEMPEL. The first of these (period about six years), found in 1867, was observed at the returns in 1873 and 1879, passing its perihelion on both occasions in May; but it has not since been seen, and M. GAUTIER has shown that the period has been lengthened by the perturbing action of

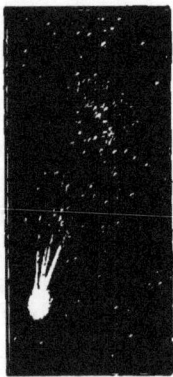

SWIFT'S COMET, 5th
APRIL 1892
(*From a photograph by*
BARNARD)

Jupiter. TEMPEL's second, discovered in 1873 (period about 5¼ years), was observed in the autumn of 1878, but escaped observation during the subsequent returns in 1883 and 1889. It was, however, re-detected near perihelion in May 1894, and the next return is due in 1899. TEMPEL's third, found in November 1869, was not recognized as a periodic comet until after it had been rediscovered in 1880 by Dr SWIFT, then of Rochester. So it is usually called SWIFT's comet; and the period being about 5½ years, an unobserved return must have taken place in 1875. The comet was due at perihelion again in 1886 and 1897 ; but, as in 1875, it was unfavorably placed, and escaped observation on both these occasions. The last observed return took place in 1891, when the comet was detected on 28th September, and passed its perihelion in the middle of November. Another return is due in 1902.

A bright comet discovered by Dr Max Wolf at Heidelberg, 17th September 1884, and moving in an elliptic orbit (period nearly seven years), returned according to prediction in the summer of 1891, and was observed again in 1898.

Mr Finlay at the Cape of Good Hope discovered a comet, 26th September 1886, which passed its perihelion on 22nd November following, and has been calculated to have a period of about 6½ years. A return duly took place in 1893 ; when the comet was rediscovered by Mr Finlay himself, 17th May, and passed its perihelion 16th June.

A comet discovered by De Vico at Rome, 22nd August 1844, and calculated to have a period of about 5½ years, was thought to be identical with one observed by La Hire at Paris in 1678, between which time and 1844 thirty periods would have elapsed. It was not seen again for more than fifty years, excepting that Goldschmidt obtained a single observation of a comet, 16th May 1855, which may have been De Vico's, though the identity is not proved. A small comet discovered by Mr E. Swift on Echo Mountain, California, 20th November 1894, moves in an orbit very similar to that of De Vico's comet. Dr Schulhof's researches show clearly the identity of the two bodies, with fluctuations in brightness ; also that the last return was delayed by the attraction of Jupiter, which was moving near the comet in 1885–86, and still nearer in 1897.

Mr BARNARD discovered at Nashville, Tennessee, July 1884, a faint comet with a period of nearly 5½ years. It was not seen at the returns of 1889 and 1895.

A comet discovered by Mr E. HOLMES at Islington 6th November 1892, has a period of about 6¾ years, and was just visible to the naked eye for a few days. As it passed perihelion 20th June, nearly five months before discovery, a return is due in 1899. Its orbit is more nearly circular than that of any other comet.

Very remarkable seems to have been the career of the comet of 1770, commonly called LEXELL'S, which was a truly unfortunate body in the way its motions were disturbed by the giant planet Jupiter. While very near the Earth it was discovered by MESSIER, 14th June 1770; and a few days afterward it approached within a distance of little more than seven times that of the Moon. On calculating its orbit, LEXELL found that it was then moving in an ellipse with a period of about 5½ years; but that had not long been the case, for in 1767 it had approached Jupiter within one sixtieth part of the radius of his orbit, and the influence of his attracting mass was so powerful that the comet's course was completely changed. At the return of 1776, its position was such that it could not become visible; and in 1779, before another return, it approached Jupiter again, much closer even than before, coming indeed within the distance of his fourth satellite. This again completely changed the comet's orbit, and made the period very much longer than LEXELL's calculations had determined it to be in 1770. On 6th July 1889, Mr BROOKS, of Geneva, New York, discovered a comet moving in an ellipse of short period. Dr CHANDLER showed that this body also made a very near approach to Jupiter in 1886, and suggested its identity with LEXELL'S comet of 1770; and the subsequent investigations of Dr POOR have corroborated this idea. As the period of BROOKS'S comet is seven years, it was visible again in 1896.

The above comprise all comets of short period known to return.[42] One discovered by PIGOTT in No-

[42] For convenient reference, the returns of many periodic comets are here tabulated, from 1898 onward: —

RETURNS OF PERIODIC COMETS

Date of Return to Perihelion	Periodic Time	Name of Comet	Apparitions Previously Observed
	Years		
1898 April	5.818	PONS-WINNECKE	6
1898 May	3.303	ENCKE	27
1898 June	8.534	SWIFT (1889 VI)	1
1898 June	6.821	WOLF (1884 III)	2
1898 September	6.507	TEMPEL$_1$ (1867 II)	3
1899 January	8.687	DENNING (1881 V)	1
1899 March	33.178	TEMPEL$_4$ (1866 I)	1
1899 April	6.309	BARNARD (1892 V)	1
1899 May	13.760	TUTTLE (1858 I)	4
1899 May	6.909	HOLMES (1892 III)	1
1899 July	5.211	TEMPEL$_2$ (1873 II)	3
1900 February	6.622	FINLAY (1886 VII)	2
1900 July	5.800	DE VICO-SWIFT	4
1900 October	5.398	BARNARD (1884 II)	1
1901 January	5.456	BRORSEN (1846 III)	5
1901 August	7.480	DENNING (1894 I)	1
1901 August	3.303	ENCKE	28
1902 December	5.534	TEMPEL$_3$ SWIFT	3
1903 September	7.566	FAYE (1843 III)	8
1903 November	7.073	BROOKS (1889 V)	2
1903 November	6.691	D'ARREST (1851 II)	6
1910 June	76.37	HALLEY	9
1955 March	71.48	PONS-BROOKS	2
1960 April	72.63	OLBERS-BROOKS	2
1985 ———	123.0	SWIFT (1862 III)	1

For the most part these are inconspicuous objects, and some of them are never visible without the telescope. They can be

vember 1783 was thought to be moving in an ellipse with a period of about five years; but this was uncertain, some thinking the period ten years, and, at any rate, the comet does not seem to have been observed either before or since. Also, one that was discovered by Mr DENNING of Bristol, 4th October 1881, has a period of about nine years; but the comet has not been seen since, probably owing to perturbations, as it approaches the paths of several planets.

Other remarkable comets, possibly seen at more than one return, cannot yet be classed as periodic bodies. Great similarity has been noticed between the elements (so far as they could be determined from descriptions of its course) of the splendid comet of 1264 and those of the fine comet observed in 1556, about the time of the abdication of the Emperor CHARLES the Fifth. This was first pointed out by DUNTHORNE in 1751, and attention was afterward specially called to it by HIND, who thought it extremely probable that these were two consecutive appearances of the same comet, and that it had returned after a sojourn of about three hundred years in the depths of space. The effect of planetary perturbation would delay another return, which, on the whole, was thought most likely to take place about 1860. Neither then, however, nor at any time since, has the comet put in an appearance. The cause of this is as yet unknown;

found by means of ephemerides, or tables of their positions among the stars, published in advance in the technical journals. Generally one or more telescopic comets may be seen at any time, and a telescope of ten inches aperture would have shown no less than seven comets visible at one time in November 1892. The total number of comets always present in the solar system has been estimated by KLEIBER (on certain hypothetical conditions) to be nearly 6,000. — *D. P. T.*

but it may be remarked that while the aphelion of HALLEY's comet is at about the distance of Neptune, a comet with a period of three hundred years would pass far beyond this outer planet of our system.

A comet observed by TYCHO BRAHÉ and others in 1596 appears to have elements similar to those of the third comet of 1845 ; and the two may be identical, with a period of about 250 years.

THE GREAT COMET OF 1843

(On 28th February, the day after its perihelion passage this comet was distinctly visible to the naked eye at noon. Its approach to the Sun was the nearest ever made by any known comet, portions of its coma being less than 50,000 miles from the Sun's surface. Its velocity at this time was at the rate of about 1,280,000 miles per hour, and the length of its tail 150,000,000 miles)

In the autumn of 1882 there was much discussion of the question whether the great comet then visible had any connection with the fine comets of 1843 and 1880, or these with each other, or with any seen in bygone centuries. It is an undoubted fact that all the orbital elements of the comets of 1843, 1880, and 1882, are very similar to each other; and it is probable that the comet of 1668, though very imperfectly observed, had similar elements also. When at perihelion, all these comets made remarkably close approaches to the Sun, coming within a distance of 700,000 miles of his centre, or about 300,000 miles of his surface. Hence it was suggested that the tremendous attractive force exerted by the Sun at so small a distance might greatly shorten the period at each return, and lead before long to the comet's absorption into the Sun, producing an incalculable outburst of solar heat. On the other hand, similarity of orbit in two or more bodies does not prove identity of those bodies, it being quite possible that two or more comets may move along the same orbit at great distances from each other. Moreover, the best determinations of the orbit in which the comet of 1882 was actually moving agree in assigning about 750 years as its period, and it is probable that all these comets are travelling along the same orbit, in about that same period.[43] An addi-

---

[43] This singular object, the most recent great comet on record, was discovered early in September 1882, by many observers in the southern hemisphere. Also Dr COMMON at Ealing, England, on the forenoon of the 17th, saw it as a bright object near the Sun; and so brilliant was it that Mr FINLAY and Dr ELKIN at the Cape of Good Hope actually followed it up to the very limb of the Sun itself. Here it utterly vanished, as if behind the solar disk; but in fact it passed in transit between the Earth and our central luminary, and became visible on the opposite side of the Sun the following day. Although it had passed within 300,000 miles of the solar surface, still there was no appreciable acceleration due to a supposed resisting medium; and Dr KREUTZ, who discussed the observations of its position, found that it is moving in a very elliptic orbit, with a period of between 800 and 1000 years. The extraordinary brilliancy, and the long period of visibility of this comet, permitted the successful application of photography to the de-

GALE'S COMET PHOTOGRAPHED AT THE LICK OBSERVATORY, 3d May 1894

(*By Professor* BARNARD *with the* 6 *in. Willard lens, exposure* 2¼ *hours. The long star-trails indicate an unusually rapid motion of the comet among the stars. The comet's tail was about* 10° *long, and the original plate showed its tendency to split into several narrow filaments*)

tional member of this comet family was discovered,
18th January 1887, which was conspicuous for a few
days in the southern hemisphere, though never visible

lineation, not only of the comet's features, but also of the lines
in its spectrum. At one time its tail was single and nearly
straight; at another, there were two tails slightly curved. But
the most curious phenomena were those exhibited at its head,
which, during recession from the Sun, underwent a remarkable subdivision into three or four, and, according to some observers, six or eight, distinct cometary masses. Stranger still
was the anomalous sunward extension of the coma, or sheath,
in a vast envelope 4,000,000 miles in diameter. All told, the
great comet of 1882 is perhaps the most remarkable on record,
and the continued eccentricities of its physical development
were, no doubt, the legitimate product of disturbances brought
about by its near approach to the Sun.

But even more vivid in memory, doubtless, is the great comet
of the year preceding, discovered by Mr TEBBUTT in New
South Wales, 22d May 1881, and regarded by some observers
as a more striking object than COGGIA's comet of 1874. Moving rapidly northward, it shone as a brilliant object in the
northern heavens through the month of June, and was the
first comet satisfactorily recorded by photography,— DRAPER
in New York, and JANSSEN in Paris securing successful pictures almost simultaneously. Dr MEYER measured with great
care the positions of stars traversed by this object, and found
evidence of refraction of the stellar rays by the comet, although
in no instance was there any perceptible absorption of the star's
light. Celestial photography has been of the greatest service
in cometary astronomy, — the long exposures regulated to con-
formity with the comet's motion making it possible to secure
most of the physical features satisfactorily. When a comet's
own motion is rapid, and its exposure long, all the stars appear
as trails, instead of bright points, as shown in the photograph
of GALE's comet (opposite page). Also a comet has been first
discovered with the assistance of photography, by Dr BARNARD
at the Lick Observatory, 12th October 1892. His photographs
of BROOKS's comet of 1893 show rapid and violent changes in
the tail, as if shattered by an encounter with a swarm of
meteors. Also a comet was discovered by Mr CHASE of the
Yale Observatory, on several plates exposed for the Leonids
in November 1898.— *D. P. T.*

in the northern. It should, however, be remembered that the few and uncertain observations of the comet of 1668 scarcely admit any decided conclusion with regard to its path. The tail only was visible in Europe; and our knowledge of the comet's orbit is chiefly derived from a map of its course in the heavens, laid down from some very rough observations made in the East Indies, extending over an interval of less than a fortnight. But they indicate that this object is probably moving in the same orbit as the comets of 1843, 1880, 1882, and 1887 are.

In modern times comets which would otherwise have eluded detection have been ' caught,' in passing near the Sun while the latter was totally eclipsed. A photograph of the eclipse of 17th May 1882, taken by Dʳ Schuster in Egypt, clearly depicted such a stranger in the rays of the outer corona (shown in the illustration on page 86). Its momentary detection was due to the obscuration of the Sun's light by the Moon when the photograph was taken. As this body was not visible either before or afterward, nothing is known of its orbit, and probably its motion was very rapid. This comet was called ' Tewfik,' after the then Khedive of Egypt. Another, much fainter and more difficult to recognize, was photographed in the corona during the total eclipse of 16th April 1893, by Mʳ Schaeberle in Chile, and by other expeditions in South America and West Africa.[44]

---

[44] The earliest comet detected during a total eclipse of the Sun was recorded by Seneca; and during a similar obscuration, probably total, slightly south of Constantinople, 19th July, A. D. 418, another was discovered. On a sketch of the corona of 18th July 1860, by Winnecke at Pobes, Spain (*Mém. Acad. Impériale, St Pétersbourg*, vii. Série, t. iv. 1), is a curious structure, which, according to Ranyard, may have been due to a

Of comets moving in elongated ellipses approaching in form to parabolas, many require several thousand years to complete a whole revolution, and they are observable for so small a proportion of their whole course that its length cannot be determined with accuracy. The revolution of the great comet of 1858 (known as Donati's) is accomplished in about 2,000 years; the splendid comet of 1811 has a period of more than 3,000 years; the fine comet discovered by M. Coggia at Marseilles in 1874 (the sheaths of the coma of which are illustrated on page 205) has been computed to be moving in an orbit of more than

DONATI'S COMET (1858 VI)

(*Drawn by* G. P. Bond, *4th October. The bright object near the comet's head is Arcturus, over which the comet passed the following day, without appreciably dimming the star's light*)

comet; also, on the Indian photographs of the total eclipse of 12th December 1871, a comet is thought to have been recorded (*Monthly Notices Royal Astronomical Society*, xxxiv. 365). It is a singular coincidence that the four modern eclipses during which comets were either seen or suspected are separated by intervals of about eleven years. Among other comets seen at one apparition only, 1847 VI is interesting from the circumstances of its discovery, at Nantucket, 1st October, by Miss Mitchell, to whom a gold medal was awarded by Christian the Eighth, king of Denmark. Madame Rümker also discovered it independently at Hamburg ten days later. A recent discussion of its motion by Miss Margaretta Palmer of the Yale Observatory assigns to it a hyperbolic orbit. — *D. P. T.*

10,000 years' period; and others have orbits of even greater duration. All the comets of short period move round the Sun in the same direction as the planets; [45] but many of those (including HALLEY's)

[45] The chemical composition of comets has been investigated by means of the spectroscope, — this instrument having been first used for this purpose in 1864 by DONATI, who applied it to TEMPEL's comet of that year, disclosing the fact that these bodies are in large part self-luminous. HUGGINS found the spectrum of WINNECKE's comet (1868 II) to coincide with that of olefiant gas in a vacuum tube; but it was not until the appearance of COGGIA's comet (1874) that the full capabilities of the new method could be tested. This brilliant object revealed the five bands of the complete hydrocarbon spectrum shading off on one

MARIA MITCHELL (1818–1889)

side; but there was also a relatively faint continuous spectrum, showing that while the gaseous portions of the comet were largely compounded of carbon and hydrogen, its light was in part reflected from the Sun. Researches on cometary spectra were continued by VOGEL, HASSELBERG, YOUNG, and others. The comet of 1881, known as TEBBUTT's, or (1881 III), was the first one whose spectrum was photographed, by DRAPER in New York, and HUGGINS in London, disclosing bands in the violet due to carbon compounds, and showing an identity with hydrocarbons burning in a Bunsen flame. Then came the two comets of 1882, approaching very near the Sun, and displaying the bright lines due to sodium, whose brightness increased as the distance of the comets from the Sun dimin-

which travel in long elliptic orbits, and many whose paths are parabolas, have retrograde motion. TEMPEL'S comet of 1866 (the comet of shortest known period combined with retrograde motion) travels round the Sun in rather more than 33 years, and is intimately associated with the stream of November meteors, or Leonids, as will be related in the next chapter.

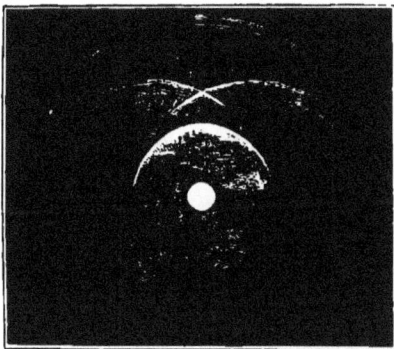

ished. Near perihelion the h y d r o c a r b o n bands were very faint or wholly invisible; and this behavior is regarded by Dr SCHEI

THE HEAD OF COGGIA'S COMET

13th July 1874 (BRODIE)

NER as proving that the intrinsic light of these comets had its origin in disruptive discharges of an electric nature in their interior. Also Dr COPELAND, now Astronomer Royal for Scotland, detected in the spectrum of the September comet of 1882 five other bright lines in the yellow and green, due to the vapor of iron. No bright comet has since appeared. It is interesting to note, in the observations just related, a partial confirmation of the theories of Dr BREDICHIN, previously stated, according to which the length and curvature of comets' tails depend upon an electrical repulsion varying with the molecular weight of the repelled gases. The comets of 1881 and 1882, largely made up of hydrocarbon compounds, presented the second type of tail, as indicated by theory. Whether the long, straight tails of the first type are hydrogenous in origin, and the shortest bushy tails of the third type are due to the vapor of iron and other heavy substances, can be decided only when bright objects of these separate classes shall have made their appearance in the future. — *D. P. T.*

BESSEL, Several papers in *Astron. Nachrichten*, xiii. (1836).

HERSCHEL, J. F. W., 'Halley's Comet,' in *Cape of Good Hope Astronomical Observations* (London 1847).

HIND, *The Comets: a descriptive treatise* (London 1852).

WATSON, *A Popular Treatise on Comets* (Ann Arbor 1860).

HOEK, 'Comet Systems,' *Recherches astron. de l'Observatoire d'Utrecht* (The Hague 1861–64).

BOND, G. P., 'Donati's Comet,' *Annals Harvard College Observatory*, iii. (Cambridge, U. S. 1862).

CARL, *Repertorium der Cometen-Astronomie* (Munich 1864).

HERSCHEL, J. F. W., *Familiar Lectures on Scientific Subjects* (New York 1866), p. 91.

HUGGINS, *Philosophical Transactions*, clviii. (1868), 556.

WATSON, *Theoretical Astronomy* (Philadelphia 1868), 536, 638.

TAIT, *Proceedings Royal Society Edinburgh*, vi. (1869), 553.

v. OPPOLZER, *Bahnbestimmung*, 2 vols. (Leipzig 1870–1880).

ZÖLLNER, *Über die Natur der Cometen* (Leipzig 1872).

v. ASTEN, *Theorie Encke'schen Cometen* (St Petersburg 1872).

KIRKWOOD, *Comets and Meteors* (Philadelphia 1873).

GYLDÉN, *Calcul des Perturbations des Comètes* (Stockholm 1877).

GUILLEMIN, *The World of Comets* (London 1877).

HOUZEAU, *Annuaire de l'Observatoire Royal* (Brussels 1877).

BREDICHIN, *Annales de l'Observatoire de Moscou*, iii. (1877), et seq. ; *Copernicus*, i. (1881) ; ii. (1882) ; iii. (1884).

NORTON, 'Physical Theory,' *Am. Jour. Science*, cxv. (1878), 161.

WILSON, 'Comets of 1880–83.' *Publ. Cincin. Obs.*, Nos. 7 and 8.

WINLOCK, W. C., 'Comet of 1882,' *Wash. Obs.* 1880, App. I.

PEIRCE, *Ideality in the Physical Sciences* (Boston 1881).

COPELAND and LOHSE, *Copernicus*, ii. (1882), 225.

YOUNG, BOSS, WRIGHT, and others, 'Comet 1881 b,' *American Journal of Science*, cxxii. (1881).

MEYER, 'Comet of 1881,' *Archives Sciences Genève*, viii. (1882).

HUGGINS, *Proc. Royal Institution*, ix. (1882) ; x. (1883), 1.

BACKLUND, 'Encke's Comet,' *Mém. Acad. Impériale Sciences St Pétersbourg*, xxxii. (1884) ; xxxiv. (1886) ; xxxviii. (1892).

GILL, '1882,' *Ann. Roy. Obs. Cape of Good Hope*, ii. (1885), pt. I.

WEISS, 'Bericht,' *Viertel. Astron. Gesell.* xx. (1885), 287.

NEWTON, 'Biela's Comet,' *Amer. Jour. Science*, cxxxi. (1886), 81.

UNTERWEGER, *Denk. Akad. der Wissenschaften Wien* (1887).

PROCTOR, *Popular Science Monthly*, xxxi. (1887), 50.

HARKNESS, 'Orbit Models,' *Sidereal Messenger*, vi. (1887), 329.

THOLLON, *Annales de l'Observatoire de Nice*, ii. (1887).

BOSS, 'Comet Prize Essay,' *Hist. Warner Obs.* (Rochester 1887).

LANGLEY, *The New Astronomy* (Boston 1888).
BERBERICH and DEICHMÜLLER, 'Brightness of Encke's Comet,' *Astronomische Nachrichten*, cxix. (1888); cxxxi. (1893).
CHAMBERS, *Descriptive Astronomy*, vol. i. (London 1889).
HIRN, *Constitution de l'Espace Céleste* (Paris 1889).
PLUMMER, 'Comet Groups,' *Observatory*, xiii. (1890), 263.
KIRKWOOD, 'Age,' *Publ. Ast. Soc. Pacific*, ii. (1890), 214.
PETERS, C. F. W., *Himmel und Erde*, ii. (1890), 316.
TISSERAND, 'Origin,' *Bulletin Astronomique*, vii. (1890), 453.
CALLANDREAU, *Ann. Obs. Paris, Mémoires*, xx. (1890).
DENNING, 'Comet-seeking,' *Telescopic Work* (London 1891).
NEWTON, 'Capture of Comets by Jupiter,' *Am. Jour. Science*, cxlii. (1891); *Memoirs National Academy of Sciences*, vi. (1893).
BARNARD, 'Classification,' *Astronomical Journal*, xi. (1892), 46.
CLERKE, *History of Astronomy* (London 1893), 109, 392, 417.
WILSON, '1889 v,' *Popular Astronomy*, i. (1893), 151.
SCHEINER and FROST, 'Spectra of Comets,' *Astronomical Spectroscopy* (Boston 1894).
GALLE, *Verzeichniss der Elemente der bisher berechneten Cometenbahnen* (Leipzig 1894).
LYNN, 'History of Encke's Comet,' *Nature*, li. (1894), 108.
OLBERS, *Sein Leben und Seine Werke* (Berlin 1894).
SPITALER, *Denk. Akad. Wissenschaften Wien*, lxi. (1894).
PLUMMER, *Jour. Brit. Astr. Association*, iv. (1894), 89.
HOLETSCHEK, *Denk. Akad. Wissenschaften Wien*, lxiii. (1896).
MÜLLER, *Photometrie der Gestirne* (Leipzig 1897).
LYNN, *Remarkable Comets* (London 1898).
PAYNE, 'Comet Families,' *Popular Astronomy*, vi. (1898).

The keen popular interest in comets is responsible for a vast literature which is excellently represented by the long lists of titles in the *Indexes* of POOLE and FLETCHER. The later volumes of these invaluable indexes contain also many titles of scientific papers; but the fullest acquaintance with them will be greatly facilitated by reference to HOUZEAU's *Vade Mecum de l'Astronome*, and his *Bibliographie Générale*, ii. (Brussels 1882). Comet literature since that date is best summarized in *Bulletin Astronomique*, a Paris monthly, begun in 1884; also the concise notes in the annual reports of the Smithsonian Institution are helpful. Current information concerning comets is given in the *Annuaire du Bureau des Longitudes*, the astronomical notes in *Nature*, and in *Knowledge, The Observatory*, and *Popular Astronomy*. — *D. P. T.*

CHAPTER XIII

METEORIC BODIES

COMETS have been for about two hundred years recognized as cosmical bodies, often travelling in regular orbits round the Sun; but that swarms of meteoric bodies move in a similar manner has been known during only the last sixty years.[46]

[46] The true cosmic nature of luminous meteors has, however, been recognized a full century; earlier they were regarded purely as phenomena of the upper atmosphere, in origin as well as manifestation. While HALLEY and others seem to have had a vague notion of the insufficiency of this view, the German philosopher CHLADNI, celebrated for investigations of the laws of sound, seems to have first expressed the true conception of meteoric matter, having published in 1794 the view, essentially unmodified to-day, that interplanetary space is tenanted by shoals of moving bodies, chiefly iron, exceedingly small in mass and dimension as compared with the planets. Multitudes of these minute bodies, then, collide with the Earth while it is

CHLADNI (1756–1827)

The first definite establishment of the relation of such a swarm to the solar system is principally due to the labors of Professor NEWTON of Yale University, in the case of the November meteors. The recurrence of the shower as seen in the United States, 13th November 1833, on about the same day of the year that a like shower had been witnessed by HUMBOLDT and BONPLAND in South America 34 years earlier (12th November 1799), led to a careful examination of previous accounts, and to the discovery of evidence that a shower radiates from the same point in the heavens about that time every year.[47] It happens later by

travelling through a shoal of them; and by their impact with the upper regions of our atmosphere, and by friction in passing swiftly through it, are heated to incandescence, thus presenting the luminous phenomena commonly known as shooting stars. For the most part they are consumed or dissipated before reaching the solid surface of our planet. The fact is determined from theory and well established from observation that more meteors are seen during the morning hours (from 4 to 6 A. M.) than at any other nightly period of equal length, because the sky is then nearly central around that general direction in which the Earth is moving in its orbit about the Sun; and to the meteors which would fall upon the Earth if it were at rest must be added those overtaken by reason of our own motion. Also it is now recognized that from July to January, while the Earth is passing from aphelion to perihelion, more meteors are seen than during the first half of the year; but this seems to be chiefly because the rich shoals of August and November are then encountered. — *D. P. T.*

[47] This point is situate in the constellation Leo, and the meteors of this shower are therefore known as Leonids. They have fallen at 33-year intervals since A. D. 902. Their average velocity on reaching the Earth is about 44 miles per second. Meteors when encountered by our atmosphere are moving through space in lines essentially parallel; a fact of the first order of significance, which TWINING and OLMSTED in 1834 were the earliest to recognize. Indeed, they actually suggested

about two days each century; but it is especially
brilliant and conspicuous at the end of each interval

the cometary character of the November shower. The apparent phenomena of a great shower are well illustrated in the
opposite chart of a fall of August meteors, or Perseids, in which

DENISON OLMSTED (1791–1859)

the apparent path of each meteor in the heavens is carefully
laid down, and the seeming radiation from a central point,
wholly due to perspective effect, is made clear. These apparent paths are the projected sections of orbits described by the
meteors round the Sun, and intersecting our orbit. If a meteor
were met by the Earth head on, it would of course be charted as
a mere point at the centre, or radiant; those meteors nearest to
this position have their luminous courses most foreshortened, as

of 33 or 34 years. Professor NEWTON showed that
the bodies must move in one of five well defined

represented by the shortest arrows; while those farthest from
the radiant are visible through the longest apparent course
among the stars. Accurate observation of the tracks of meteors is very difficult ; and on every shower-chart will be found a
few arrows whose prolongation does not pass near the radiant.

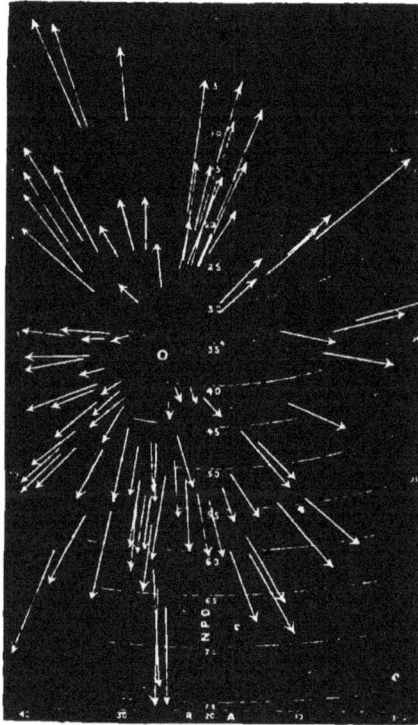

THE APPARENT RADIATION OF METEORS
(DENNING)

*(A shower of Perseids, the arrows showing the vis-
ible path of each meteor observed, and their prolongation backward defining the position of the
radiant in the sky)*

orbits to satisfy this delay; but the investigation of ADAMS, of Cambridge, England, showed that one only of these would explain all the circumstances of the

Stray or sporadic meteors are often thus indicated, as well as the effects of imperfect observation.

Dr MAX WOLF of Heidelberg, having exposed a plate an hour for the stars of the Milky Way, 7th September 1891, found

APPARATUS FOR PHOTOGRAPHING TRAILS OF METEORS

(An inclined or polar axis driven by clockwork, so as to follow the stars accurately. Upon this axis are mounted several cameras which can be so pointed as to cover a selected portion of the sky. Then, the clockwork being started, and the caps removed, each luminous meteor makes its own exposure, and its trail is impressed among the stars upon the plate with absolute precision)

on development a fine, dark, nearly uniform line crossing it, due to a meteor of uneven brightness in its flight; and this is the first meteor trail ever photographed (see also page 359). During the showers of 1894, Professor NEWTON and Dr ELKIN of New Haven brought into successful service at the Yale Observatory the novel piece of apparatus pictured in the ac-

WINNECKE (1835–1897)

(FRIEDRICH AUGUST THEODOR WINNECKE was an astronomer of unusual talent and exceptional versatility, as shown by his extensive researches in both theoretic and observational astronomy, especially relating to the comets (p. 193). He was for a time Vice-director at Pulkowa)

NEWTON (1830–1896)

(HUBERT ANSON NEWTON, late Professor of Mathematics at Yale, besides his classic researches on the connection between comets and meteors, proved in 1875 the origin of comets extraneous to the solar system; and he showed how planets, Jupiter especially, capture these bodies)

case. This true explanation is that the swarm of meteors, of which the November shooting-stars form a portion, moves round the Sun in a regular elliptic orbit, with a period of about 33¼ years ; and that the perihelion of that path, where the meteors are of course nearest to the Sun, very nearly touches the Earth's orbit, as indicated by the accompanying diagram.

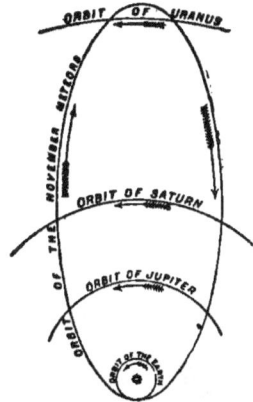

The Earth, passing through that point of its orbit in the middle of November, encounters meteoric matter on every such occasion ; but the meteors being much more closely aggregated in a certain portion, called the 'gem of the ring,' the most abounding showers occur only every thirty-third year.

PATH OF THE NOVEMBER METEORS AND OF TEMPEL'S COMET (1866 I)

(*In relation to the planetary orbits, to which it is inclined in space about 17°*)

As, however, the 'gem' extends a great distance along the line of the meteoric orbit (about, in fact, $\frac{1}{15}$ · part of its whole length), fine displays are generally seen on several years in succes-

companying illustration ; and by means of it the trails of bright meteors have been located among the stars impressed on photographic plates, with a precision wholly unattainable by the purely optical methods of the past. These investigations are conducted at the charge of a trust fund of the National Academy of Sciences, provided by the late Dʳ J. LAWRENCE SMITH, whose researches added very greatly to our knowledge of the meteorites. The principal of this fund, about $8000, was derived from the sale of Dʳ SMITH's splendid collection of meteorites to Harvard University. — *D. P. T.*

sion. The orbital motion of these meteors, like that of many comets, is in a reverse direction to that of the planets; and this of course greatly increases the relative velocity with which a portion of them enter the Earth's atmosphere, while the larger part of the mighty stream sweeps on nearly toward the part of space which the Earth has left. The aphelion of their path is a little beyond the orbit of Uranus. As a result of these investigations, another grand display of meteors was predicted for 14th November 1866; and this was strikingly verified.[48]

[48] Like displays are expected, 14th November 1899 and 1900. Sometimes astronomers when sweeping the sky for comets encounter small meteoric bodies in their search. A remarkable flight of faint, telescopic meteors was thus observed by Professor BROOKS, 28th November 1883, as depicted in the accompanying illustration. They were very small, and most of them left a faint train, visible in the telescope for one or two seconds. Mr DENNING estimates that the telescopic meteors are twenty-fold more numerous than the naked-eye objects of this nature. At rare intervals telescopic views of very large

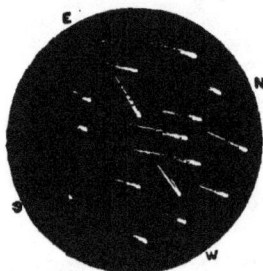

FLIGHT OF TELESCOPIC ME-
TEORS, 28th November 1883
(BROOKS)

meteors, usually called fireballs, or aerolites, have been caught, and all these physical phenomena are doubtless of the same general character, although variously represented by the different astronomers who have been favored with the opportunity of making such observations. Generally the individual particles are seen to be balloon-shaped.

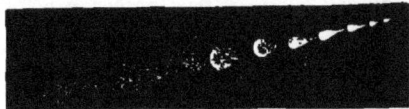

METEOR OF 28th DECEMBER 1888 (DENNING)

Above is an excellent picture of the disintegration of a slow

Not long afterward the meteors seen about 10th August were subjected to a similar examination, with the result that they too move in an elliptic orbit round the Sun in a reverse direction to that of the Earth and planets; but their path is much more eccentric than that of the November meteors, and though at perihelion it meets the Earth's orbit, at aphelion it stretches far beyond Neptune.

These discoveries were speedily followed by another, priority in which belongs to Professor SCHIAPARELLI of Milan. It is that each of these meteoric orbits is identical with that of a comet, which bodies must therefore be regarded as sustaining very close and intimate connection with the meteors. The orbit of the

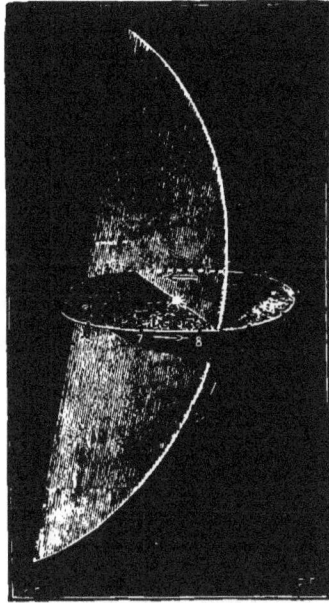

PATH OF THE AUGUST METEORS

(*In relation to the Earth's orbit, which is represented by the small ellipse. Around it are the numbers of the months. Only the perihelion portion of the meteoric path is shown, and its intersection with our orbit is at the point where the Earth is about 10th August. The radiant of this shower is in the constellation Perseus*)

moving meteor in its aerial flight, as drawn by Mr DENNING, of Bristol, England, who ranks among the foremost of living authorities on meteoric astronomy. At the last phase the pear-shaped body spread into a wide stream of fine ashes and disappeared. Such celestial displays are most likely in April, August, November, and December. — *D. P. T.*

November meteors [49] has the same elements as that

[49] Só systematically have meteoric phenomena been watched and recorded during the past quarter century, particularly by the Italian astronomers, that nearly 300 distinct showers are now recognized. Following is a selected list of those observed in recent years by Mr DENNING at Bristol : —

Name of Shower	Position of Radiant Point		Dates of Observation
	Right Ascension	Declina- tion	
	h m	°	
Quadrantids	15 20	52 N.	January 2–3
ζ Cancrids	7 56	16 N.	January 2–4
κ Cygnids	19 40	53 N.	January 14–20
θ Coronids	15 32	31 N.	January 18–28
α Draconids	14 4	69 N.	February 1–4
α Serpentids	15 44	11 N.	February 15–20
β Leonids	11 40	10 N.	March 13–15
β Ursids	10 44	58 N.	March 24
ζ Draconids	17 32	62 N.	March 28
β Serpentids	15 24	17 N.	April 17–25
Lyrids	18 0	32 N.	April 17–20
η Aquarids	22 28	2 S.	April 29–May 6
α Coronids	15 24	27 N.	May 7–18
η Aquilids	19 36	0	May 15
η Pegasids	22 12	27 N.	May 29–June 4
δ Cepheids	22 20	57 N.	June 10–28
Vulpeculids	20 8	24 N.	June 13–July 7
α Cygnids	20 56	48 N.	July 11–19
α–β Perseids	3 12	43 N.	July 23–August 4
δ Aquarids	22 36	12 S.	July 27–29
λ Andromedes	23 20	51 N.	July 30–August 11
Perseids	3 0	57 N.	*August 9–11
κ Cygnids	19 28	53 N.	August 5–16
o Draconids	19 24	60 N.	August 21–25
ε Perseids	4 8	37 N.	August 21–September 21
γ Pegasids	0 20	10 N.	August 25–September 22
β Piscids	23 4	0	September 3–8
η Aurigids	4 52	42 N.	September 12–October 2
Lyncids	6 36	43 N.	September 21–25
ε Arietids	2 40	20 N.	October 11–24
Orionids	6 8	15 N.	October 17–20
δ Geminids	7 4	23 N.	October
ε Taurids	3 40	9 N.	November 2–3
Leonids	10 0	23 N.	November 13–14
Leo Minorids	10 20	40 N.	November 13–28
ε Taurids	4 12	22 N.	November 20–28
Andromedes	1 40	43 N.	November 26–27
Geminids	7 12	33 N.	December 1–14
α Geminids	7 56	29 N.	December 7–12
κ Draconids	12 56	67 N.	December 18–29

* Probably, Mr DENNING says, the Perseids are observable more than a month, during which period their radiant is con-

of a small comet discovered by TEMPEL at Marseilles, 19th December 1865, which passed its perihelion, 11th January 1866, and was found to have a
period of 33¼ years. The Chinese observed a comet
in the year corresponding to A. D. 1366, which, it is
thought, was probably moving in an orbit identical
with TEMPEL'S, seen just 500 years afterward (equal to
15 of its calculated periods). Farther, the orbit of
the meteors of 10th August coincides with that of the
third comet of 1862 (discovered by Professor SWIFT,
15th July, and visible to the naked eye during part of
August and September), which has been calculated to
revolve round the Sun in about 120 years. The dispersion of meteoric particles along the cometary
orbit must be much more complete in the case of
the August than of the November meteors, since
the abundance of the former is far more nearly
uniform.[50]

tinually shifting eastward and northerly among the stars.
Many of the showers in the above list are as yet incompletely
determined, and scattering observations by amateurs are well
worth recording. It will readily be inferred that there are few
clear, moonless nights when meteors can not be seen with a
little patient watching. — *D. P. T.*

[50] A word concerning the origin of meteors. Many theories
have been advanced — that they came from the Sun, from the
Moon, from the Earth as a product of volcanic action, and so
on ; but the difficulties barring the acceptance of these and many
other hypotheses are very great. On the other hand, the theory
that all meteors were originally parts of cometary masses is one
that may be accepted without much hesitation. Comets in the
past have been known to disintegrate, — for example, that of
BIELA, already illustrated on page 192, in its double form ; and
its story has been admirably told by Professor NEWTON. Apparently the process went on so rapidly that in 1885 farther
subdivision had taken place, and it is now exceedingly unlikely
that any part of the comet will ever be seen again, as a comet.

Another interesting fact bearing upon this subject must be mentioned here. On 27th November the

During the shower of the Biela meteors, 27th November 1885, however, a piece of this comet fell upon the Earth, at Mazapil, Mexico; and on subsequent occasions we may expect to capture still other fragments, as the Bielids enter the atmosphere with a velocity of only ten miles a second. The strange disintegration of the nucleus of the great comet of 1882 was watched and depicted by numerous reliable observers; and as it is a member of a cometary family, the comets of 1843, 1880, and 1887 being fellow voyagers along the same orbit through space, it is not unreasonable to assume that all

BROOKS COMET OF 1890

(*In process of disintegration*)
(BARNARD)

four of these bodies may in past ages have been a single comet of unparalleled proportions. Other comets, too, have been recognized as fragmentary, but only a single additional instance need be cited, — that of BROOKS's comet of 1890, the separate nuclei of which are shown in the above illustration.

Many comets, then, having been known and seen to disintegrate under the repeated action of solar forces, which forces are exerted in greater or less degree upon all bodies of this class, the theory of the secular disruption of all comets is reached as an obvious conclusion. From several independent lines of inquiry, it is known that the age of the solar system is exceedingly great, probably many hundreds of millions of years; we should therefore expect to find almost the whole mass of a few comets (perhaps the older ones upon which the disrupting forces had been operant for indefinite ages) already shattered into small pieces; so that instead of the original comet, we should have an orbital ring of opaque meteoric bodies, travelling round the Sun in the original cometary path, each body too small to reflect an appreciable amount of solar light, and only visible as a meteor when in collision with our atmosphere. Already the identity of four such comet-orbits

Earth crosses the orbit of BIELA'S comet, which body, it will be remembered, has ceased to appear according to calculation; and about that time showers of meteors have been seen to radiate from a point in Andromeda, from which a body moving in the orbit of the comet would seem to approach. A very conspicuous display was seen on that date in 1872, when the Earth crossed the comet's orbit about three months after the body itself should have passed that point; and it was even suspected that a portion of BIELA'S comet was afterward seen receding in the opposite direction to that of the radiant point of the meteors. But this remains very uncertain, as it was difficult to reconcile the motion of that body with the theoretical motion of the comet. Another fine display of meteors was seen 27th November 1885, when the principal part of the comet was even nearer the Earth than in 1872. Doubtless there is some connection between these meteors and the comet, and the radiant will continue to be watched with great interest. A considerable shower of meteors was noticed to emanate from this region of Andromeda, 23d November 1892, somewhat earlier in the month than previous appearances; and it seems probable that this was a sporadic group, also that a larger shower may be expected near the end of November 1905.

and meteor streams has been established beyond a doubt, giving rise to the well-known showers of 20th April, 10th August, 14th and 27th November; and there no longer seems to be any substantial reason for doubting a like origin for all other meteoric bodies. Meteors, then, belonged originally to comets, which in the lapse of ages have trailed themselves out, so that discrete particles of the primal mass are liable to be encountered anywhere along the original cometary paths. — *D. P. T.*

CHAPTER XIV

METEORIC BODIES (*continued*)

WITHIN very recent years, meteoric bodies have been seen to fall upon our globe during periods of well known showers; and these objects, captured by the Earth and now in relatively small numbers in our cabinets and laboratories, are to be regarded as actual fragments of celestial bodies which were originally comets, but which have become disintegrated by the action of solar forces at their returns to perihelion. The recent recognition of this connection between meteorites and the showers of April, August, and November, breaks down the last objection to the theory (toward which astronomers have long been drifting, and which may now be accepted without reserve) that all luminous meteors, heretofore classified and subdivided as aerolites, meteoroids, shooting stars, bolides, siderolites, fireballs, meteorites, and the like, belong to a single class, and have a community of origin. Their points of difference relate only to size, and to composition, chemical and physical.

Only two general terms, then, are necessary. All those celestial bodies colliding with the Earth and producing sudden and characteristic luminous phenomena of the sky are (1) *meteors*. Sometimes they are very large, perhaps tons in weight; or their composition may be peculiar; or their encounter with the atmosphere may take place near a minimum velocity: for those and other reasons, meteors which resist that intense disruptive action due to the friction of the atmosphere and the heat due to atmospheric impact, and fall down upon the surface of the Earth, are known as (2) *meteorites*. Of the last there are many classes, representing in all about 400 different falls, few of which were, however, actually seen to fall. The most celebrated collections of meteorites are at Vienna, the British Museum, Paris (illustrated below), Harvard University (collected by LAWRENCE SMITH), Amherst College (collected by SHEPARD), Yale University, and Berlin.

THE PARIS COLLECTION OF METEORITES
(*In the Muséum d'Histoire Naturelle, begun by* CORDIER *and largely augmented by* DAUBRÉE)

Although the descent of meteoric bodies from the sky was generally discredited till the beginning of the present century, still such falls have been recorded from the earliest times. Usually regarded as prodigies or miracles, such stones have commonly been objects of worship among the oriental peoples; for example, the Phrygian stone, the ' Diana of the Ephesians

which fell down from Jupiter,' the famous stone built into the
Kaaba at Mecca, and to-day revered by Mohammedans as a holy
relic.   The earliest known meteoric fall is historically recorded
in the Parian Chronicle as having occurred in the island of Crete,
B. C. 1478.   We must pass over the numerous falls of meteoric
bodies first brought together and discussed by CHLADNI
(chief among which was the Pallas or Krasnoiarsk iron, an
irregular mass weighing about ¾ of a ton, found in 1772 by
the celebrated traveller PALLAS near Krasnoiarsk, Siberia, and
the greater part of which is now preserved in the Imperial
Museum at Saint Petersburg), and devote a few words to a
' shower of stones ' in the department of Orne, France, early in
the present century.   BIOT, the distinguished physicist and
academician, who was directed by the Minister of the Interior
to investigate the subject, reported that about 1 P. M. on Tues-
day, 26th April 1805, there had been a violent explosion in the
neighborhood of L'Aigle, heard for a distance of 75 miles.
round, and lasting five or six minutes.   A rapidly moving fire-
ball had been seen from several adjoining towns in a sky generally clear, and there was absolutely no doubt that on the
same day many stones fell in the neighborhood of L'Aigle.
They were scattered over an elliptical area 6¼ miles long, and
2½ miles broad, and BIOT estimated their number between two
and three thousand.   Thenceforward the descent of meteoric
matter upon the Earth from interplanetary space has been
recognized as an undoubted fact.

Many similar phenomena since investigated show (as would
be expected from their cosmic origin) that meteorites fall upon
our planet without reference to latitude, or season, or day and
night, or weather.   Their temperature on entering our atmosphere is probably not far from 200° centigrade below zero, and
their velocities range from 10 to 45 miles a second.   So' great
is the atmospheric resistance to their flight that on traversing
the whole of this protecting envelope, their velocity is vastly
reduced, and very suddenly.   On the basis of experiments by
Professor A. S. HERSCHEL, it was found that the velocity (at
ground impact) of the meteorite which fell at Middlesborough,
Yorkshire, 14th March 1881, was only 412 feet a second; and
at Hessle, Sweden, 1st January 1869, several stones fell on ice
only a few inches thick, and rebounded without either breaking
it or being themselves broken.   The flight of a meteor through
the atmosphere is only a few seconds in duration, and owing to
the sudden reduction of velocity, it will continue to be lumin-

ous throughout the upper part of its course only; on the average, visibility is found to begin at an elevation of about 70 miles, and end at about half that altitude. The work done by the atmosphere in suddenly reducing the meteor's velocity appears in considerable part as heat, fusing its exterior to incandescence. But conduction of heat in stony meteoric masses is slow; and notwithstanding the excessively high temperature of the exterior, during luminous flight, there is good reason for believing that all large meteorites, if they could be reached at once on striking the Earth, would be found to be cold ; because the smooth, black crust or varnish, which invariably encases them as a result of intense heat, is always thin.

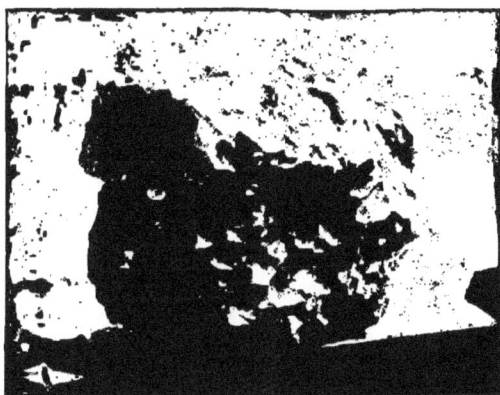

METEORIC IRON FROM ORANGE RIVER, SOUTH AFRICA
IN THE SHEPARD COLLECTION OF AMHERST COLLEGE

(*Extreme length* 19 *inches, weight* 326 *lbs.*)

The Orange River meteorite of the Amherst College collection, here photographed, is an excellent typical specimen. Particularly are the pittings or thumb-marks, generally covering the exterior of iron meteorites, well shown in this specimen ; they are due probably to the resistance and impact of the minute columns of air which impede its progress, and to the unequal conduction and fusibility of the surface material. Meteorites are never regular in form or spherical ; and the Orange River meteorite exhibits about average irregularities of exterior. These indicate a fragmentary character and lack of uniform structure in the original mass. Often there is an explosion, which is in part accounted for by the sudden heat to which the exterior is exposed, and the impact of the air column in front of the swiftly moving meteor, which is so densely compressed that it acts almost like a solid. Fragments are sometimes found miles apart which fit perfectly together.

The surface of our Earth containing three times as much water as land, a large proportion of all falling meteorites are necessarily lost in the ocean. The processes of deep sea dredging have, indeed, in some instances brought to light an appreciable amount of meteoric matter. Of meteorites which fall without being seen, in regions where ordinary climatic conditions prevail, by far the greater part soon disappear because of atmospheric action, the irons gradually oxidizing, and the stones rapidly disintegrating because of their porous structure. In desert climates, however, their lifetime is greatly extended; and in such regions the search for meteorites is best rewarded.

Although meteorites are classified as *metallic* and *stony*, the division is far from abrupt. According to their structure and composition, the metallic meteorites are subdivided into siderites, siderolites containing 80 to 95 per cent of iron, and pallasites, the first being homogeneous masses of iron and nickel combined, while the last are spongy masses of iron filled with olivine, or chrysolite. About eleven twelfths of the stony meteorites (which are greatly in excess of the metallic ones) are termed *chondrites*, by Rose, because their composite structure is that of rounded grains (chondri), with nodules of iron generally scattered through their mass. Professor Newton has carefully investigated the history of about 265 observed falls, represented by specimens in existing collections, and he has found that their motions about the Sun were direct, not retrograde. He concludes also that the larger meteorites moving in the solar system are allied much more closely to the group of short period comets than to the comets whose orbits are nearly parabolic.

The largest fall actually heard or seen occurred in Emmett County, Iowa, 10th May 1879, in a shower of stones, the most massive weighing 437 pounds. The largest meteoric stone ever discovered weighs 647 pounds; it fell in Hungary, in 1866, and is now in the Vienna Museum. The iron meteorites are often much larger, a famous one found in Tuczon, Arizona, and now in the National Museum at Washington, being in the shape of a ring weighing about 1000 pounds, and one found in Texas, now in the Yale collection, weighing 1635 pounds.

There are numerous others in different parts of the world, both in collections and still in the field, whose meteoric character is perfectly established, the largest of which, more than six tons in weight, is at Bahia, Brazil. The iron meteorites actually seen to fall are, however, remarkably few, as the following table shows. The largest is known as the Nedjed iron, and its weight is 130 pounds.

IRON METEORITES SEEN TO FALL

Name of place (or meteorite)	Name of country (or state)	Date of Fall	Where preserved	Weight Grams
Agram	Croatia, Hungary	26th May 1751	British Museum	282.3
Charlotte	Tennessee, U.S.	1st August 1835	British Museum	77.5
Braunau	Bohemia	14th July 1847	British Museum	553.2
Tabarz	Saxony	18th Oct. 1854	British Museum	9.0
Victoria West	South Africa	1862	British Museum	158.5
Nedjed	Central Arabia	Spring, 1865	British Museum	59,420.0
Nedagalla	Madras, India	23d Jan. 1870	British Museum	4,379.7
Marysville	California, U.S.	1873		
Rowton	Shropshire, Eng.	20th Apr. 1876	British Museum	3,109.0
Mazapil	Mexico	27th Nov. 1885		
Cabin Creek	Arkansas, U.S.	27th Mar. 1886		

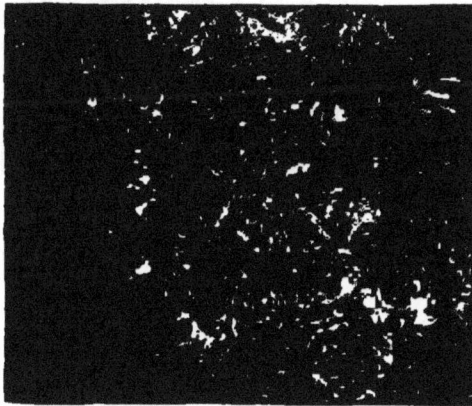

WIDMANNSTÄTTIAN FIGURES

(*Meteoric iron of Texas — Impression taken from the iron*)

S & T — 15

One of the first tests applied to a mass of iron suspected to be meteoritic is that known as the 'Widmannstättian figures,' from VON WIDMANNSTÄTTEN of Vienna, who in 1808 polished a piece of the Agram meteorite and etched it with acid. These figures have a general resemblance in all meteoric irons, well illustrated on p. 225, and they are sections of planes of cleavage, along which chemical change of some sort has taken place. Their form and relative position are determined by the laws of crystallography. The test of the Widmannstättian figures is not, however, regarded by modern investigators as infallible, because certain terrestrial substances exhibit them in some degree, particularly the famous Ovifak masses discovered by NORDENSKIÖLD on Disko Island. The largest of these dis-

PSEUDO-METEORIC IRON DISCOVERED AT OVIFAK, GREENLAND, IN 1870
(*Length,* 6½ *feet ; weight,* 20 *tons. Now in the Hall of the Royal Academy, Stockholm*)

puted meteorites is shown in the annexed engraving. For a long time these irons were thought to have had an extraterrestrial origin, but the opposite opinion is now generally held, due chiefly to the examinations of LAWRENCE SMITH and RAMMELSBERG.

Chemical analysis of meteorites reveals the presence of all

those elements most commonly recognized in the Earth's crust, — iron, nickel, sulphur, carbon, oxygen, calcium, with less of hydrogen, nitrogen, chlorine, sodium, cobalt, manganese, copper, and a trace of bromine, lead, and strontium. Iron is generally alloyed with nickel, and phosphorus nearly always combined with both nickel and iron. Cobalt is very generally associated with the nickel. Hydrogen and nitrogen are present as occluded gases, and carbon usually appears as graphite, in a few cases as diamond.

All investigations hitherto made of the question whether meteorites bear any testimony as to the existence of living organisms in other worlds fail to show such existence. Chemical compounds new to mineralogy have been brought to light, but no new elemental substance has been discovered in analyzing meteorites, and the study of these bodies has become a department of mineralogy rather than astronomy.

The Zodiacal Light. — After twilight in the western sky on clear moonless nights from January to April, may be seen the zodiacal light, a faintly luminous and ill-defined triangular area, expanding downward obliquely from the Pleiades to the northwest horizon. The illumination of this region is not uniform, the central portions being the brightest and outshining the white luminosity of the Milky Way with a slightly yellowish tint. First recognized by CHILDREY about the middle of the 17th century, CASSINI observed it carefully from 1683 to 1688, establishing its form and position so critically that no observations at the present day would appear to indicate any change. In the latter half of the present century this strange object has been carefully observed by SCHMIDT, HEIS, MAXWELL HALL, JONES, SERPIERI, and others; but there is still much divergence of view as to what may be the cause of the phenomenon. Dr WRIGHT, of Yale University, regards it as due to the reflection of solar rays from innumerable small interplanetary bodies, crowded more densely about the mean plane of the solar system, and in parts extending outward from the Sun far beyond the orbit of the Earth. It may be conceived as a vast cloud of meteoric dust, approximately lens-shaped in figure, still in slow process of formation from the *débris* of comets. That the zodiacal light is reflected sunlight has been abundantly proved by the spectroscope in the hands of LIAIS, COPELAND, and SMYTH, who find a short continuous spectrum similar to that of the Sun, possibly crossed by dark Fraunhofer lines, but without a trace of bright ones; and by the polariscope deftly manipulated by Dr WRIGHT, who in 1873–74 found the light strongly polarized. This fact proves that it is reflected ; and the especial manner in which it is polarized indicates the Sun as its only source.

The Gegenschein. — In this connection should be mentioned the zodiacal counterglow, or 'gegenschein,' first described in 1854 by BRORSEN, who gave it this name. A very faint and evenly diffused nebulous light, it is seen almost exactly opposite the Sun, and undergoes various changes of diameter (from about 3° to 15°), and of shape (from circular to elliptical). Even the proximity of bright stars or planets renders it quite invisible ; and it can never be seen in June and December, because it is then crossing the Milky Way. It is best seen in September and October, in the constellations Sagittarius and Pisces. Mr BARNARD, whose eye has the advantage of long training upon comets and nebulae, has been very successful in seeing this faint and difficult luminosity of the midnight sky, and his observations with others show that it does not lie in the ecliptic, though always close to it, being generally displaced somewhat north. Apparently connected with the gegenschein is often seen, especially in the autumn, a zodiacal band, six or eight times the breadth of the Moon, lying along the ecliptic and crossing the entire heavens. Whether the gegenschein is due to abnormal refraction by our atmosphere, or to a zone of small planets, or to cosmic conditions connected with meteoric matter, is not yet satisfactorily made out.

CHLADNI, *Ursprung der von* PALLAS *entdeckten Eisenmasse* (Leipzig 1794).

BIOT, *Mémoires de l'Institut National de France*, vi. (1806), pt. i.

CHLADNI, *Ueber Feuermeteore* (Vienna 1819).

OLMSTED, 'Fall of 1833,' *American Journal Science*, xxv. (1834), 363; xxvi. (1834), 132 ; xxix. (1836), 376.

BENZENBERG, *Die Sternschnuppen* (Hamburg 1839).

COULVIER-GRAVIER, *Les étoiles filantes* (Paris 1847–54–59).

HEIS, *Die Periodischen Sternschnuppen* (Cologne 1849).

SCHMIDT, *Resultate aus zehnjahrigen Beobachtungen ueber Stern-schnuppen* (Berlin 1852).

SMITH, Numerous papers in *American Journal Science* (1855–83), and *Comptes Rendus* (1873–81).

JONES, 'Zodiacal Light,' *Report U. S. Japan Expedition*, iii. (Washington 1856).

GREG, HERSCHEL, GLAISHER, and others, *Reports British Association Advancement Science* (1860 and onward).
BUCHNER, *Die Meteoriten in Sammlungen* (Leipzig 1863).
ROSE, *Beschreibung und Eintheilung der Meteoriten* (Berlin 1864).
NEWTON, 'Shooting Stars,' *American Journal Science*, lxxxviii. (1864), 135; *Memoirs Nat. Acad. Sciences*, i. (1866), 291.
SCHIAPARELLI, 'Origine probabile delle stelle meteoriche,' *Bull. Met. dell' Osservatorio del Collegio Romano*, v. (1866), 8, 10, 11, 12.
PHIPSON, *Meteors, Aerolites, and Falling Stars* (London 1867).
KIRKWOOD, *Meteoric Astronomy* (Philadelphia 1867).
ADAMS, 'Orbit November Meteors,' *Mon. Not. Royal Astron. Society*, xxvii. (1867), 247.
GRAHAM, *Proceedings Royal Society*, xv. (1867), 502.
DAUBRÉE, 'Meteorites,' *Smithsonian Report* for 1868.
RAMMELSBERG, *Die Chemische Natur der Meteoriten* (Berlin 1870-79).
MASKELYNE, 'Meteorites,' *Nature*, xii. (1875), 485, 504, 520.
TSCHERMAK, 'Die Bildung der Meteoriten', *Sitz. kk. Akad. Wissenschaften Wien*, lxxi. (1875), 661.
HEIS, *Publ. Royal Observatory Münster*, i. and ii. (1875-77).
SERPIERI, 'La Luce Zodiacale,' *Mem. Soc. Spettr. Ital.*, v. (1876), Appendix.
v. KONKOLY, *Spectres de 140 étoiles filantes* (Budapest 1877).
NEWTON, 'Relation of Meteorites to Comets,' *Nature*, xix. (1879), 315, 340.
BALL, 'Source of Meteorites,' *Nature*, xix. (1879), 493.
LEWIS, 'Zodiacal light and Gegenschein,' *American Journal Science*, cxx. (1880), 437.
HERSCHEL, 'Meteor spectroscopy,' *Nature*, xxiv. (1881), 507.
NEWTON, 'Meteors,' *Encyclopædia Britannica*, 9th ed. xvi.(1883).
LEHMANN-FILHÉS, *Die Bestimmung von Meteorbahnen* (Berlin 1883).
TSCHERMAK, *Die mikroskopische beschaffenheit der meteoriten* (Stuttgart 1883-85).
MEUNIER, 'Météorites,' *Encycl. Chimique*, tome ii. (Paris 1884).
JOULE, 'Shooting Stars,' *Scientific Papers*, i. (London 1884), 286.
SEARLE, A., 'Zodiacal light,' *Proc. Am. Academy Arts and Sciences*, xix. (1884), 146.
BREZINA, *Die Meteoritensammlung des kk. mineralog. Hofkabinetes in Wien* (1885).
FLETCHER, *Study of Meteorites* (British Museum 1886).

DAUBRÉE, 'Origin and structure of Meteorites,' *Popular Science Monthly*, xxix. (1886), 374.

NEWTON, 'Meteorites, Meteors, and Shooting Stars,' *Proc. Am. Assoc. Adv. Science*, xxxv. (1886), 1.

WENDELL, 'Meteor Orbits,' *Astron. Nachr.*, cxiv. (1886), 285.

SCHIAPARELLI, *Le Stelle Cadenti* (Milan 1886).

NEWTON, 'Biela Meteors,' *American Journal of Science*, cxxxi. (1886), 409; 'Meteorite orbits,' cxxxvi. (1888), 1; also, cxlvii. (1894), 152.

DENNING, 'Distribution of meteor streams,' *Month. Not. Royal Astron. Society*, xlvii. (1887), 35.

FLIGHT, *A Chapter in the History of Meteorites* (London 1887

HUNTINGTON, 'Catalogue of all recorded meteorites,' *Proc. Am. Acad. Arts and Sciences*, xxiii. (1888), 37.

DENNING, 'August Meteors,' *Nature*, xxxviii. (1888), 393.

DENNING, In CHAMBERS'S *Astronomy*, i. (London 1889), bk. v.

WRIGHT, 'Zodiacal Light,' *The Forum*, x. (1890), 226.

LOCKYER, *The Meteoritic Hypothesis* (London 1890).

DENNING, 'Catalogue of 918 radiants,' *Month. Not. Royal Astron. Society*, l. (1890), 410.

PLASSMANN, *Verzeichniss von Meteorbahnen* (Cologne 1891).

DENNING, *Telescopic Work for Starlight Evenings* (London 1891).

BARNARD, 'Gegenschein,' *The Astronomical Journal*, xi. (1891), 19; *Popular Astronomy*, i. (1894), 337.

BALL, *In Starry Realms* (London 1892).

BREDICHIN, *Mélanges Math. et Astron.* vii. (St Petersburg 1892).

KIRKWOOD, 'Leonids and Perseids,' *Astronomy and Astro-Physics*, xii. (1893), 385, 789; 'Andromedes,' xiii. (1894), 188.

BACKHOUSE, DENNING, and others, *Memoirs British Astron. Association*, i. (1893), 17; iii. (1894), 1.

DENNING, 'How to Observe Shooting Stars,' *Popular Astronomy*, i. (1893-94).

SCHULHOF, *Bulletin Astronomique*, xi. (1894), 126, 225, 324, 406.

MONCK, 'Radiants,' *Jour. Brit. Astron. Assoc.*, v. (1895), 253.

RAMSAY, 'Argon and Helium,' *Nature*, lii. (1895), 224.

FARRINGTON, *Cat. of Collection*, Field Museum (Chicago 1895).

HARVEY, *Trans. Roy. Soc. Canada*, 1896, § iii. 91.

WÜLFING, *Meteoriten in Samml. u. Literatur* (Tübingen 1897).

MIERS, 'Ancient and Modern Falls,' *Sci. Progress*, vii. (1898), 349.

WILSON, 'November Shower,' *Pop. Astron.* vi. (1898), 502.

PICKERING, W. H., 'Leonids 1897,' *An. Harv. Col. Ob.* xli. (1898), v.

Consult also POOLE'S *Indexes*, under *Meteors* and *Meteorites*.

CHAPTER XV

THE distribution of the stars visible in the northern hemisphere into groups, or (as they are usually called) constellations, was made in very ancient times, probably by the Babylonian star-gazers. Several less important or conspicuous groups have been added since ; and the principal constellations near the south pole, visible only in the southern hemisphere, were formed and named by the navigators of the 16th century, especially PIERRE DIRCKSZ KEYSER (sometimes called PETRUS THEODORI), chief pilot of a Dutch fleet which sailed to the East Indies in 1595. Others were added by LACAILLE when observing at the Cape of Good Hope in the middle of the 18th century.

About one hundred of the brighter stars have special names (Sirius, Arcturus, etc.) given them by Greek and Arabian astronomers or star-gazers ; but with these exceptions, the only way of referring to any particular star, until comparatively modern times, was by means of its place in the imaginary figure or animal which the constellation containing it was supposed to resemble, — for example, the star in the head of Andromeda, in the tip of the tail of Ursa Major, etc.

The suggestion to designate each star by a letter of the alphabet placed before the name of the constellation was first made by ALEXANDER PICCOLOMINI, an Italian ecclesiastic and amateur astronomer, who pub-

lished at Venice in 1559 a small work on the fixed
stars, in which he gave some rough diagrams of the
principal constellations, with Roman letters affixed to
the most conspicuous stars in each. The suggestion,
however, was not generally adopted until JOHN BAYER
of Augsburg published in 1603 his well-known series
of maps of the constellations, in which he designated
each star therein marked by one of the letters of the
Greek alphabet; and these have been universally
adopted, Roman letters being used when the Greek
alphabet is exhausted. Following the letter is the
name of the constellation in the genitive case ; for
example, α Ursæ Majoris, β Cygni, etc. Numbers
also (following the lists in FLAMSTEED'S catalogue) are
employed in the same manner, taken in the order of
right ascension of the stars in each constellation, —
being used exclusively for the fainter stars which have
no letters, and in addition to them for the brighter
stars which have.[51]

[51] The tracing of constellations, fascinating as it is to the
popular mind, is of slender importance to the modern astronomer, because he is accustomed to designate most of the stars
which he observes, not so much by their names as by their
numbers in standard catalogues of their positions relatively to
the imaginary circles described in the next paragraphs. The
ability to recognize at sight about 100 of the principal stars of
the firmament, is, however, well worth the little exertion it
costs ; and the helps to acquire this knowledge are abundant in
number and simple in use. Better for the beginner than the
star charts and atlases are the planispheres (p. 240), for reasons
which will become apparent on using them ; and PROCTOR'S
Easy Star Lessons has an advantage over both, in reiterating
the asterisms in their varying seasonal relations to vertical
circles and the horizon. To the observer of luminous meteors
and of variable stars an intimate acquaintance with constellation
figures and stellar names is more important than in other lines
of astronomical inquiry. — *D. P. T.*

It should be noticed that in lettering the principal stars BAYER did not (as some have supposed) make the sequence of the letters follow throughout the order of apparent brightness. The stars in each constellation follow the old division into orders or classes of magnitude, and the letters given to the stars in each class were arranged rather according to the form of the figure represented than the alphabetic succession of the letters. This is clearly seen in the seven principal stars of Ursa Major, all rated of the second magnitude. Of these, α and β form the 'pointers,' or two stars pointing toward the Pole Star; γ, δ, are the two other stars of the so-called 'wain,' and ϵ, ζ, η, are the three in the tail of the bear, or the three horses of the imaginary chariot; but every one must have noticed that, of these seven stars, δ is much the faintest, although taken like the others as of the second magnitude. It should also be mentioned that, from occasional touching or overlapping of the figures, a given star has sometimes, in BAYER'S original nomenclature, a name by a different letter in two constellations: thus, β Tauri is the same star as γ Aurigæ; α Andromedæ is identical with δ Pegasi, etc. To avoid confusion, astronomers have in each case dropped the secondary and retained only the primary designation. LACAILLE, in affixing letters to the southern stars, made their order correspond in most cases more nearly to the successive gradations of brightness than BAYER had done with the great mass of the constellations lettered by him.

Full description of all the constellations would require many maps; but a few remarks may be made upon some of the more noticeable stars and groups, especially those in and around the zodiac.

The plane of the Earth's equator, extended to the heavens, traces out a circle called the celestial equator or equinoctial, everywhere distant 90°, or a quarter circumference from both celestial poles. The north pole of the heavens is indicated in the sky by a star of the second magnitude very near it, called from this circumstance ' [Stella] Polaris,' or the Pole Star. Similar prolongation of the plane of the Earth's orbit to the heavens traces out a circle called the ecliptic which is inclined to the equinoctial 23° 28', or the angle between the Earth's equator and its orbit. These two great circles intersect at two opposite points called the equinoxes, or equinoctial points.

A zone or band extending 8° on each side of the ecliptic is called the zodiac; within this the Moon and all the large planets are always found, as the inclinations of all their orbits to that of the earth are less than 8°. The stars distributed along and around this zone have been formed into 12 constellations, proceeding eastward in the following order, — Aries, Taurus, Gemini, Cancer, Leo, Virgo, Libra, Scorpio, Sagittarius, Capricornus, Aquarius, Pisces; and the line of the ecliptic itself is divided into twelve signs (as they are called), each having taken its name from the constellation nearest to it. At the time of the earliest astronomical writings, the equinoctial points were in the constellations Aries and Libra, and they are still said to be at the first or initial points of these signs; although in consequence of precession of the equinoxes (page 22) they are now in the constellations Pisces and Virgo, and two thousand years hence will be in Aquarius and Leo respectively.

As the vernal equinox (where the Sun crosses the equator when moving northward) is situate in Pisces, it is evident that six constellations of the zodiac (Pisces, Aries, Taurus, Gemini, Cancer, and Leo) occupy a more northern position in the heavens than the other six (Virgo, Libra, Scorpio, Sagittarius, Capricornus, and Aquarius), so that the former are visible more of the time from the northern hemisphere of the Earth than the latter are.

The following considerations will show when these constellations are above the horizon at night. Although the Sun's light, diffused in our atmosphere, generally renders the stars around him invisible during the day, the place of the Sun in the heavens is always in the ecliptic, and therefore in the middle of the zodiac. On 20th March he reaches the vernal equinox, so called because it is the one passed by the Sun in the spring ; but, as previously mentioned, while that point is still called the first point of the sign Aries, as in ancient times, its present position is at the beginning of the constellation Pisces. The Sun is not therefore at the beginning of the group called Aries till near the end of April, and he passes the middle of it in May. Hence it follows that Libra, at the opposite side of the zodiac, will at that season be on the meridian, or due south at midnight, and the rest in order ; so that the annexed table gives the zodiacal constellation on or near the meridian at the beginning of successive months : —

January	. . . .	GEMINI.	July	. . . .	SAGITTARIUS.
February	. . . .	CANCER.	August	. .	CAPRICORNUS.
March	. . . . .	LEO.	September	.	AQUARIUS.
April	. . . . .	VIRGO.	October	. .	PISCES.
May	. . . . . .	LIBRA.	November	.	ARIES.
June	. . . . .	SCORPIO.	December	. .	TAURUS.

Before midnight each of these constellations will be to the east, and after it to the west, of the meridian during the month against which it is here placed. The winter Sun being south, and culminating low down, the portions of the zodiac on the meridian at night in winter are much higher in the heavens than those which are similarly visible in summer; just as it is matter of common observation that the full Moon in winter runs high and in summer low.

The most beautiful of the zodiacal constellations is Taurus, which has several very bright stars (one of the first magnitude called Aldebaran, or a Tauri) and contains the splendid cluster known as the Pleiades, also the more diffused cluster called the Hyades, in which Aldebaran is situate. Gemini is remarkable for two very bright stars near together, named Castor and Pollux, — the former being also designated a, and the latter β, Geminorum. Leo is a fine group, and has perhaps (what few have) a little resemblance to the figure indicated by the name. The forepart of Leo has something the shape of a sickle, and contains a star of the first magnitude, a Leonis, or Regulus. Virgo has only one bright star, to which the name Spica

THE PLEIADES

(Ordinary naked-eye view. The telescope and photographic plate have revealed nearly 2,000 stars in this region, hundreds of which are shown in the enlarged view on the opposite page)

Virginis was attached by the ancients; and it is now known either as Spica or a Virginis. The only other very remarkable star in the zodiac is the beautiful red object known as Antares (Cor Scorpii), the brightest star in the constellation Scorpio.

With regard to the other constellations, the most

THE PLEIADES — CLUSTER AND NEBULA

(*From photographs by the Brothers* HENRY *of the Paris Observatory*)

splendid is Orion, partly in the northern and partly in the southern celestial hemisphere. The three bright stars in a straight line above the middle of it (com-

THE GREAT NEBULA IN ORION

(Photographed by ROBERTS, *4th February* 1889. *Exposure,* 3 *hours* 25 *minutes. This nebula is faintly visible to the naked eye surrounding the middle star in the Sword of Orion. See page* 260. *Among the most successful observers of this magnificent object, before the days of its photographic portraiture, were the* BONDS [WILLIAM CRANCH, *and* GEORGE PHILLIPS, *father and son*], *successively Directors of the Harvard Observatory. The latter's drawing of the Great Nebula is unsurpassed)*

monly called Orion's belt, but by astronomers δ, ε, ζ Orionis) point in an easterly direction somewhat above Sirius, the principal star of Canis Major, which is the brightest star of the entire heavens. Of the stellar objects near the north pole, the seven which form part of Ursa Major are best known. As already mentioned, a line prolonged through β and α points nearly toward the pole star. Another (carried to a considerably greater distance), through ζ, η, passes very near to Arcturus, the principal star in Boötes; and one extended in the reverse direction through δ, α, will similarly indicate the position of Capella, the principal star in Auriga, and the brightest of the northern hemisphere. It is thought that this object has somewhat increased in brightness of late years, and that Vega was formerly the brightest.

On the opposite side of the pole star from Ursa Major, and at about an equal distance from it, is the fine group called Cassiopeia, which contains several bright stars arranged somewhat in the shape of an irregular W. Next in brightness to Capella is Vega, the principal star of the small constellation Lyra, which, at the average latitude of places in the United States, passes very near the zenith at its upper culmination. To the west of Lyra is the large and scattered group known as Hercules; to the east of it, Aquila, the three most conspicuous stars in which are near together and almost in a straight line, the brightest of the three, α Aquilæ, being in the middle. Southward from Cassiopeia is Andromeda, whose three principal stars (with the three brightest in Pegasus at one end and the largest star in Perseus at the other) form a configuration resembling the seven principal stars of Ursa Major. That star in Andromeda (α) nearest to Pegasus was regarded by the ancients, and marked by BAYER, as belonging to both constellations; and the square formed by it and *a*, *β*,*γ* Pegasi, is called the 'square of Pegasus.' Nearly between Aquila and Pegasus is Cygnus, resembling a Latin cross; while between Aquila and the southern part of Hercules is the northern part of Ophiuchus, a constellation extending into both hemispheres, sometimes called Serpentarius (serpent-holder), the same in signification as Ophiuchus, the latter word being derived from the Greek.

W. C. BOND (1789–1859)

IDELER, *Ursprung und die Bedeutung der Sternnamen* (Berlin 1809).
HIGGINS, *Names of the Stars and Constellations* (London 1882).
GORE, *An Astronomical Glossary* (London 1893).
WEST, ' Pronunciation of Arabic Star-names,' *Popular Astronomy*, ii. (1895), 209.

WHITALL, *Movable Planisphere* (Philadelphia).
PHILLIPS, *Planisphere for the Latitude of England* (London).
GOLDTHWAITE, *Planisphere for Latitude 40° N.* (New York).
HARRINGTON, *Planisphere for Latitude of the U.S.* (Ann Arbor).
POOLE, COLAS, *Celestial Planisphere* (Chicago).
DUNKIN, *The Midnight Sky* (London 1869).

PROCTOR, *Half-hours with the Stars* (London 1878).
PROCTOR, *Easy Star Lessons* (New York 1882).
COLBERT, *The Fixed Stars* (Chicago 1886).
CLARK and SADLER, *The Star-guide* (London 1886).
JEANS, *Handbook for finding the Stars* (London 1888).
YOUNG, 'Uranography,' in his *Elements of Astronomy* (New York 1890).
HILL, R., *The Stars and Constellations* (New York 1891).
UPTON, 'Constellation study.' *Popular Astronomy*, i. (1893–94).
SERVISS, *The Popular Science Monthly*, xlv.–xlvii. (1894–95).
Also the constellations are accurately and artistically figured in *The Century Dictionary* (New York 1889–91).
CLARKE, *How to Find the Stars* (Boston 1893) ; for use in connection with his convenient astronomical lantern. Also Mr BAILEY'S 'Astral Lantern' will be found very helpful in learning the stars and constellations.

ARGELANDER, *Uranometria Nova* (Berlin 1843).
MITCHEL, O. M., *Atlas* (Cincinnati 1849), accompanying his edition of BURRITT.
BURRITT, *Geography of the Heavens*, with *Atlas* (N. Y. 1856).
CHACORNAC, *Atlas Écliptique* (Paris 1862).
ARGELANDER, *Atlas des Nörd. Gestirnten Himmels* (Bonn 1863).
HEIS, *Atlas Coelestis Novus* (Cologne 1872).
PROCTOR, *New Star Atlas* (London 1874).
NEWCOMB, Star maps in his *Popular Astronomy* (N. Y. 1878).
HOUZEAU, *Uranométrie Générale* (Brussels 1878).
GOULD, *Uranometria Argentina* (Buenos Ayres 1879).
PETERS, C. H. F., *Celestial Charts made at the Litchfield Observatory* (Clinton, N. Y. 1882).
JOHNSTON, GRANT, *School Atlas of Astronomy* (New York 1884).
ESPIN, *Elementary Star Atlas* (London 1885).
SCHÖNFELD, *Bonner Sternkarten* (Bonn 1886).
SCHURIG, *Himmels-Atlas* (Leipzig 1886).
MESSER, *Stern-Atlas für Himmelsbeobachtung* (St Petersburg 1888).
KLEIN, McCLURE, *Star Atlas* (London 1888).
COTTAM, *Charts of the Constellations* (London 1889).
PECK, *Handbook and Atlas of Astronomy* (New York 1891).
WEISS, *Bilder-Atlas der Sternenwelt* (Esslingen 1891).
BALL, *An Atlas of Astronomy* (New York 1892).
ROHRBACH, *Sternkarten in gnomon. Projection* (Berlin 1894).
UPTON, *Star Atlas* (Boston 1895).
PECK, *The Observer's Atlas of the Heavens* (London 1898).

CHAPTER XVI

THE COSMOGONY

TO inquire into the origin and development of the heavenly bodies is the object of the Cosmogony, or the science of the formation of the world.

Have the planets, the Sun, and the stars always existed, as they now appear in the sky; or was there a time when the present beauty and order of the Cosmos was supplanted by a state of chaos? To this question the cosmogony of the present time answers that the heavenly bodies have been gradually developed from a diffused or nebular condition, just as animals and plants are developed on the earth; but that it has taken countless ages for the Sun and stars to attain their present state of evolution. It is shown that in the beginning the nebular condition prevailed, and the law of universal gravitation caused the matter to gather into masses called nebulæ. The process then is the following: as these gaseous masses condense, there arises a motion of rotation around some centre, so that the nebula begins to rotate on an axis. The particles of the nebula are acted upon by two forces: (1) The gravitation of the mass; (2) The centrifugal force of rotation. As the mass condenses, the rotation becomes more rapid, owing to the conservation of the moment of momentum, and the contraction of the radius of

gyration ; so that at length the centrifugal force becomes equal to gravity, and the particles on the
equator of the mass cease to fall toward the centre.
Thus a ring (or lump) of nebulous matter is left

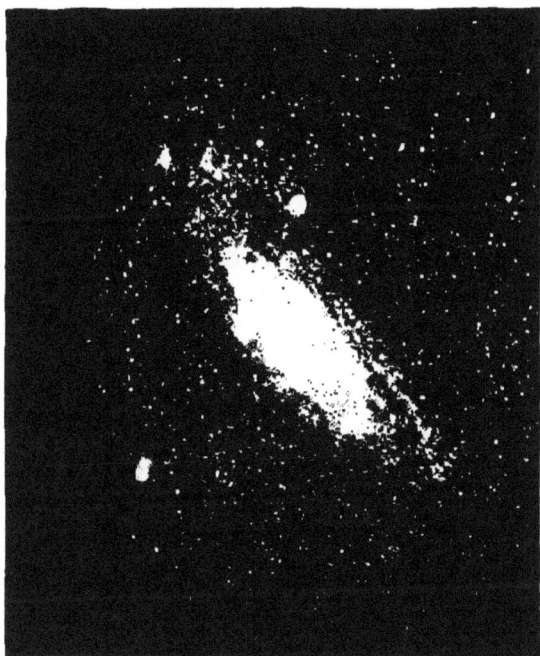

THE GREAT NEBULA IN ANDROMEDA
(*Photographed by* ROBERTS, 1888 — *Exposure, four hours*)

behind, revolving around the central mass in an
elliptic orbit, so well illustrated in the above engraving of the great nebula in Andromeda. The body
thus separated will in the course of ages condense
into a satellite or planet, or double star, as observed
in space.

Having stated this essential idea of celestial evolution, the history of the subject will now be briefly surveyed. Cosmogony has engaged the attention of philosophers from time immemorial; but it is only in the last century and a half that substantial progress has been rendered possible by the general advance of all related sciences. ANAXIMANDER, ANAXAGORAS, DEMOCRITUS and other ancient philosophers held that the world had arisen from the falling together of diffused matter in a state of chaos; and proceeding on this supposition, they endeavored to predict its future career and ultimate destiny. Some of the Greek philosophers, ARISTOTLE, for example, and PTOLEMY, while admitting the perishable character of all terrestrial things, maintained that the world beyond the orbit of the Moon is imperishable and eternal. Other philosophers followed ANAXIMANDER and ANAXIMENES in maintaining that, as the world had arisen in time, so it would in time pass away; and thus the universe had undergone and would continue to undergo alternate renewals and destructions.

In like manner the still more ancient cosmogony of *Genesis* declares that in the beginning the Earth was 'without form and void,' and implies that the visible world has arisen by a process of gradual evolution. These theories of the ancients are of course mere speculations supported by broad analogies of nature, but as they are in general accord with modern results, they are of interest in the history of the cosmogony.

Cosmogonic speculation of a scientific character began with KANT, who was the first of modern philosophers to advance a definite mechanical explanation of the formation of the heavenly bodies, and particularly of the bodies composing the solar system.

The views of KANT, however, do not seem to have received much scientific recognition until after LA PLACE'S independent formulation of the nebular hy-

IMMANUEL KANT (1724–1804)

pothesis. LA PLACE advanced the nebular hypothesis as a result of his prolonged study of the mechanics of our system, and the sound dynamical conceptions underlying his explanation secured for it immediate recognition. The work of Sir WILLIAM HERSCHEL lent observational support to LA PLACE'S ideas, and the observations of Sir JOHN HERSCHEL strengthened the evidence gathered by his illustrious father.

But about 1850, when Lord ROSSE'S great reflector

La Place (1749–1827)

showed the discontinuous nature of some of the objects then classed as nebulæ, the question arose whether with sufficient power all nebulæ might not be re-solved into discrete stars. Fortunately, the invention of the spectroscope by KIRCHHOFF about 1860, and Dʳ HUGGINS's application of it to the heavenly bodies at once answered this question in the negative, by showing that many of the nebulæ are masses of glowing gas in the process of condensation. It then became a matter of great scientific interest to investigate the formation of the heavenly bodies. The principle of the conservation of energy and the mechanical theory of heat, which VON HELMHOLTZ was

LORD ROSSE'S GREAT SIX-FOOT REFLECTING TELESCOPE
(*Birr Castle, Parsonstown, Ireland*)

the first to apply to the nebular contraction of the Sun, and LANE'S researches on condensing gaseous masses, together with the researches of Lord Kelvin on the Sun's age and heat, have each marked important epochs in the development and confirmation of the nebular hypothesis, as now maintained and generally accepted by astronomers. The importance of Lane's work consisted in showing that a gaseous mass condensing under its own gravitation might rise in temperature, and thus La Place's assumption of an original high tempera-

ture became unnecessary, as the heat might be developed by the falling together of cold matter. La Place supposed that the planets and satellites resulted from the condensation of rings which were successively shed by the contracting nebula, but to explain how a ring would become a planet has always been somewhat difficult. Yet prior

LORD ROSSE (1800–1867)

to the researches of Professor G. H. Darwin, the 'ring-theory' was never seriously questioned, at least in regard to planetary evolution. But the course of thought was greatly changed when he showed that the Moon probably separated from the Earth not in the annular form supposed by La Place, but in the form of a globular mass, which had been driven away to its present distance by the tidal reaction of the Earth. The tides here alluded to are not surface oceanic tides, but tides in the body of the molten globe, which were especially high when the two bodies were close to-gether. As the matter composing our planet is fric-tional, the tides lag, and the tidal protuberance in the Earth therefore points in advance of the Moon, and exerts on it a small force tending to accelerate its motion, with the result that the distance increases ; while the reaction of the Moon against the protube-rance retards the rotation of our globe on its axis.

In this way Professor DARWIN explains the expan-sion of the Moon's orbit from age to age, and shows that the working of tidal friction would also render it more and more eccentric. Also he sought to apply this remarkable theory to the development of the planetary system as a whole, but it was found that elsewhere in our system the effects of tidal friction had been comparatively unimportant.

The general result of Professor DARWIN's work on the cosmogony of the solar system is to leave LA PLACE's theory without any material change except in the system of Earth and Moon. If it were possi-ble to overcome the mathematical difficulty of ex-plaining how a ring could condense into a planet, the regularity in the motions of the planets and satellites and the circularity of their orbits might be explained. The too rapid revolution of the satellite Phobos round the planet Mars can be accounted for as a result of tidal friction ; but it is difficult to explain in a satisfac-tory manner the retrograde motion of the satellites of Uranus and Neptune, though the suggestions of M. FAYE are ingenious and possibly of lasting value.

Recently Dᵣ T. J. J. SEE, of the University of Chi-cago, has taken a farther and very important step in the development of the theory of cosmical evolution, and incidentally he has applied it to the planetary system. In investigating the origin of double stars, he found that their orbits are generally very eccentric, as compared with the orbits of the planets and satellites, and it occurred to him that the cause of this remarkable phenomenon might be the secular action of tidal friction. His orbit of α Centauri shown here is a typical binary with about the average eccentricity.

It appears that double stars have arisen from the breaking up of double nebulæ by a process of division resembling fission among the protozoans, and that their orbits were originally nearly circular. According to Dʳ SEE the average eccentricity is about 0.45, while in the planetary system the average eccentricity is only 0.0389, or one twelfth of that found among binary stars. Another very important point in his researches lies in the connection pointed out between double nebulæ and the dumb-bell-shaped figures of equilibrium discovered by DARWIN and POINCARÉ.

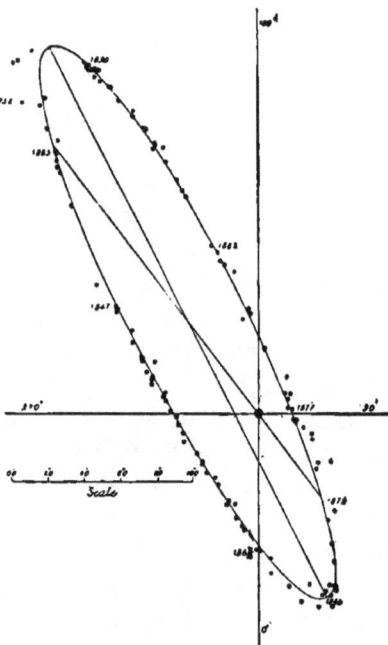

ORBIT OF ALPHA CENTAURI

(*Computed and projected by Dʳ* SEE)

These eminent mathematicians have substantially proved that it is possible for a rotating nebula to break up, under its own contraction, into an hourglass figure exactly resembling the double nebulæ observed in space. On these pages are reproduced the figure of POINCARÉ, and a typical double nebula, to show the process of division.

The masses resulting from the figures of DARWIN and POINCARÉ are much too large relatively to render the results applicable to our own system, where the attendant bodies are always very small. Just such masses, however, exist among double nebulæ and double stars; and Dr SEE therefore concludes that if such large relative masses are not found in our system they are found in abundance elsewhere in space. He points out that the double stars are remarkable for two fundamental facts: (1) The large mass-ratios of the components, so that the masses are often nearly equal and always comparable; and (2) the high eccentricities of their orbits.

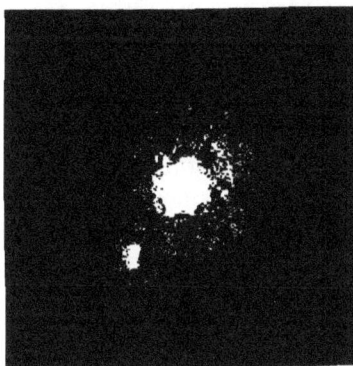

TYPICAL DOUBLE NEBULA IN VIRGO

(*No* 1202 *in Sir* JOHN HERSCHEL's *Catalogue*)

(*No.* 61 *in* MESSIER's *Catalogue*)

Among all known systems, that of the Sun is remarkable for the great number and small masses of the attendant bodies, and for the near approach to circularity in the orbits. Nearly all the mass of the

original solar nebula is embraced in the Sun, which has 750 times the mass of all the attendant planets and satellites combined. Our system is essentially single — all the mass in the Sun — while in binary stars the systems are essentially double.

Applying the law of binary evolution to the planetary system, D^r SEE concludes that

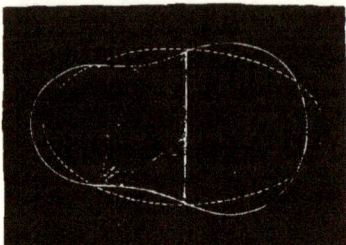

THE FIGURE OF POINCARÉ

the planets were separated not in the form of rings, as LA PLACE supposed, but in the form of lumps or masses, which would easily condense into planets and satellites. In this manner he escapes the necessity of explaining how rings would condense into single masses; indeed, he maintains that rings would not condense at all, but become swarms of small bodies, like those which make up the rings of Saturn and the small planets between Mars and Jupiter.

It is still too early to pronounce an opinion as to the ultimate value of these later researches in cosmogony. The weak point in the older theories lies in the deduction of all cosmical laws from our own system, whose formation appears to have been somewhat anomalous. But as D^r SEE's theory of cosmical evolution has been deduced from the study of other systems in space, and is in harmony with the known facts of Nature, it may fairly be regarded as the most advanced step yet taken in the science of the formation of the heavenly bodies.

KEPLER, *Epitome Astronomiae Copernicanae* (Frankfort 1635).

SWEDENBORG, *De Chao Universali Solis et Planetarum* (1734, and London 1845).

WRIGHT, *A New Hypothesis of the Universe* (London 1750).

* KANT, *Naturgeschichte und Theorie des Himmels* (Königsberg 1755).

LAMBERT, *Cosmologische Briefe die Einrichtung des Weltbaues* (Augsburg 1761).

HERSCHEL, ' Nature and Construction of Sun and Fixed Stars,' *Philosophical Transactions for* 1795.

LA PLACE, *Exposition du Système du Monde* (Paris 1796); and *Annuaire du Bureau des Longitudes* for 1867.

COMTE, ' Cosmogonie Positive,' *L'Institut*, iii. (1835), 31.

MAYER, *Dynamik des Himmels* (Heilbron 1848).

RANKINE, ' Reconcentration of the Mechanical Energy of the Universe,' *Philosophical Magazine*, iv. (1852), 358.

THOMSON, *Mechanical Energies of the Solar System* (Edinburgh 1854).

KIRKWOOD, ' On the Nebular Hypothesis,' *Am. Jour. Science*, xxx. (1860), 161.

TROWBRIDGE, ' On the Nebular Hypothesis,' several papers in *American Journal of Science* for 1864 and 1865.

PROCTOR, ' LA PLACE'S Nebular Theory,' in his *Saturn and its System* (London 1865), 201.

ROCHE, *La Constitution et l'Origine du Système Solaire* (Montpellier 1875).

NYRÉN, ' Ueber die von EMANUEL SWEDENBORG aufgestellte Kosmogonie,' *Viertel. Astron. Gesell.*, xiv. (1879), 80.

CLIFFORD, ' The First and the Last Catastrophe,' in his *Lectures and Essays*, i. (London 1879).

PEIRCE, ' Cosmogony,' and ' From Nebula to Star,' in his *Ideality in the Physical Sciences* (Boston 1881).

DARWIN, Papers in *Proceedings* and *Philosophical Transactions Royal Society*, 1878–1881.

HELMHOLTZ, ' On the Origin of the Planetary System,' *Popular Lectures*, 2d series (New York 1881).

NEWCOMB, ' The Cosmogony,' in his *Popular Astronomy* (New York 1885), 503.

FAYE, *Sur l'Origine du Monde* (Paris 1885).

FAYE, ' Sur la Formation de l'Univers et du Monde Solaire,' *Annuaire du Bureau des Longitudes* for 1885.

WOLF, *Les Hypothèses Cosmogoniques* (Paris 1886).

PARKES, *Unfinished Worlds* (New York 1887).

BRAUN, *Ueber Cosmogonie vom Standpunkt Christlicher Wissenschaft* (Münster 1887) ; CLERKE, *Nature*, xxxviii. (1888), 365.

JANSSEN, ' L'Age des Étoiles,' *Annuaire du Bureau des Longitudes* for 1888.

COAKLEY, ' On the Nebular Hypothesis of LA PLACE,' *Papers American Astronomical Society* (Brooklyn 1888).

CROLL, *Stellar Evolution* (New York 1889).

HIRN, *Constitution de l'Espace Céleste* (Paris 1889).

BALL, *Time and Tide* (London and New York 1889).

LOCKYER, *The Meteoritic Hypothesis* (London 1890).

GREEN, *The Birth and Growth of Worlds* (London 1890). Bibliography.

TUCKERMAN, ' References to Thermodynamics,' *Smithson. Misc. Collections*, No. 741 (1890).

KLEIN, *Kosmologische Briefe* (Leipzig 1891).

KEELER, *Astronomy and Astro-Physics*, xi. (1892), 567, 768.

CLERKE, *The System of the Stars* (New York 1892).

DARWIN, ' Cosmogonic Speculations founded on Tidal Friction,' Article ' Tides,' *Encyclopædia Britannica* (9th ed.), 1892.

GORE, *The Visible Universe* (London and New York 1893).

SEE, *Die Entwickelung der Doppelstern-Systeme* (Berlin 1893).

BALL, *The Story of the Heavens*, Chapter xxvii. (London 1893).

STANLEY, *Notes on the Nebular Theory* (London 1895).

LOCKYER, *Evolution of the Heavens and the Earth* (London 1895).

NOLAN, *Satellite Evolution* (Melbourne 1895).

SEE, *Researches on the Evolution of Stellar Systems* (Lynn 1896).

GERLAND, in VALENTINER'S *Handwörterbuch der Astronomie* (Breslau 1897).

SEE, *The Atlantic Monthly*, lxxx. (1897), 484.

LIGONDÈS, *Formation Mécanique du Système du Monde* (Paris 1897).

CHAMBERLIN, ' Climatic Changes,' *Jour. of Geology*, v. (1897), 653.

MOULTON, *Popular Astronomy*, v. (1898), 508.

LA FOUGE, *Formation du Système Solaire* (Châlons-sur-Marne 1898).

DARWIN, *The Tides and Kindred Phenomena in the Solar System* (Boston 1898).

Consult also under *Nebulæ* the references in POOLE'S *Indexes* cited on page 315 of this volume. Also in *The Westminster Review* (July 1858) is an able discussion by HERBERT SPENCER. MATSON'S *References for Literary Workers* (Chicago 1892) contains at pp. 388–9 a bibliography of popular articles.— *D. P. T.*

CHAPTER XVII

THE FIXED STARS

IT remains to give a short account of the motions and distances of those far more remote heavenly bodies which come under the general designation of fixed stars, devoting also a few words to the investigations concerning the probable motion of our own system, as a whole, among them.

Those fixed stars which are bright enough to be visible to the naked eye have been watched from the most remote antiquity; but, excepting the changes of brightness to which a few are subject, and the occasional appearance of a temporary star never seen before, no other knowledge of the fixed stars than that of their comparative brightness and apparent distribution in the sky was obtained before the invention of the telescope, or indeed could be. With regard to the temporary stars, it is not in all cases easy to say whether the accounts handed down to us really refer to a new star or to a comet. The most ancient record of such an object relates that in B. C. 134, a new star burst out with a brightness sufficient to render it visible in the day-time, and attracted the attention of HIPPARCHUS, leading him to draw up a catalogue of stars (the first ever made), with a view of enabling future ages to trace with certainty any subsequent appearances or disappearances. The Chinese annals show that the astronomers of that country also noticed this star, which seems to have been situate in the southern hemisphere and in the constellation Scorpio. They also speak of one said to have been seen in the constellation Centaurus in a year corresponding to A. D. 173.[52]

[52] Cataloguing the stars is the most tedious labor of the modern practical astronomer. A high degree of accuracy in the positions is demanded, and the latest catalogues embrace many thousand faint stars, most of them invisible to the naked eye. Stellar catalogues nearly always contain the magnitudes of the stars, given according to an arbitrary scale of numbers expressing their brightness. Twenty of the brightest stars of the firmament, ten in the northern celestial hemisphere and ten in the southern, are rated of the first magnitude. Then follow about

65 of the second magnitude,
200 of the third,
500 of the fourth,
1400 of the fifth,
5000 of the sixth,

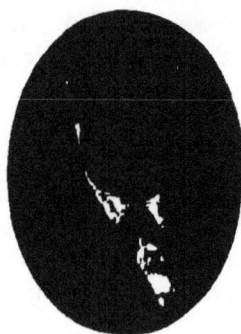

HEIS (1806–1877)

these last being so faint that they are just visible to a keen eye on clear moonless nights. The number increases rapidly with each fainter magnitude, nearly in geometric proportion; so that, if we include stars of the tenth magnitude, the faintest included in catalogues, there are about a million in all. Estimates of the uncounted hosts of uncatalogued stars still fainter and fainter, down to the 17th magnitude, make the total number far in excess of 100 millions. Professor BARNARD's fine photograph opposite shows how thickly they are strewn in many parts of the celestial vault. Until recent years the magnitudes of the brighter stars were for the most part determined by mere eye estimates, and HEIS, the German astronomer, who spent many years observing paths of meteors, was

MILKY WAY IN SAGITTARIUS (R. A. = 18^h 10^m. DECL. = S. 20°)

(Photographed by Professor BARNARD *at the Lick Observatory. Exposure,* 4¼ *hours)*

Another star, as bright at one time as the planet
Venus, is said to have appeared in the year A. D. 389

an exceedingly accurate observer of stellar magnitudes as well.
But the most precise results have been obtained by means of
instruments called photometers, in which every star is repeatedly compared with a standard light of known intensity.

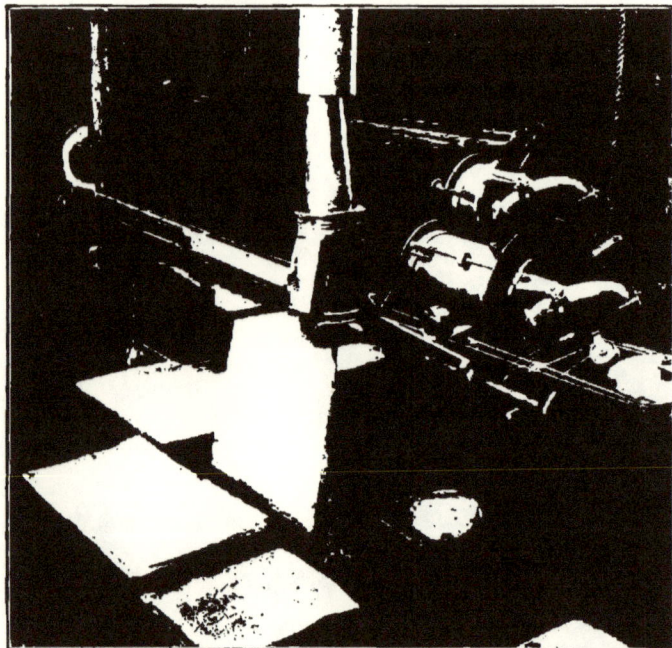

THE MERIDIAN PHOTOMETER OF HARVARD COLLEGE OBSERVATORY
(*Devised by Professor* EDWARD C. PICKERING, *and used under his direction in
determining the magnitude of all the bright stars in the entire sky*)

Stellar photometers are of varied designs. PRITCHARD used
a so-called wedge photometer, in which a thin strip of colored
glass, with faces slightly divergent, was attached to the eye-
piece, and that position of the glass wedge was recorded at

(in the reign of the Emperor THEODOSIUS) : but from
the accounts given by the ecclesiastical historians, it is

which the star's light was extinguished. Professor PICKERING
has successfully employed a type of his own devising, and
named the meridian photometer, shown opposite, and so
called because each star, when near the meridian, is directly
compared with Polaris, itself an invariable standard. The
oblique mirrors enable this to be done, no matter what the

PHOTOGRAPHIC EQUATORIAL TELESCOPE

*(Built by Sir HOWARD GRUBB for the Royal Observatory at
Cape Town. The lower tube is the photographic telescope,
with a 13-inch object-glass; the upper one is an optical or
guiding telescope with a 9-inch objective. The tubes are 12
feet long. Similar instruments by the same maker, for the
International Astrographic Survey, are mounted at Greenwich, Oxford, Melbourne, Sydney, and Tacubaya, in Mexico)*

evident that a remarkable comet was seen at the time
to which this appearance is referred, so there can be

altitude of the star. With this instrument, as mounted at
Cambridge, Massachusetts, and later at Arequipa, Peru, the
magnitudes of all the brighter stars of both northern and
southern hemispheres have been determined with a degree of
accuracy unsurpassed. Attempts have been made in the hope
of increased precision in magnitude by means of photography,
but as yet without much encouragement. Photography, however, has already revolutionized the construction of star catalogues, because the highly sensitive plate records in an hour
as many stars as an astronomer could observe the position of
in many weeks. Foremost in these highly significant researches is Dʳ DAVID GILL, the earliest to propose an international congress of astronomers, which first met at Paris in 1887,
and undertook an astrographic survey and photographic map of
the entire heavens. This is now in progress, with 18 instruments
like the one on page 259, six of which are located in the southern hemisphere. Nearly all the great nations originally joined
in this important work, including the observatories at Paris,
Algiers, Bordeaux, Toulouse ; San Fernando (Spain), Catania,
and Rome (Vatican) ; La Plata, Rio de Janeiro, and San-
tiago, in South America; Helsingfors and Potsdam in Germany; in addition to the observatories named on the preceding
page. Some, however, have dropped out, from lack of funds.
The total number of plates exposed at the conclusion of this
comprehensive survey will be more than 20,000, and the total
expense for all nations will aggregate not far from $100 for
each plate. Then the plates must be carefully measured, and
the resulting data converted into arcs of the celestial sphere,
so that the right ascension and declination of every star may
be given as in catalogues constructed in the old-fashioned way
by means of the meridian circle (see page 368). Dʳ KAPTEYN
of Groningen is foremost in the work of deriving star-positions
from photographs, and he has already measured and catalogued
nearly half a million stars from the Cape Town plates. At
Paris the work is progressing rapidly under the direction of
Mᶫᶫᵉ KLUMPKE. But the degree of precision attainable by the
photographic method does not displace the necessity for the
labors of the astronomer with the telescope alone. The long-
tested methods of slow observation with meridian instruments

little doubt that this report of a new star is simply due to a misunderstanding of the description of the

must still be employed when the last degree of precision is sought. The name of ARGELANDER, a pupil of the great BESSEL, stands out pre-eminently for persistent faithfulness in

F. W. A. ARGELANDER (1799–1875)

this class of observation. His famous *Durchmusterung* of the northern heavens embraces no less than 324,198 stars, set down with some precision, both in catalogues and in a series of stellar charts, which are a prime essential in all working observatories. In 1865 the *Astronomische Gesellschaft* of Germany undertook

comet, which moved in the heavens from near the place where Venus then was to the constellation Ursa

a supplementary work of vast magnitude, the re-observation of about one third of ARGELANDER's stars, with the highest degree of accuracy attainable. These *Gesellschaft* zones, observed at 13 observatories, European and American, are now nearly complete. What ARGELANDER did for the northern sky, GOULD

BENJAMIN APTHORP GOULD (1824–1896)

did for the southern heavens, by residing for fifteen years at Cordoba, Argentine Republic. Together with his assistants, he observed nearly 75,000 stars, extending to the $9\frac{1}{4}$ magnitude; published in 1879 the *Uranometria Argentina*, an elaborate series of charts of the southern heavens; in 1886 the Argentine General Catalogue of 32,000 stars; and in 1897, shortly after his lamented and accidental death, Dr CHANDLER brought out the last great work of GOULD, being a complete discussion of about 40 clusters of stars in the southern sky, with positions derived

Major. A similar remark applies to new stars which are said to have appeared in the northern heavens in A. D. 945 and 1264. In all probability both were comets; in the latter year (that of the battle of Lewes) a splendid comet is known to have been seen.

But it is quite certain that a very conspicuous and remarkable new star *did* appear near the constellation Cassiopeia in 1572. It was noticed in other parts of Europe several days before it was seen by TYCHO BRAHÉ; but he made a series of careful observations of the star, and left an elaborate account of it. He first saw it while residing at his uncle's house in Scania (then included in the kingdom of Denmark, but united to the rest of Sweden in the following century), on 11th November 1572. Its brightness even then rivalling Sirius, and increasing until it surpassed that of Jupiter, the star was for a short time visible in broad daylight; but it soon began gradually to fade away, and by March 1574 had totally disappeared, after having been seen, with more or less brilliancy, during a period of about 16 months.

from measurements upon photographs, which GOULD began at Cordoba in 1872 as the pioneer. In 1889 Miss CATHERINE W. BRUCE of New York donated to Harvard College Observatory the sum of $50,000 for a great photographic telescope, of 24 inches aperture. This has been constructed by ALVAN CLARK & Sons, from designs by Professor PICKERING, and it is now mounted in Peru, where the more interesting regions of the southern heavens are in process of permanent record. This telescope and its unusual type of mounting are illustrated on page 264, and opposite is a reproduction of a part of one of the star charts taken with it. On the entire original glass, 14 X 17 inches square, were 400,000 stars by actual count. Probably the Bruce telescope is the most powerful optical instrument in existence. Certainly it can photograph the faint glimmer of thousands of stars which no human eye can ever expect to see in any telescope. — *D. P. T.*

VICINITY OF ETA CARINAE (ARGUS) R. A. 10ʰ 41ᵐ, DECL. 59°S.

(*Photographed by Professor* BAILEY *with the Bruce telescope, exposure 4 hours*)

[*Engraving from* TODD'S '*New Astronomy*,' *by special permission of the American Book Company, Publishers*]

Another temporary star appeared in the right foot of the constellation Ophiuchus in the year 1604, and was observed by KEPLER and others. At one time as bright as Venus, it continued visible for more than a year, disappearing about the end of 1605. On 20th June 1670 Father ANTHELME, a French Carthusian of Dijon, noticed another new star of the third magnitude, very near the head of Cygnus. Observed both by HEVELIUS and by CASSINI, it underwent several remarkable changes in brightness during two years, then completely disappeared, and has not since been seen. Another star in the same constellation (now known as 34 Cygni) underwent some remarkable fluctuations of light in the 17th century, but has remained unchanged in brightness since, being just visible to the naked eye.

Within the last fifty years several temporary stars have appeared quite suddenly, whose previous existence is unrecorded. One of these was in 1848, when HIND noticed, on 28th April, a star of the fifth magnitude (easily visible to the naked eye), in a part of the constellation Ophiuchus, where he was certain that, so recently as the 5th of that month, no star so bright as the ninth magnitude was visible; nor is there any record of a star having been observed there at any previous time. From the date of its discovery it began continuously to diminish in brightness, and it is now visible only through very powerful telescopes. A still more remarkable, as well as more recent case of the appearance of a temporary star (temporary at least as regards its visibility to the naked eye) is that of a star in the constellation Corona Borealis, which seems to have burst out quite suddenly, 12th May 1866, when it was first noticed by BIRMINGHAM in Ireland. Of the second magnitude at the time of discovery, it diminished so rapidly as to be invisible to the naked eye by the end of the month. In the autumn of the same year it increased somewhat in brightness again, — but not sufficiently to become visible without a telescope,— and afterward sank down to what must be considered its normal brightness. A third instance of the kind occurred in 1876, when SCHMIDT, of Athens, noticed on 24th November, a new star of the third magnitude in the constellation Cygnus, which afterward gradually faded away, ceasing to be visible to the naked eye in about a month, and is now to be seen only with the aid of a very powerful telescope. Toward the end of August 1885 another new star appeared in the midst of the great nebula in Andromeda ; but it never became bright enough to be visible to the naked eye. The most recent case is that of a star in Auriga, which was first noticed at the end of January 1892, but was afterward found to have been registered by photography in the preceding December. It faded away in March 1892, after having been for a few weeks visible to the naked eye.[53]

[53] Nova Aurigae has the most singular history of all the temporary stars. Dʳ ANDERSON, an amateur of Edinburgh, was its first discoverer, his meagre equipment consisting of only a star atlas and a pocket telescope magnifying ten times. But examination of the Harvard stellar photographs at once showed that the star had many times been recorded by its own light on plates exposed between 10th December 1891 and 20th January 1892. Measures of them showed that Nova Aurigae had rapidly attained its maximum brightness unobserved on 20th December, when it was of the 4.4 magnitude, and therefore plainly visible without a telescope. By the 1st of April the star had become invisible, except with a very large telescope, but in August 1892 a temporary brightening brought its magnitude up from the 13th to the 10th. Through

268 *Stars and Telescopes*

The above are special cases, but the number of
stars which may be called regularly variable, being

1893 and 1894 it remained between the 10th and 11th magnitude, and Professor BARNARD's observations with the 36-inch
Lick telescope from 1892 to 1895 show that Nova Aurigae had
become a small, bright nebula with a star-like nucleus. Professor CAMPBELL, Dʳ VON GOTHARD and others, who critically
examined the star's spectrum, showed that it was practically
identical with the spectra of planetary nebulæ. Parallax
observations, similar to those described later in this chapter,
showed that the star was as remote from the solar system as
the average of the stars whose distances are known. The
outburst which produced the sudden rise in its brightness
must, therefore, have taken place on a scale inconceivably
grand. Just what caused it cannot be said to be known; but
the complex character of the star's spectrum in February 1892
indicates a probable collision, or at least a near approach, of
two vast gaseous bodies travelling through interstellar space,
with a relative motion exceeding 500 miles a second.

A swarm of meteorites swiftly traversing a remote gaseous
nebula would account for most of the peculiarities of the
spectrum of Nova Aurigae. But a wholly satisfactory theory
is not likely until many other new stars appear in the sky, and
their constitutions are studied by means of spectra photographed at short intervals. That lynx-eyed watch of the whole
sky, both northern and southern, now kept by means of photography, enables us to say that new stars are not infrequent
objects, as formerly supposed. Mⁿ FLEMING in 1893 found
one of the seventh magnitude in the southern constellation
Norma, with a spectrum practically identical with that of Nova
Aurigae (indicating hydrogen and calcium), and which is now
nebular also. And the same keen observer has since found
many other objects of this highly interesting character, the
most noticeable of which appeared in 1895 in Carina, a part
of the wide southern constellation Argo. Nova Carinae fell
in three months from the eighth magnitude to the 11th, and
exhibited all the essential features of the new stars in Norma
and Auriga. Still another was found by her the same year
in Centaurus, which had a spectrum different from its prede-
cessors, but like them eventually became a gaseous nebula.
It is significant that nearly all the Novae have gleamed in or

subject to regular changes in brightness of various degrees, is at present known to be very large, and is from time to time increased by fresh discoveries. Usually these are subject only to smaller mutations of brightness. Of the few stars which undergo remarkable irregularities of this kind, the two most interesting cases are o Ceti and η Argûs. The former, sometimes called Mira Ceti, was first noticed as being subject to change in 1596, by DAVID FABRICIUS, who saw it of the third magnitude in August, and perceived in October that it had ceased to be visible. BAYER saw it in 1603, and PHOCYLIDES in 1638, the latter noticing (which BAYER did not) that it was the same star, thus shown to be variable in brightness. The observations of HEVELIUS between 1648 and 1662 established its period, which is about 331 days in length; but the changes of brightness during this interval are themselves of a very variable kind, nor is the period quite constant in length. The star is usually visible to the naked eye for about six months, and invisible during the succeeding five.

CLUSTER ω CENTAURI

(J. F. W. HERSCHEL)

HALLEY was the first to suspect changes of brightness in that remarkable star in the southern hemisphere, η Argûs, which he had observed at Saint Helena

near the Milky Way; and in the case of Nova Aurigae at least, the striking display of 1892 must really have taken place as long ago as the beginning of the 19th century. In the latter part of 1898, a star-like condensation was reported in the centre of the nebula in Andromeda. — *D. P. T.*

in 1677 as of the fourth magnitude. LACAILLE in 1751 saw it of the second ; and its changes from time to time since have been exceedingly irregular. In 1838, and again in 1843, it surpassed for a while all other stars in brightness, excepting only Sirius. After the latter date, it slowly but steadily diminished, and ceased to be visible to the naked eye in 1867, and it so remains, though its brightness has slightly increased in the last few years.

SCHÖNFELD (1828–1891)

Of the variable stars of short period, the most remarkable is Algol or β Persei. Its usual magnitude is about the second, and as such it shines constantly and regularly for a period of about two and a half days ; then it fades gradually almost down to the fourth magnitude, afterward recovering its ordinary brightness by a similar gradual increase. The whole amount of change takes place in about nine hours, so that the complete length of a period is not quite three days ; more accurately, 2^d 20^h 48^m $55^s.4$. Spectroscopic observations have proved that these variations are caused by the regular interposition of an opaque body revolving round Algol, and cutting off at each revolution a portion of its light in the manner of a partial eclipse.

The above are the most conspicuous and easily detected changes of brightness in the fixed stars ; but about 150 stars are now known to be variable, and the application of photometry as an accurate means of measuring the amount of light of a star is constantly increasing this number. In many cases the variability is confined within narrow limits; in others, the stars, even when at their greatest brightness, are scarcely, or not at all, visible to the naked eye.[54]

[54] So rapid has been the recent progress in our knowledge of the variable stars that the total number now known embraces about 1,000, if we include the rather remarkable discoveries of Professor BAILEY, who, beginning in 1896, has found about 500 variables in the dense globular clusters photographed with the Harvard telescopes. Their changes of magnitude are marked within a few hours. In ω Centauri alone (the cluster shown on page 269) 125 variables were discovered. Dr CHAND-LER of Cambridge has published several catalogues of variables in which about 500 of these objects are carefully classified. They are distributed all the way round the heavens, and for the most part are included in a belt or zone tilted about 18° to the celestial equator. Variables are best observed by means of Father HAGEN's excellent Atlas. Following are a few of the variables :—

VARIABLE STARS

STAR'S NAME	Position for 1900.0		VARIATION	
	R. A.	Decl.	Period	Range
	h m	° '	d	
Omicron Ceti	2 14	− 3 26	331	1.7 to 9.5
β Persei	3 2	+40 34	2$\frac{7}{8}$	2.3 to 3.5
ζ Geminorum	6 58	+20 43	10$\frac{1}{2}$	3.7 to 4.5
R Leonis	9 42	+11 54	313	5.2 to 10
δ Librae	14 56	− 8 7	2$\frac{1}{3}$	5 to 6 2
α Herculis	17 10	+14 30	90±	3.1 to 3 9
X Sagittarii	17 41	−27 48	7	4 to 6
β Lyrae	18 46	+33 15	12$\frac{7}{8}$	3.4 to 4.5
δ Cephei	22 25	+57 54	5$\frac{1}{3}$	3.7 to 4.9

Many of the newly discovered variables being faint stars, they have no regular name except their number in some catalogue of positions. They are, therefore, designated by the

The determination of the parallaxes, and thereby the distances of the fixed stars, constitutes a problem which completely baffled astronomers until little more than fifty years ago; for so vast is the distance of these bodies, that, although we use the diameter of the Earth's orbit as a base, by comparing observations made at opposite parts of the year, nevertheless the angle at the star subtended by this diameter is generally so small as to be scarcely distinguishable from inevitable errors of observation. The half of this angle is the star's annual or heliocentric parallax.

Early in the present century BRINKLEY thought his observations at Dublin indicated that he had found

---

letters R, S, T, and so on, in the constellation to which they belong. The number of variables of the Algol type is now about 20, two of the most pronounced being U Cephei and W Delphini, and the total variation in the magnitude of each is about 2.5. W Delphini is never visible to the naked eye, but remains for about four days at its full brightness of about the 9th magnitude, when in seven hours it falls to the 12th magnitude, so that a 5-inch telescope will barely show it. Algol, the type star, is best visible in the northeastern heavens in the evenings of early autumn, and nearly overhead in winter. The minima are given in many almanacs, and such astronomical journals as *The Observatory* and *Popular Astronomy*. Its full decline takes place in rather more than four hours, and it pauses at minimum brightness only 15 or 20 minutes. Variables are differently classified by various investigators, and SCHÖNFELD of Bonn (page 270), a pupil of ARGELANDER, devoted much time to these puzzling bodies in his best years. Many variables are quite anomalous in their fluctuations, having no regular period, and still others present several mutations of brightness in every complete period; as, for instance, β Lyrae, whose complex spectrum shows a wealth of helium bands and hydrogen lines not yet unravelled. The shortest known variable is ω Centauri 91, whose period is $6^h 11^m$; and other variables in this interesting cluster exhibit many exceptional peculiarities. — *D. P. T.*

parallaxes amounting to 2″ or 3″ for several of the brighter stars ; but POND (who succeeded MASKELYNE as Astronomer Royal in 1811) showed by more accurate Greenwich observations that these conclusions were untenable. The first really satisfactory determi-

FRIEDRICH WILHELM BESSEL (1784-1846)

nation was made long afterward by BESSEL, in the case of a star only just visible to the naked eye, known as 61 Cygni. Attention was directed to it as being probably much nearer us than others, in consequence of its unusually large proper motion in the heavens, which carries it though an arc of about 5″ every year. 61 Cygni is itself double ; and taking advantage of the circumstance that there are two other small stars in its close neighborhood (which, not sharing in this motion, are probably much farther off), BESSEL, in 1837, began observations at Königsberg, with the view of determining the parallax of this star relatively to the small objects in the same field. As their parallax might fairly be considered insensible, the result would practically be the parallax of 61 Cygni, or of the two stars composing it. In 1840 he was able to announce that this quantity was measurable, amounting to about 0″.35, a value which later observations by other astronomers have essentially confirmed. The resulting distance of 61 Cygni from the solar system is over 50,000,000,000,000 of miles. Beginning with those of HENDERSON (made about the same time as BESSEL's of 61 Cygni), observations of a bright star in the southern hemisphere, α Centauri, show that its parallax is the greatest of all the fixed stars yet measured ; and the best determination, made by Dr GILL and Dr ELKIN at the Cape of Good Hope in 1881–83, amounts to 0″.75, indicating a distance equal to about 25,000,000,000,000 of miles. Several other stellar parallaxes have since been measured : that of Sirius is 0″.38 ; one small star in LALANDE's catalogue

C. A. F. PETERS (1806–1880)

(No. 21,185) has yielded a parallax of about 0″.5, and that of another (No. 21,258) is about 0″.3. Also many, smaller in amount, have been determined. Only by the rapid propagation of light is it possible to express the distances of the enormously remote fixed stars in small figures. In a year light travels nearly 6,000,000,000,000 of miles ; and this inconceivable distance, now generally adopted as the unit of length in expressing stellar distances, is termed a 'light year.' Of α Centauri, for example, it is customary to say that its distance is about 4⅓ light years, and in like manner the distance of 61 Cygni is about seven light years.[55]

BRÜNNOW (1821–1891)

[55] Stellar distances tax to the utmost the skill and patience of the practical astronomer. So minute are the quantities to be measured and calculated that finding a star's distance may be likened to the problem which would confront a prisoner who possessed instruments of the utmost precision, and was told to measure the distance of a mountain twenty miles away, but was given liberty only to observe the mountain from the window of his cell. In this case, the orbit of the Earth, 186,000,000 miles across, would be represented by the largest hoop he could place in the window; while our globe itself, on the same scale, would shrink to a speck requiring a microscope to render it visible. The steadiest atmosphere, instruments of perfect type and construction, and observers of consummate training can alone cope with the obstacles encountered at every step. Following BESSEL came the painstaking PETERS, who, about the middle of the 19th century, attempted the more difficult undertaking of ascertaining, not the relative, but the absolute parallaxes of certain fixed

In speaking of 61 Cygni, the expression 'double star' was used, and it is essential to distinguish between

stars, among them Polaris, Capella, Arcturus, and Vega. His method employed the meridian circle (page 368) by which he was skilful enough to detect very delicate changes in the apparent direction of these stars with the change of season. About 1870 BRÜNNOW, then Astronomer Royal for Ireland,

HUGO GYLDÉN (1841–1896)

began to devote his attention to these critical researches, determining many stellar parallaxes with high precision, and his work was ably continued at Dunsink for several years by Sir ROBERT BALL. Their measures were made with the micrometer (page 343), and the mathematical formulæ requisite in calculating such observations are presented in elegant form by BRÜNNOW in his *Handbook of Spherical Astronomy* (London, 1865). Also GYLDÉN (from 1871 to 1896 Director of the Royal Observatory at Stockholm), and Dʳ AUWERS, of

Dʳ GILL'S HELIOMETER AT CAPETOWN

(*Aperture*, 6 *inches. Built by* REPSOLD *of Hamburg in* 1888, *and chiefly used in determining distance of Sun and stars. Instruments of this type in the hands of trained observers have given results of the highest precision yet reached in astronomical measurement*)

the two classes of stars which come under that designation.   A stellar object which appears single to the naked eye, but when viewed with a telescope is seen

Berlin, and Dʳ OTTO STRUVE, late Director at Pulkowa (pp. 130 and 133), have greatly advanced our knowledge of the distances of the stars.   But the favorite method at the present day is that originally employed by BESSEL — measures by the heliometer, which is a telescope of medium size, mounted equatorially, as shown on page 277, and provided with a variety of intricate appliances for facilitating the astronomer's use of the instrument, and enhancing the accuracy of his measures with it.   But the cardinal peculiarity of the heliometer is its divided object-glass, as in the opposite figure, one half of the glass being mounted in a sliding frame, which enables the observer to move it by a definite amount relatively to the stationary half. As each portion makes a perfect image in the field of view, any lateral displacement of the half-objective will produce a double image.   In the case of the sun, the two images of his disk are then brought just tangent to each other; and so his diameter has been most accurately measured (heliometer meaning sun-measurer).   The Yale Observatory possesses the only instrument of this type in America, mounted in 1882, and it has been successfully employed by Dʳ ELKIN in determining stellar parallaxes.   Following is a table including many of the stars whose distances are best known : —

DISTANCES OF FIXED STARS

Star's Name (In order of Distance)	Magnitude	Position for 1900.0		Parallax	Distance in —		Proper motion
		R. A.	Decl.		Light-years	Trillions of miles	
		h m	° ′	″			″
α Centauri	−0.1	14 33	− 60 25	0.75	4⅓	25	3.67
61 Cygni	5.1	21 2	+38 15	0.45	7⅖	43	5.16
Sirius	−1.4	6 41	−16 35	0.38	8½	50	1.31
Procyon	0.5	7 34	+ 5 29	0.27	12	71	1.25
Altair	0.9	19 46	+ 8 36	0.20	16	94	0.65
o² Eridani	4.4	4 7	− 7 7	0.19	17	100	4.05
Groomb. 1830	6.5	11 7	+38 32	0.13	25	147	7.65
Vega	0.2	18 34	+38 41	0.12	27	158	0.36
Aldebaran	1.0	4 30	+16 18	0.10	32	191	0.19
Capella	0.1	5 9	+45 54	0.10	32	191	0.43
Polaris	2.1	1 23	+88 46	0.07	47	276	0.05
Arcturus	0.2	14 11	+19 41	0.02	160	950	2.00

to consist of two so near as not to be separated by the unaided vision, is naturally called a double star. But the fact that there are two stars apparently in such close proximity does not of itself prove any connection between them ; they may, indeed, be merely optically double, that is, appearing double only because the individual stars happen to be nearly in the same line of sight, without reference to their actual *distance*. But Sir WILLIAM HERSCHEL'S persevering survey of the heavens revealed to him so large a number of these apparently double stars that there was in many cases great probability of physical connection between the

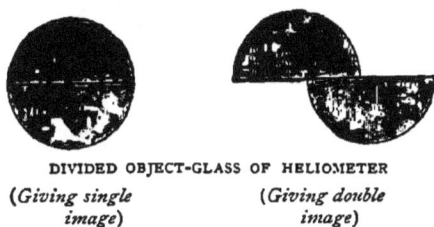

DIVIDED OBJECT-GLASS OF HELIOMETER
(*Giving single image*)   (*Giving double image*)

Of all the stars whose distances have been measured, per-haps 50 are regarded as satisfactorily known, and many of them are shown approximately in the diagram on page 280, adapted from calculations by RANYARD and GREGORY, in which the solar system is at the centre and the concentric circles are drawn at intervals of five light-years. Any star just outside the outer circle would have a parallax of 0″. 1. Within recent years photography has rendered great assistance in find-ing stellar parallaxes, chiefly because plates once exposed can be critically measured and re-measured at any time. The pho-tographs of PRITCHARD and RUTHERFURD have contributed conspicuously to this desirable end. GYLDÉN, from an inves-tigation of the parallaxes and proper motions of about half of the first magnitude stars, finds their average distance repre-sented by 40 light-years or 230 trillions of miles, a result con-firmed by Dr ELKIN's careful researches. — *D. P. T.*

components, as suggested several years before by
Michell in the *Philosophical Transactions* for 1767.
Herschel, in 1782, began the practice of carefully
observing and recording the relative positions and

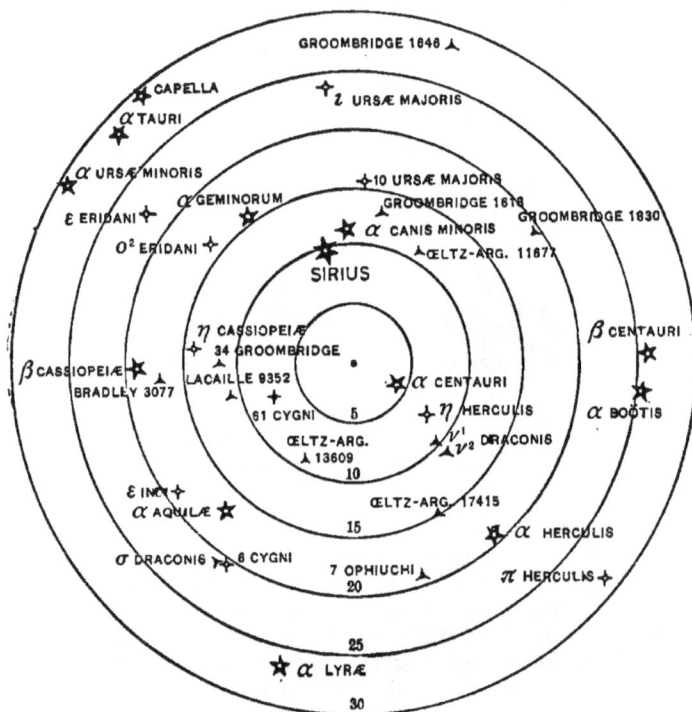

GROOMBRIDGE 1846

CAPELLA
α TAURI
ι URSÆ MAJORIS

α URSÆ MINORIS
10 URSÆ MAJORIS
ε ERIDANI
α GEMINORUM
GROOMBRIDGE 1618
GROOMBRIDGE 1830
O² ERIDANI
α CANIS MINORIS
ŒLTZ-ARG. 11877
SIRIUS

η CASSIOPEIÆ
34 GROOMBRIDGE
β CENTAURI
β CASSIOPEIÆ
LACAILLE 9352
α CENTAURI
BRADLEY 3077
61 CYGNI   5
α BOÖTIS
η HERCULIS
ŒLTZ-ARG.
ν¹ν² DRACONIS
13609   10
ε INDI
α AQUILÆ
ŒLTZ-ARG. 17415
15
α HERCULIS
σ DRACONIS   δ CYGNI
7 OPHIUCHI
π HERCULIS
20

25
α LYRÆ
30

KNOWN DISTANCES OF STARS FROM THE SUN IN LIGHT-YEARS

(*From* Todd's ' *New Astronomy,*' *by special permission of the American
Book Company, publishers*)

distances of the stars which he had noticed as being
apparently closely double; and when, after an interval of some years, these values were determined again,
he at once discovered that in many cases the compo-

nents of a double star have a distinct motion with reference to each other.

In HERSCHEL's first paper on the subject, contained in the *Philosophical Transactions* for 1803, he remarks that his observations distributed over a period of twenty-five years, and there recounted, prove that many of the stars whose mutual positions and distances had been measured by him, 'are not merely double in appearance, but must be allowed to be real binary

β *Orionis*   γ *Leonis*   *Polaris*   γ *Virginis*

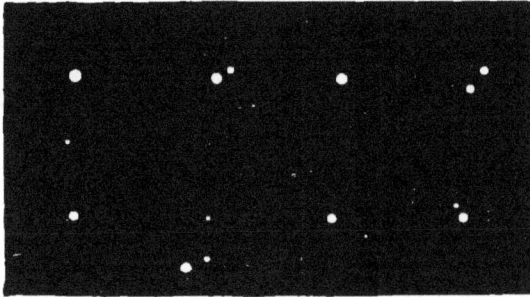

δ *Cygni*   γ *Arietis*   γ *Andromedæ*   δ *Serpentis*

EIGHT TYPICAL DOUBLE STAR SYSTEMS

(*As seen in an inverting telescope*)

combinations of two stars, intimately held together by the bond of mutual attraction.' These, then, are called binary stars, to distinguish them from such as are only optically double ; and the discoveries of binary systems, or stars physically double, is constantly increasing. The high interest attaching to this subject has led many skilful astronomers to devote much attention to it since HERSCHEL's time. The elder STRUVE was one of the first ; and a catalogue of no fewer than

3,134 double stars was published at Saint Petersburg
in 1837, containing the results of his observations at
Dorpat and Pulkowa.   Many of the binary stars have
a regular orbital revolution,
the most interesting cases
being ζ Herculis, ξ Ursæ
Majoris, and 70 Ophiuchi,
whose periods of revolution
have been determined quite
accurately to be 34, 61, and
94 years respectively. The
shortest known, those of κ
Pegasi and δ Equulei, are
only about eleven and a
half years.   Much longer

F. G. W. STRUVE (1793–1864)

than any of these, and therefore less precise, are the
periods of those interesting and long-known double
stars, γ Virginis, γ Leonis, and Castor, the first of
which amounts to nearly 200, the second to about
400, and the third to nearly 1,000 years.

Accurate investigation of the proper motions of the
larger stars has disclosed, by means of irregularities
in these motions, more than one instance of the disturbance of a star by an unseen companion ; so that
the binary character of the star was only discovered
in this way.   Such was the case with the brightest of
all the fixed stars, Sirius ; and the disturbing companion was afterward detected as a small star near it,
visible only with a powerful telescope.   Procyon, the
principal star in Canis Minor, is also subject to a similar irregularity of proper motion, and one which would
seem to result from orbital motion round a companion
possessing so feeble a luminosity that it was long supposed to be opaque ; and its probable perception has

been quite recent. The period of revolution in this case is about 40 years.[56]

[56] More than 10,000 double stars are now known, measured, and catalogued, a number only reached by the faithful enthusiasm of many gifted astronomers — chief among them the STRUVES, father and son, DAWES, Professor SCHIAPARELLI, Baron DEMBOWSKI, and Professor BURNHAM. Few stellar objects will better repay the amateur than 'doubles.' Even a small telescope will show many of the wider and brighter pairs, a few of which, illustrated on page 281, are readily located by ordinary star charts. The easiest is perhaps γ Virginis, discovered by BRADLEY in 1718; and it has now been measured through so large a portion of its binary arc that

W. R. DAWES (1799–1868)

its period is pretty well established. The diagram on the next page exhibits the orbit and the observations upon which it is based. The real orbit has a high eccentricity, surpassing that of every other known binary. Of yet greater interest is Sirius, the orbit of whose companion, a star of the tenth magnitude, is also shown. The existence of this companion was first predicted by Dᵣ AUWERS, and verified independently in 1862 by ALVAN GRAHAM CLARK, the distinguished optician, on the completion of the 18½-inch telescope now belonging to the Dearborn Observatory at Evanston. The accompanying diagram of the companion's orbit is by Professor BURNHAM, from observations between 1862 and 1896. There were no observations between 1890 and 1896, because the companion was then so near the blazing central star as to be lost in its rays, and wholly invisible even with the Lick 36-inch telescope. It is now visible again, and will remain so till 1942, when it will disappear once more. The full period of the companion of Sirius is 51.8 years and its mass equals that of the Sun itself. The nearest known fixed star, α Centauri, is also a binary

When stellar positions observed at considerable intervals of time are compared, nearly all the stars

system with a period of 81 years (page 250). The component stars, very nearly equal, have each about the mass of the Sun. When at their nearest (*periastron*) they are about as far apart as Saturn is from the Sun, and at their farthest (*apastron*) their distance from each other is equal to 36 astronomical units, or one fifth greater than Neptune's distance from the Sun. About 200 binary systems are now known, of which Dʳ SEE regards 40 as well ascertained. The longest known binary is ζ

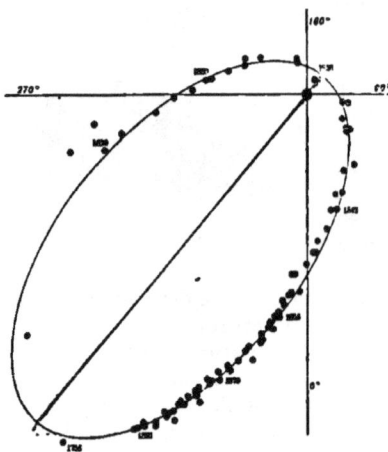

APPARENT ORBIT OF THE
COMPANION OF SIRIUS
(BURNHAM)

APPARENT ORBIT OF γ VIRGINIS (Σ 1670)
(*The length of the major axis is nearly 7″*)

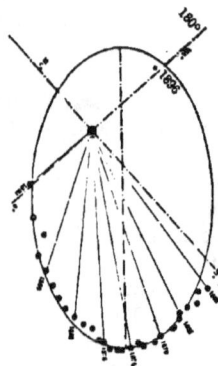

Aquarii, with a period not less than 1,500 years; and among the very short ones not already mentioned are 82 Ceti (16 years), and β 883 (Lalande 9091), the shortest of all, with a period of 5½ years.

A great gap still exists between even these short periods, and the times of revolution of companion stars recently discovered by means of the spectroscope, and therefore called spectroscopic binaries; for their periods are not reckoned in years, but in

are found to be endowed with regular though slow
changes of place in the heavens, called their proper

days, or even hours. If the orbit of a close binary is seen
edge on, twice in every revolution the light of the two stars
will coalesce; half way between which the two stars will be
moving, one toward and the other from the Earth, also twice.
Now photograph the star's spectrum at each of these four

ALVAN GRAHAM CLARK (1832–1897)

critical points: in the first pair, the lines appear sharply
single; in the second pair, they are found to be double. This
singular phenomenon is readily explained by DOPPLER'S important and most useful principle (page 56), according to
which the lines in the spectrum of a star receding from us are
displaced toward the red, while if the star is coming toward
us, the spectral lines are displaced toward the violet. At con-

motions. Presently we shall consider whether these are in all cases wholly due to actual motion of the

junction, then, as the two stars are passing athwart each other the lines in their combined spectrum will be single, but double at the quadratures, or intermediate between the conjunctions. Mizar (ζ Ursae Majoris) was the first discovered spectroscopic binary, by Professor EDWARD C. PICKERING in 1889. This moderately bright star in the middle of the Dipper's handle has a mass exceeding that of the Sun forty-fold, and the period

JOHANN CHRISTIAN DOPPLER (1803–1853)

of its invisible companion is 52 days. β Aurigæ is another, with a much smaller mass and a period of four days. Also, α^1 Geminorum, the larger of the two stars composing the brilliant double star Castor, proves to be a rapid binary with a period of only three days. But the case of μ^1 Scorpii is the most remarkable of all, with the exceedingly short period of $34^h 42^m 30^s$. On the same photographic plate, side by side, are the spectra of μ^1 and its companion μ^2, the lines of the latter

stars in space; but certainly a considerable part of the proper motions which are exceptionally large must be so. It has already been mentioned that 61 Cygni was thought to be probably nearer us than most of the fixed stars, on account of its very large proper motion of about 5″ in a year. Two small stars, one in the northern, the other in the southern hemisphere, having no designation but numbers in star catalogues (1,830 GROOMBRIDGE, and 9,352 LACAILLE), have proper motions even larger than this, being each 7″ in amount, or nearly so; but these stars do not, like 61 Cygni, appear to be much nearer the solar system than others. Only a few stars are known to have proper motions between 4″ and 5″ in amount, and the proper motions of the great mass of stars are much smaller, and only to be recognized by comparing accurate observations widely separate in time.

When the proper motions of several stars had been tolerably well determined, the idea was soon suggested that they might be due in part, not to actual motions of these stars in space, but to translation of the solar system among them. Such movement would produce an apparent motion in the stars in the opposite direction to that of the solar motion, or a tendency on the whole to recede radially from the point in the heavens

always single and sharply defined. Those of μ^1, on the other hand, are now single and again double, thus leaving no room for doubting this most modern contribution of the spectroscope to stellar astronomy, and marvellously confirming the brilliant BESSEL's prophecy of 'the astronomy of the invisible.' By repeating these spectrum photographs and making comparative measurements upon them, the period of revolution and other data can be found for binary stars so close together as to be forever beyond the separating power of the largest telescopes. —*D. P. T.*

toward which the Sun was journeying. Approximate determinations of this point were simultaneously made in 1783 by Sir WILLIAM HERSCHEL and by PIERRE PREVOST, of Geneva, both coming to the conclusion that it was situate in the constellation Hercules. In 1805 HERSCHEL made a similar investigation, founded

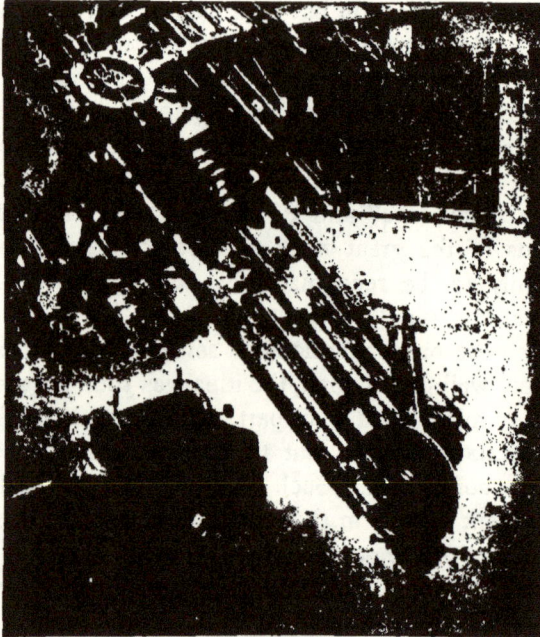

STAR SPECTROSCOPE BY BRASHEAR

(*Attached to eye end of the 36-inch* LICK *telescope. Also shows accessory apparatus for producing comparison spectra*)

on proper motions determined by MASKELYNE; and in later times others were made by different astronomers, all of whom agree in placing the apex of the Sun's motion in the same region of the heavens, either in Lyra or in the adjacent constellation Cygnus, and therefore not far from Hercules.[57]

[57] Stellar proper motions were first made out by HALLEY in 1718. M. BOSSERT, of the Paris Observatory, in 1896 published a catalogue of all the well-ascertained proper motions of stars, over 2,600 in number, in which the motion exceeds 0ˢ.01 in right ascension and 0″.1 in declination. An orange yellow star of the eighth magnitude in Pictor has a proper motion of nearly 9″, the largest known. Sir WILLIAM HUGGINS was the first to make, in 1868, that most important application of the spectroscope to sidereal astronomy, by which a star's light reveals unerringly its motion toward the Earth or from it, in accordance with DOPPLER's principle. Thus was demonstrated beyond a shadow of doubt the fact that the swarms of the stellar universe are in constant motion through space, not only athwart the line of vision as their proper motions had long disclosed, but some swiftly toward our solar system and others as swiftly from it. So the term 'fixed stars' appeared to be a double misnomer. Research of this character has also been vigorously prosecuted by Dʳ VOGEL of Potsdam, near Berlin (page 61), at the Lick Observatory by Professor CAMPBELL with the fine spectroscope shown above, and by Mʳ MAUNDER at Greenwich (page 39), where the high significance of this work led the Astronomer Royal many years ago to direct its inclusion in the programme of observational routine. Following are a few determinations of

MOTION IN THE LINE OF SIGHT

Star's Name	Position for 1900.0		Motion per Second Toward or from the Sun	
	Right Ascension	Declination	In Miles	In Kilometers
	h m	° ′		
α Arietis	2 2	+22 59	−11.7	−19
Aldebaran	4 30	+16 18	+31.1	+50
Rigel	5 10	− 8 19	+13.6	+22
Betelgeux	5 50	+ 7 23	+17.6	+28
γ Leonis	10 14	+20 21	−25.1	−40
Spica	13 20	−10 38	−10 6	−17
α Coronae	15 30	+27 3	+20.3	+33
Altair	19 46	+ 8 36	−23.9	−38

Attempts have been made to extend the theory farther, and to ascertain the region round which the

The sign minus means that the star is coming toward us, and + that it is receding. It is to be remembered that these are simply motions in the line of sight, relative to the solar system as a whole, for the effect of the Earth's motion round

OBSERVATORY OF Dᵣ ISAAC ROBERTS, F. R. S.
(Starfield, Crowborough Hill, Sussex, England)

the Sun has been eliminated. To find the actual motions of these stars through space, it is necessary to combine the above data with their distances and proper motions, a geometric operation which will usually increase the sight-line motions very considerably. Part of the research regularly pursued with the 40-inch Yerkes telescope (page 338) is work of this

solar motion is carrying the Earth and all the bodies that move obediently to the Sun. Such attempts, however, are yet premature, and can only be regarded as tentative. MÄDLER, about sixty years ago, considered it probable that the point in the heavens in question was situate in or near the cluster of stars (see next paragraph) known as the Pleiades. But it would seem that the stellar motions noticed by him were rather due to special cases of star-drift ; and M^r MAX-WELL HALL, as the result of an investigation founded on the proper motions of a few stars whose parallax and distance are approximately known, has indicated a point in the constellation Pisces as that around which the solar motion is probably performed. In the present state of our knowledge, the period of this great revolution can be little more than matter of conjecture ; and as a rough approximation, it is put down by Mr HALL at about 13,000,000 years.

Many of the fixed stars are arranged in groups or constellations which imagination has likened to the forms of animals and other objects, — some of these in strings of various twists suggesting the idea of serpents or dragons. While these groups are spread out over regions of the sky many degrees in extent, casual inspection of the heavens shows here and there a group of stars arranged close (sometimes very close) to each other, so as to form a cluster. Of these, the most remarkable is a loose cluster called the Pleiades,

character, in charge of Professor FROST ; and photography is a most helpful adjunct, because the plates accumulated in periods of clear and steady atmosphere may thereafter be measured, and the necessary results derived at leisure and with that high degree of patience and painstaking which alone render such investigations of any value. — *D. P. T.*

in Taurus (pages 236–237). It is easy to perceive
six stars in this well-known group, and some persons
with unusually acute vision can detect several more ;
but while HYGINUS and OVID say there are only six,
there is evidence that as many as thirteen were seen
before the invention of the telescope. With a moder-

20-INCH REFLECTOR AND PARALLEL GUIDING TELESCOPE
(*Used by* Dr ISAAC ROBERTS *in photographing stars and nebulæ*)

ately good instrument at least a hundred become visible, while very large telescopes reveal the existence of
more than a thousand. Near the Pleiades and in the
same constellation is a more scattered group called
the Hyades, resembling a letter V. Still another,
called Præsepe, is visible to the naked eye as a faintly

luminous spot or nebula in the constellation Cancer, and is not well seen unless the night be clear and moonless. The telescope, while showing that this and many other nebulous or cloudy-looking masses are composed of very distant or very minute stars, brings into view multitudes more not visible to the naked

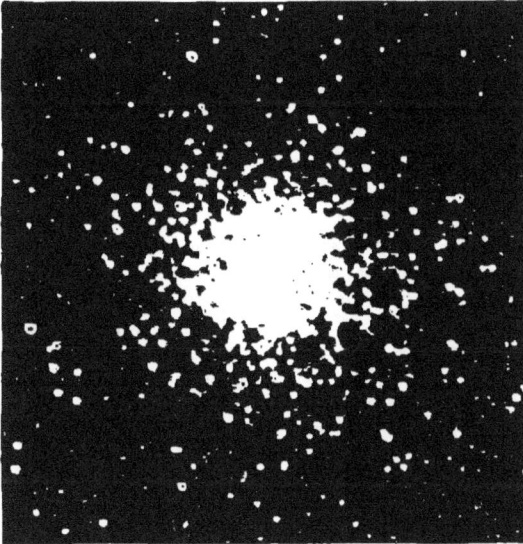

GLOBULAR CLUSTER IN PEGASUS

(Photographed by ROBERTS, *4th November* 1890. *Exposure,* 2 *hours. Central portion nebulous)*

eye at all. The close clusters are excellently illus-'trated by the reproduction above, from photographs by Dr ISAAC ROBERTS, of a typical globular cluster in Pegasus, and the magnificent double cluster in Perseus (page 297) ; and on pages 290 and 292 are D^r ROB-

ERTS's observatory, and the ingenious arrangement of telescopes with which he took these classic photographs.[58]

[58] And this is practically all that even the most powerful telescope, if unaided, can avail. With it can be found the star's exact position in the firmament, not only with reference to neighboring stars, but its distance from us also. The telescope solved the problem of *where*, but was utterly baffled by the question *what*, until supplemented by the spectroscope, with observations interpreted by KIRCHHOFF's laws (page 69).

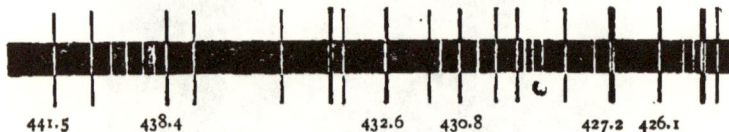

441.5   438.4   432.6   430.8   427.2 426.1

SPECTRUM OF SIRIUS COMPARED WITH THAT OF IRON (VOGEL)

Here again Sir WILLIAM HUGGINS was the pioneer, and Betelgeux and Aldebaran were the first stars whose chemical constitution was revealed to the eye of man. Many of these distant twinkling luminaries were at once found to contain calcium, hydrogen, iron, magnesium, sodium and other metals whose existence in the Sun had been demonstrated. What, for example, could be better proof of the fact of iron in Sirius than a spectrum like the one adjacent, in which are repeated coincidences of the dark (negative) iron bands (above and below) with the intermediate bright lines of the spectrum of the star itself? Only a year intervened before the spectra of stars had been examined in

PIETRO ANGELO SECCHI (1818-1878)

numbers sufficient to permit their classification. This was first satisfactorily done by Father SECCHI of Rome, whose comprehensive scheme embraced four distinct types based on optical observations solely : —

Stretching across the sky in clear nights is seen a band or zone of luminous matter, which was a great

Type I is chiefly characterized by the breadth and intensity of dark hydrogen lines; also a decided faintness or entire lack of metallic lines (see next page). Stars of this type are very abundant, and are blue or white. Sirius, Vega, Altair and numerous other bright stars belong to this type, often called 'Sirian stars,' a class embracing perhaps more than half of all the stars.

Type II is characterized by a multitude of fine dark, metallic lines, closely resembling the solar spectrum. They are yellowish, like the Sun. Capella (page 301) and Arcturus illustrate this type, often called 'solar stars,' and they are rather less numerous than the Sirians. According to recent results of Professor KAPTEYN, absolute luminous power of first type stars exceeds that of second type stars sevenfold. Stars nearest to the solar system are mostly of the second type.

Type III is characterized by many dark bands, well defined on the side toward the blue, and shading off toward the red — a 'colonnaded spectrum,' as Miss CLERKE very aptly terms it (see the next page). Orange and reddish stars and a majority of the variables fall into this category, of which α Herculis, Antares and Mira are examples.

Type IV is characterized by dark bands, often technically called flutings, similar to those of the previous type, but reversed as to shading: they are well defined on the side toward the red, and fade out toward the blue. Stars of this type are few, perhaps 50 in number, faint, and nearly all blood-red in tint. Their atmospheres contain carbon.

Type V has been added to SECCHI's classification by Professor PICKERING, and is characterized by bright lines. They are about 70 in number, and are all found near the middle of the Galaxy. From two French astronomers who first investigated objects of this class, they are usually known as WOLF-RAYET stars. They are a type of objects quite apart from the rest of the stellar universe, and many objects termed planetary nebulæ yield a like spectrum.

A classification by Dr VOGEL combines SECCHI's types III and IV into a single type, and several other classifications have been proposed since the inauguration of stellar spectrum photography. The Harvard classification embodies about 20

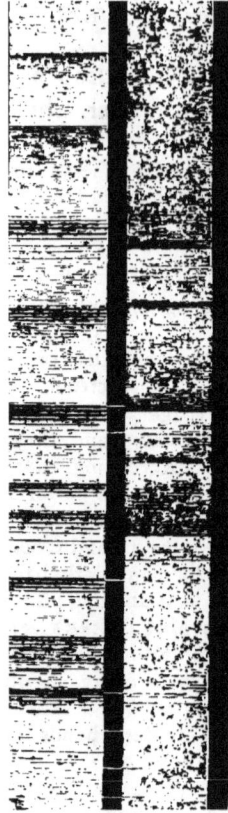

FATHER SECCHI'S FOUR TYPES OF STELLAR SPECTRA

(Dependent on visual observations)

enigma to the ancients, and received the name of the Galaxy, or Milky Way. As soon as GALILEO applied

groups. Whether these differences of spectra are due to different stages of stellar development, or whether they indicate real differences of constitution, is not yet known. Probably they are due to a combination of these causes. According to Mr ESPIN, stars having the most bands in their spectra present the greatest differences between their visual and photographic magnitude. As in the Sun, so in the stars, the girdling atmosphere must be held responsible for numerous dark lines.

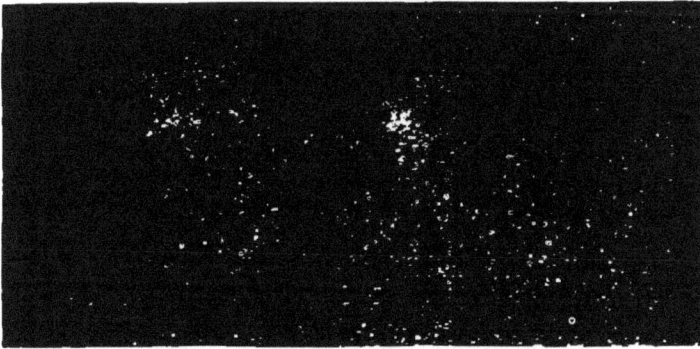

DOUBLE CLUSTER IN THE SWORD HILT OF PERSEUS

(*Photographed by* ROBERTS, *13th January* 1890. *Exposure,* 3 *hours. This cluster is visible to the naked eye as a faint spot of light in the Milky Way*)

In the case of α Aquilæ, M. DESLANDRES, the able spectroscopist of Paris, has discovered fine double bright lines traversing the middle of the dark hydrogen lines, also through the iron lines and the K-line of calcium; and he traces their origin to a chromosphere similar to that enveloping our Sun.

The first star whose spectrum was successfully photographed is Vega, by HENRY DRAPER in 1872, and our detailed knowledge of stellar spectra is in great part consequent upon his initiative. At the time of his lamented death, research upon the constitution of the stars was still in the experimental stage; but the ample funds of the Henry Draper Memorial, generously

his telescope to the heavens, he saw that this appearance is produced by millions of stars apparently scat-

maintained by Mʳˢ DRAPER, have enabled Professor PICKERING
and his efficient corps to prosecute an unparalleled line of investigations embracing the stars of both hemispheres. These
researches were at first conducted with the Bache telescope,
illustrated opposite, whose chief peculiarities are a large object
glass of short focal length, and a mounting with a forked polar

HENRY DRAPER (1837-1882)

axis. For the furtherance of this work Mʳˢ DRAPER provided
an exact duplicate of this instrument. Also, both are equipped
with great prisms (not shown), mounted in front of the objectives, thereby spreading out the stellar images into lines on the
photographic plate, instead of the mere points that appear on
chart plates. Suitable adjustment of the prism and the clock-
motion causes the broken spectral line to trail over the plate
latitudinally, so that every star down to the eighth magnitude

tered like dust along the black ground of the sky. Analogy led many to suppose that the nebulæ or

records its own spectrum, sometimes many hundred on a single plate. The advantages of this fortunate arrangement are obvious: peculiarities of spectra are continually leading to the detection of variable stars that would otherwise pass unobserved; several new or temporary stars have been discovered; and the spectra of about 50 stars in the Pleiades exhibit at a glance their close identity of chemical constitution. In 1890

THE BACHE TELESCOPE AT CAMBRIDGE

(*Aperture, 8 inches. Employed since 1885 in photographing the stars and their spectra. Now mounted at Arequipa, Peru, and in use upon the southern stars. The success of this photographic instrument led to the construction of the Bruce telescope, page 264*)

HARVARD COLLEGE OBSERVATORY — PROFESSOR EDWARD C. PICKERING, *Director.*

(*Ideal observatory construction — Instruments housed for the most part in separate small buildings*)

smaller patches of luminous matter brought into

was published the 'Draper Catalogue of Stellar Spectra,' including more than 10,000 stars; and the larger and more powerful Bruce telescope (page 264) is now continuing these researches to stars of fainter magnitudes. If a star's spectrum is required with a high degree of dispersion, the ordinary slit spectroscope is employed, which so reduces the star's light that large objectives become necessary. Dr VOGEL and Dr SCHEINER at Potsdam, and Sir NORMAN LOCKYER at South Kensington have likewise been successful in detailed photography of especially bright stars, recording many thousand lines in highly luminous objects of type II. Capella's spectrum adjacent, from a fine photograph kindly sent me by M. DESLANDRES of Paris, exhibits not only the wealth of lines, many of which are due to iron, calcium, helium and magnesium, but by displacement of bright lines in the terrestrial spectrum, shows the swift recession of Capella from our solar system. The door is now wide opened to a study of the constitution of at least the brighter stars in every detail of their chemical composition; and advance along these

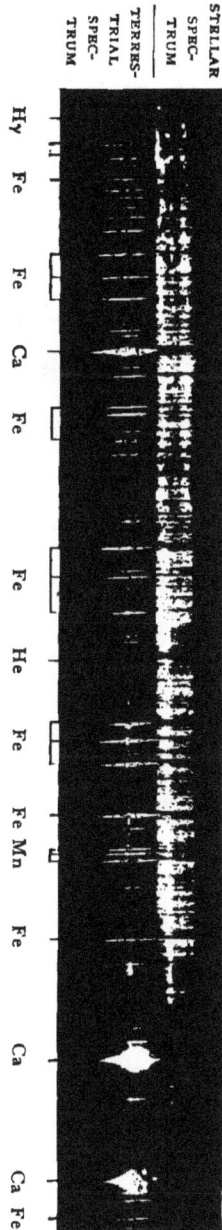

SPECTRUM OF α AURIGÆ (CAPELLA), A YELLOWISH STAR OF TYPE II (DESLANDRES)

(*The marked shift of the bright lines in the terrestrial spectrum shows that the star is receding at the rate of 43.8 kilometers or 27 miles per second*)

view by telescopes are also masses of stars, resembling, on account of their greater distances, portions of the Milky Way as seen by the naked eye, — and, indeed, many of these were resolved into stars by more powerful telescopes ; but (as just stated with regard to the great nebula in Orion) it is now known, from modern

SIR JOHN F. W. HERSCHEL (1792–1871)

observations with the spectroscope, that many of the nebulæ are wholly or in part irresolvable. The improbability of the older view had been shown by Sir JOHN HERSCHEL's observations of those remarkable celestial objects in the southern hemisphere which, from having been first noticed by MAGELHAENS in the earliest

lines cannot much longer be checked by the seemingly endless magnitude of the task. — *D. P. T.*

voyage ever made to the Pacific Ocean, are called the Magellanic clouds. These consist of two large and nearly circular masses of nebulous light, — one larger than the other, — which are conspicuous to the naked eye, and look not unlike portions of the Milky Way. The larger remains visible even in strong moonlight, which obliterates the smaller. When viewed through large telescopes these objects are seen to contain multitudes of stars, and hundreds of nebulæ which cannot be resolved into stars, all wedged together in such a way as to suggest some close and immediate connection between them, quite inconsistent with the idea that any portions are at very much greater distances from us than others.[59]

[59] Sir WILLIAM HERSCHEL, the goal of whose researches was a knowledge of the construction of the heavens, undauntedly essayed a tedious series of 'gauges' or counts of the stars visible in different parts of the sky in the field of his 20-foot reflector. Twice the breadth of the field reached across the Moon's diameter; and the field itself contained about $\frac{1}{130,000}$ part of the area of the whole celestial sphere. HERSCHEL actually counted the stars in nearly 3,500 such fields, and the number therein ranged all the way from 0 to nearly 600. These gauges were distributed all around the sky, along a galactic meridian inclined about 35° to the celestial equator, Sir JOHN HERSCHEL continuing his father's work at the Cape of Good Hope. The average number of stars in a field was as follows : —

In the Galaxy itself	122
In galactic latitude 15°	$30\frac{1}{3}$
30°	$17\frac{2}{3}$
45°	$10\frac{1}{3}$
60°	$6\frac{1}{2}$
75°	$4\frac{2}{3}$
90°	$4\frac{1}{2}$

From this very rapid increase in the number of the stellar hosts, as we approach nearer and nearer the Milky Way, HERSCHEL drew unwarrantable conclusions, which we cannot

The labors of the two HERSCHELS greatly increased
the number of known nebulæ, and many more have

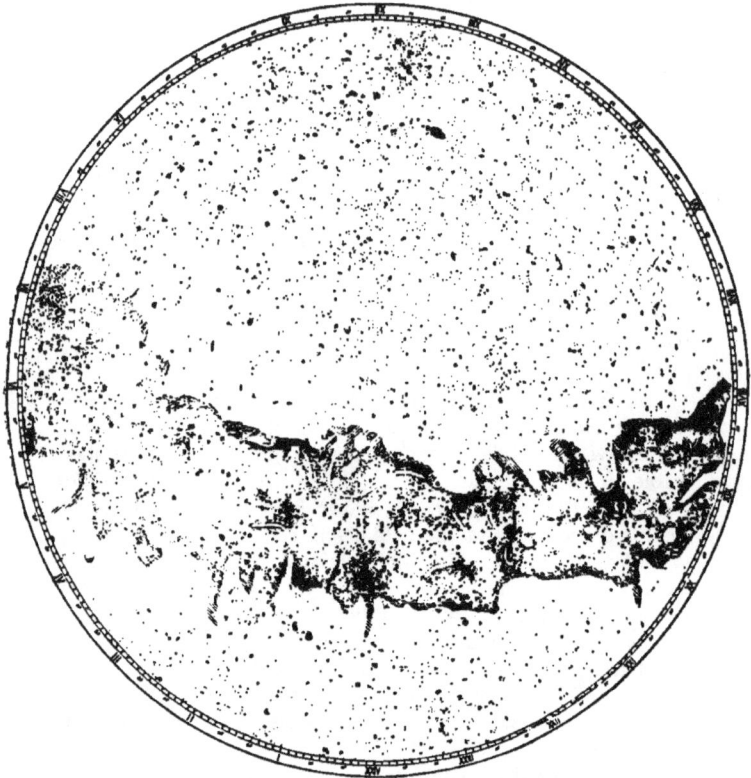

THE NEBULAE AND CLUSTERS OF THE NORTHERN HEAVENS (WATERS)

(*Milky Way according to* BOEDDICKER, *the dots in it marking star clusters. The dots
above and below it indicate nebulae, most numerous in regions remote from the Galaxy*)

here enter into; but we may summarize as follows our knowledge of this recondite subject down to the year 1890: —
The component bodies of our sidereal universe are scattered.
without regard to uniformity, throughout a vast system having
in general the shape of a watch, whose thickness is, perhaps,

been discovered since their time. About 10,000 are now catalogued, and their distribution over the celes-

one-tenth of its diameter. On either side of this, and clustered about the poles of the sidereal system, are the regions of nebulæ, thus making the entire visible universe spheroidal in general shape. The plane of the Milky Way passes through the middle of this stellar aggregation, and near its centre are our Sun and his attendant worlds. In large part, the stars are clustered irregularly, in various and sometimes fantastic forms, but without any approach to systematic order. Often there are streams of stars; and elsewhere aggregations leaf-like or tree-like in apparent structure; also,'star-sprays,' as PROCTOR termed them. This painstaking investigator laboriously charted all the stars of ARGELANDER'S *Durch-musterung*, nearly 325,000 in number, and bare inspection of his comprehensive map affords the best notion of the seeming capriciousness of stellar distribution. But he proved clearly the detailed connection between the distribution of the stars and

R. A. PROCTOR (1837–1888)

the complex branching and denser portions of the Galaxy. Sir JOHN HERSCHEL remarked many well-defined offshoots from the Milky Way, each tipped at its extremity with a fairly bright star; and the physical connection between the two may be regarded as certain. That is, the fainter stars of the Galaxy that give the effect of nebulous light are at practically the same distance as the brighter stars. PROCTOR, too, directed attention to the fact that the luminous streamers in the irregular nebulæ often coincide with streams of small stars in the same field; and in the case of Orion, spectroscopic analysis by Sir WILLIAM and Lady HUGGINS has proved the connection between wisps of the nebulous light and the stars adjacent, thus leaving little room for doubt as to the distance of the nebulæ. See also the nebulous streamers of the Pleiades (page 237). Upon such inference, in fact, our slender knowl-

tial vault — by no means uniform — manifests a marked
tendency to avoid the zone of the Milky Way, and to

edge of the distance of the nebulæ depends; for the parallax
of a nebula, although attempted, has never been measured
directly.

Since 1890 the spectroscope has come to the assistance of
the telescope in solving the intricate problem of stellar distribution. Professor PICKERING had noticed that most of the
brighter stars of the Galaxy yield spectra of the first or Sirian
type ; and Professor KAPTEYN, by combining the well-ascertained proper motions of certain stars with their classification in
the 'Draper catalogue of stellar spectra,' concludes that, as stars
having very small proper motions show a condensation toward
the Galaxy, the stars composing this girdle are in great part of
the Sirian type and lie at vast distances from the solar system.
If a star is near to us, its proper motion will ordinarily be large ;
and in the case of stars of the second or solar type, the larger
the proper motion the greater their number. So it would appear that the solar stars are aggregated round the Sun himself ;
a conclusion greatly strengthened by the fact that of stars
whose distance and spectral type are both ascertained, seven
of the eight nearest to us are solar stars.

From Professor KAPTEYN's further research he concludes
that our solar system lies, not at the centre, but slightly to the
north of the Galaxy. Returning then to the Sirian stars, he
reaches the conclusion that if we compare stars of equal brightness, those of the Sirian type average nearly three times more
distant from the Sun than do those of the solar type. The
Sirian stars may therefore be legitimately regarded as far exceeding the solar stars in intrinsic brightness. GOULD's theory
of a solar cluster receives remarkable confirmation, though the
evidence as to the figure of this cluster is far from complete ;
indeed, it may be ring-shaped. But Professor KAPTEYN's conclusion seems hard to escape, that the Galaxy itself has no
connection with our solar system ; and is composed of a vast
encircling annulus of stars, far exceeding in number those of
the solar aggregation, and everywhere more remote than the
stars composing it, as well as differing from them in physical
type. So it is the mere element of distance that reduces their
individual glow and seemingly crowds them thickly together
into that gauzy girdle which we call the Galaxy. — *D. P. T.*

aggregate north and south of it, near the poles of the galactic circle, as illustrated on page 304. This is additional proof that the nebulæ are connected with the general system of the stars, and do not form other systems, as was formerly thought, at immense distances beyond it. Sir JOHN HERSCHEL noticed that one third of the nebulæ catalogued by him, chiefly from his father's and his own observations, are congregated in a portion of the heavens occupying about one eighth of the celestial sphere, extending from Ursa Major in the north to Virgo in the south. If the irresolvable nebulæ are not placed at distances from us immensely greater than are those that have been resolved into clusters of stars, what is the alternative? Simply that the irresolvability of the former arises, not from the greater distance of the stars comprising them, but from their actually smaller size, which prevents their being seen separate, even under the highest telescopic power.

We conclude with a few words respecting three of the most interesting of the nebulæ. The most remarkable of all surrounds θ Orionis and is in that part of the constellation Orion called 'the sword.' Only barely visible to the naked eye, it appears to have been first seen by CYSAT, a Swiss astronomer, in 1618. It was first drawn and described by HUYGENS (who was apparently unaware of CYSAT's perception) thirty-seven years afterward. Although this gigantic object contains a large number of telescopic stars, spectral analysis has shown that a portion of the light received from it, and from some other nebulæ as well, is due to glowing gas (page 238).

The great nebula in Andromeda (page 243) is just visible to the naked eye, and was noticed before the

invention of the telescope, there being evidence that
AL-SÛFI, the Persian astronomer, saw it in the tenth
century of our era. SIMON MAYR first examined it
telescopically and described it in 1612. A very interesting nebula in the southern hemisphere and invisible in our northern latitudes, surrounds the remarkable
variable star η Argûs. The first to describe this object
seems to have been the French astronomer LACAILLE,
in the course of his observations, in 1750–53, at the
Cape of Good Hope, when Sir JOHN HERSCHEL in the
following century drew an elaborate representation of
it, which still serves as a standard of comparison.[60]

[60] Regarding the nebulæ as the primordial substance that
worlds are fashioned from, our ultimate knowledge of the construction of the heavens rests
upon (1) the discovery and
classification of thousands of
nebulæ, (2) the accurate determination of their positions
among the stars (repeated at
long intervals), (3) the study of
their physical appearance in
the telescope and by means of
photography, (4) their motions
in space and chemical constitution by the spectroscope, (5)
determinations of their distance
by the stars to which they are
found to be related. About
8,000 nebulæ are now recorded
in our catalogues, the most
complete of which is that by
Dr DREYER of the Armagh Observatory in Ireland. The keen-
eyed Dr SWIFT, now work-

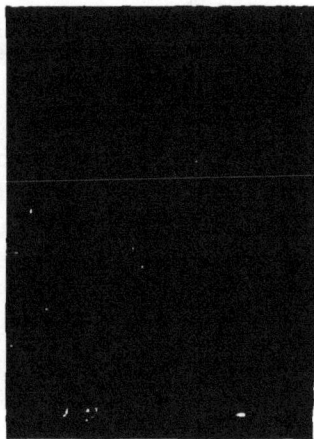

LAMONT (1805–1879)

ing in the clear skies of Southern California, is continually
adding new discoveries. Following the HERSCHELS came LA-
MONT of Munich and D'ARREST of Copenhagen, who devoted
many faithful years to observation of the nebulæ. Large tele-

scopes are now requisite for continuing this important work,
such as the 26-inch refractor of the University of Virginia, employed to good advantage in this field by Professor STONE.
The Orion nebula was the first one ever photographed, by
HENRY DRAPER in 1880, and nebular photography has been
successfully followed up by Dr VON GOTHARD in Hungary;
by Professor BARNARD who has discovered extensive masses
of diffused nebulosity outside
the Pleiades, and an optically
invisible nebula of vast extent
in Scorpius; by Professor
W. H. PICKERING, whose
plates reveal spiral filaments
outlying the nebula of Orion;
and by Dr ISAAC ROBERTS.
whose fine photographs of
star clusters, the Orion nebula
(page 238) and the La Placian
ring nebula in Andromeda
(page 243) have already been
shown. The spectrum of the
latter shows that it is not
gaseous, still no telescope has
yet resolved it. Two other
products of his remarkable
skill and patience here follow

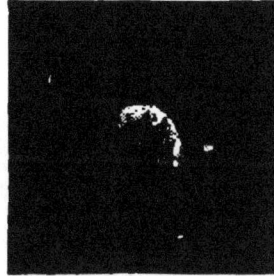

RING NEBULA IN LYRA

(*Photographed by* ROBERTS, *27th
July* 1891. *Exposure* 30 *minutes.
The central star, very obvious in
the photograph, is an exceedingly
difficult object, even with great
telescopes*)

—the ring nebula in Lyra (not a good reproduction of his original
negative), and the marvellous spiral nebula in Canes Venatici,
the characteristic whorls of which, as drawn by Lord ROSSE, had
been doubted by many astronomers till this photograph revealed
them convincingly. Here we seem to see a multiple star in
process of actual evolution, much as the double stars are
thought to have originated from the double nebulæ (page 251).
Following in the footsteps of Sir JOHN HERSCHEL, Dr GILL at
the Cape, Mr RUSSELL, Government Astronomer at Sydney,
and Professor BAILEY at Arequipa, have continued the highly
important task of photographing the southern nebulæ and
clusters with eminent success. In 1897 BAILEY photographed
a fine spiral nebula in Hydra. Wide zones of the southern
sky, however, remain yet unexplored, except the mere reconnaissance. Evidences of variability have been sought in different regions of several of the larger nebulæ; but only a single
instance of change is satisfactorily made out—in the great

irregular nebula enveloping η Argûs (page 265), a conspicuous part of which, recorded near the centre of HERSCHEL'S drawing, is wholly absent from Mᵣ RUSSELL'S photograph of 1890, in which the same space is occupied by a great dark oval. The facts of highest significance concerning nebulæ recently brought to light depend upon observations with the spectroscope by Professor KEELER, now director of the Lick Observatory. Sir WILLIAM HUGGINS in 1864 first found bright lines in their spectra, evidencing a community of chemical composition, and due to glowing gas, mostly hydrogen. Recently helium has been added, and still other lines are due to substances as yet unrecognized on the Earth. The character of the lines indicates exceedingly high temperatures, or a state of strong electric excitement. Temperature and pressure both increase toward the nucleus. Not all the nebulæ yield bright lines; and this may be due to gas under extreme pressure, or to aggregations of stellar bodies.

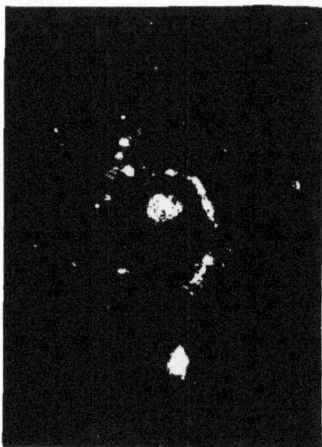

SPIRAL NEBULA IN CANES VENATICI

(*No. 51 in* MESSIER'S *Catalogue. Photographed by* ROBERTS, *29th April* 1889. *Exposure,* 4 *hours. The most marked of the spiral nebulae*)

There are about forty lines in the photographic spectra of nebulæ, the chief line, according to Professor KEELER, failing of identity with the magnesium fluting. His careful measurements upon the spectra of the Orion nebula are perhaps the most significant of all, for they prove that the distance between the nebula and our solar system is increasing at the rate of eleven miles in every second of time. However, neither this nor any other nebula has been discovered to partake of proper motion, although the bright nebula in Draco is coming toward us at the rate of forty miles per second. All the evidence so far seems to show that the nebulæ are in motion through interstellar space with velocities comparable with those of the stars themselves. —*D. P. T.*

STRUVE, W., *Stellarum Compositarum Mensurae Micrometricae* (Saint Petersburg, 1837). See also *Dun Echt Observatory Publications*, i. (Aberdeen, 1876).

STRUVE, W., *Études d'Astronomie Stellaire* (Saint Petersburg, 1847).

SMYTH, W. H., 'Colors of Multiple Stars,' *Sidereal Chromatics*, (London, 1864).

HUGGINS, 'Motion in Sight-line,' *Proceedings Royal Society* xvi. (1866), 382.

D'ARREST, *Siderum Nebulosorum Observationes Havnienses* (Copenhagen, 1867).

WOLF, RAYET, 'Bright-line Stars,' *Comp. Rend.* lxv. (1867), 292.

HUGGINS, 'Résumé of Spectrum Analysis,' *Report British Association*, 1868, pp. 140–165.

MONTIGNY, 'Scintillation,' *Bulletins de l'Académie de Belgique*, 1868, *et seq.*

PROCTOR, *Universe and Coming Transits* (London, 1874).

KNOBEL, 'Bibliography of Stars and Nebulæ,' *Month. Not. Roy. Astr. Soc.* xxxvi. (1876), 365.

NEWCOMB, *Popular Astronomy* (New York, 1877).

KNOBEL, 'Chronology of Star Catalogues,' *Mem. Roy. Astr. Soc.* xliii. (1877), 1.

HALL, M., 'Sidereal System,' *Memoirs Royal Astronomical Society*, xliii. (1877), 157.

HALL, A., 'Double Stars,' *Washington Obs.* 1877, App. vi.

HOLDEN, 'Nebula of Orion,' *Washington Obs.* 1878, App. i.

SECCHI, *Les Étoiles* (Paris, 1879).

CROSSLEY, GLEDHILL, WILSON, *Handbook of Double Stars* (London 1879), Notes, etc. (1880). Bibliographies.

ROSSE, 'Observations of Nebulæ and Clusters, 1848–78,' *Trans. Roy. Soc. Dublin,* ii. (1880).

PICKERING, 'Dimensions of Stars,' *Proc. Am. Acad. Arts and Sciences*, viii. (1881), 1.

PLUMMER, 'Sidereal System,' *Copernicus,* ii. (1882), 45.

NEWCOMB, 'Catalogue of Standard Stars,' *Astr. Papers Am. Ephemeris*, i. (1882), 147.

VOGEL, MÜLLER, 'Stellar Spectroscopy,' *Publ. Astrophys. Obs. Potsdam*, iii. (1883).

PLUMMER, 'Sun's Motion in Space,' *Mem. Roy. Astron. Society*, xlvii. (1883), 327.

GRANT, 'Catalogue of 6415 Stars,' *Glasgow Univ. Obs.* (1883).

GILL, ELKIN, 'Parallaxes in So. Hemisphere,' *Mem. Roy. Astr. Soc.* xlviii. (1884), 1.

YARNALL, FRISBY, 'Catalogue of 10,964 Stars,' *Washington Observations*, 1884, Appendix I.

DRAPER, 'Stellar Spectrum Photography,' *Proc. Am. Acad.* xi. (1884), 231.

GILL, 'Future Problems in Sidereal Astronomy,' *Proc. Roy. Institution*, xi. (1884), 91.

PRITCHARD, *Uranometria Nova Oxoniensis* (Oxford, 1885).

PICKERING, 'Harvard Photometry,' *Annals Harv. Coll. Obs.* xiv. (1885); xxiv. (1890). BAILEY, xxxiv. (1895).

PICKERING, 'Photographic Photometry,' *Ann. Harv. Coll. Obs.* xviii. 1888, No. VII.

GORE, J. E., *Planetary and Stellar Studies* (London, 1888).

LOCKYER, 'Classification,' *Proc. Roy. Society*, xliv. (1888), 1.

DREYER, 'Catalogue of Nebulæ and Clusters,' *Mem. Roy. Ast. Soc.* xlix. (1888), 1 ; li. (1895), 185.

PICKERING, 'Index to Observations of Variables,' *Annals Harvard College Observatory*, xviii. (1889), No. VIII.

CHANDLER, 'General relations of variables,' *The Astronomical Journal*, ix. (1889), 1.

PRITCHARD, *Researches in Stellar Parallax by the Aid of Photography* (Oxford, 1889–92). History, iv. (1892).

CHAMBERS, *Descriptive Astronomy*, iii. (Oxford, 1890).

CLERKE, *The System of the Stars* (New York, 1890).

BOSS, 'Solar Motion,' *The Astronomical Journal*, ix. (1890), 161.

PICKERING, 'Draper Catalogue of Spectra,' *Ann. Harv. Coll. Obs.*, xxvii. (1890) ; MAURY, xxviii. (1897).

RUSSELL, *Description of the Star Camera* (Sydney 1891).

SCHEINER, 'Photographic Photometry,' *Astronomische Nachrichten*, cxxviii. (1891), 113.

HUGGINS, 'Celestial Spectroscopy,' *Nature*, xliv. (1891), 372.

RAMBAUT, 'Binary Orbits by Spectroscope,' *Month. Not. Roy. Astron. Society*, li. (1891), 316.

BOEDDICKER, *The Milky Way* (London and New York, 1892).

PROCTOR, RANYARD, *Old and New Astronomy* (London, 1892).

VOGEL, NEWCOMB-ENGELMANN'S *Populäre Astronomie* (Leipzig, 1892).

VOGEL, 'Motion in Sight-line,' *Astronomy and Astrophysics*, xi. (1892), 203.

PORTER, 'Proper Motions,' *Publ. Cincinnati Obs.* xii. (1892).

BALL, *The Story of the Heavens* (London, 1893).

GORE, J. E., 'Sun's Motion in Space,' *The Visible Universe*, p. 193 (New York, 1893).

EASTON, *La Voie Lactée* (Paris, 1893).

ROBERTS, *Photographs of Stars and Nebulæ* (London, 1893).
BÉLOPOLSKY, 'β Lyrae,' *Mem. Spettroscopisti Ital.* xxii. (1893).
CLERKE, *History of Astronomy during the XIXth Century.* (London, 1893).
LOCKYER, ' Spectra of the brighter stars,' *Phil. Trans.* clxxxiv. (1893), 675.
BALL, ' Solar Motion,' *In the High Heavens* (London 1893).
VOGEL, ' β Lyræ,' *Sitz. Akad. Wiss. Berlin* 1894 (1), 115.
JANSSEN, ' Photographic Photometry,' *Smithsonian Report* 1894, p. 191.
SCHEINER, FROST, *Astronomical Spectroscopy* (Boston and London, 1894). Bibliography.
CAMPBELL, ' Wolf-Rayet Stars,' *Astronomy and Astrophysics,* xiii. (1894), 448.
GORE, J. E., *The Worlds of Space* (London, 1894).
MÜLLER, KEMPF, ' Photometry,' *Publ. Obs. Potsdam,* ix. (1894).
FLAMMARION, GORE, *Popular Astronomy* (New York 1894).
MÄDLER, BURNHAM, ' Double Stars,' *Himmel und Erde,* vii. (1894), 41.
BURNHAM, ' Double Stars,' *Publ. Lick Observatory,* ii. (1894).
KEELER, ' Spectra of Nebulæ,' *Publ. Lick Obs.* iii. (1894); *Astron. and Astrophys.* xiii. (1894), 476.
GIBERNE, *Radiant Suns* (New York, 1894).
CHAMBERS, *The Story of the Stars* (New York, 1895).
D'ENGELHARDT, ' Nebulæ and Clusters,' *Observations Astronomiques* (Dresden, 1895).
SAMTER, ' Milchstrasse,' *Himmel und Erde,* vii. (1895), 508, 544.
VALENTINER, *Handwörterbuch der Astron.* (Breslau 1895-98).
CHANDLER, ' Catalogue of Variables,' *The Astronomical Journal,* xvi. (1896), 145.
RADAU, ' Solar Motion,' *Bulletin Astron.* xiii. (1896), 169.
EVERETT, ' Poles of Binary Orbits,' *M. N. Roy. Astr. Soc.* lvi. (1896), 462.
SEE, *Researches on the Evolution of Stellar Systems* (Lynn, 1896).
CERASKI, ' Variables,' *Annales Obs. Moscou,* iii. (1896), livr. ii.
CHASE, ' Cluster in Coma Berenices,' *Trans. Yale Obs.* i. (1896).
MOULTON, ' Spectroscopic Binaries,' *Pop. Astr.* iii. (1896), 337.
CLARKE, H. L., ' Life-history of Star Systems,' *Pop. Astr.* iii. (1896), 489.
RUSSELL, ' Southern Circumpolars,' *Pop. Astr.* iv. (1896), 61.
BURNHAM, ' Binary Systems,' *Popular Astr.* iv. (1896), 169.
GILL, KAPTEYN, ' Photographic Durchmusterung,' *Annals Cape Observatory,* iii. (1896); iv. (1897).

YENDELL, 'Variable Stars,' *Popular Astr.* iii.–v. (1896–97).
SCHIAPARELLI, 'Color of Sirius,' *Rubra Canicula* (1897).
LOCKYER, 'Celestial Eddies,' *Nature*, lv. (1897), 249.
FOWLER, 'Chemistry of the Stars,' *Knowledge*, xx. (1897), 77.
MONCK, 'Spectra of Binaries,' *The Observatory*, xx. (1897), 389.
CLERKE, 'β Lyræ,' *The Observatory*, xx. (1897), 410.
SCHEINER, *Die Photographie der Gestirne* (Leipzig, 1897). Atlas and bibliography.
GOULD, 'Clusters,' *Cordoba Photographs* (Lynn, 1897).
WHITNEY, 'Solar Motion,' *Popular Astronomy*, v. (1897), 309.
WILSON, 'Ring nebula in Lyra,' *Popular Astr.* v. (1897), 337.
CLERKE, 'New Class of Variables,' *Observatory*, xx. (1897), 52.
MÜLLER, *Die Photometrie der Gestirne* (Leipzig, 1897). Bibliography.
HALL, M., 'Sidereal system revised,' *M. N. Roy. Astr. Soc.* lvii. (1897), 357; lviii. (1898), 473.
PANNEKOEK, 'β Lyræ,' *Kon. Akad. Wetenschap. Amsterdam*, v. (1897).
HUGGINS, 'Celestial Spectroscopy,' *The Nineteenth Century*, xli. (1897), 907.
LOCKYER, *The Sun's Place in Nature* (London, 1897).
PANNEKOEK, 'Galaxy,' *Jour. Brit. Ast. Assoc.* viii. (1897–98).
GORE, J. E., 'Sidereal Heavens,' *Concise Knowledge Library* (New York, 1898).
DUNÉR, SCHEINER, 'Stellar Evolution,' *Popular Astr.* vi. (1898), 85.
MYERS, 'System of β Lyræ,' *The Astrophysical Journal*, vii. (1898), 1; *Popular Astronomy*, vi. (1898), 268.
KEELER, 'Physiological Phenomena, and Spectra of Nebulæ,' *Publ. Astron. Society Pacific*, x. (1898), 141.
KAPTEYN, 'Solar Motion,' *Astron. Nachr.* cxlvi. (1898), 97.
EASTON, 'Theory of Universe (PROCTOR),' *Knowledge*, xxi. (1898), 12.
EASTON, 'New Theory of Galaxy,' *Knowledge*, xxi. (1898), 57.
MAUNDER, McCLEAN, 'Photographs of Stellar Spectra,' *Observatory*, xxi. (1898), 163.
SCHWEIGER-LERCHENFELD, *Atlas der Himmelskunde* (Vienna, 1898).
DESLANDRES, 'Motion in Sight-line,' *Bulletin Soc. Astron. France*, September 1898.
McCLEAN, 'Stellar Spectroscopy,' *Proc. Roy. Soc.* lxiv. (1898).
DOUGLASS, 'Stellar Bands in Zodiac,' *Pop. Astr.* v. (1898), 511.
EASTMAN, *Catalogue of 5,151 Stars* (Washington 1898).

DESLANDRES, 'Motion in Sight-line,' *Bulletin Astron.* xv. (1898), 49.

YOUNG, *General Astronomy*, revised edition (Boston, 1898).

HAGEN, *Atlas Stellarum Variabilium* (Berlin, 1898).

References to the popular literature of the stars and nebulæ are to be found in the volumes of POOLE's *Index*, on the pages indicated below : —

Volume of Index, and Years	References to Nebulæ	References to Stars
I (1800–81)	p. 903	p. 1243
II (1882–86)	308	417
III (1887–91)	298	407
IV (1892–96)	396	545
Annual Literary Index, 1897	89	122
Index to General Literature, 1893	203	274

Most of the earlier and more important scientific papers are classified and titled in Sir ROBERT BALL's *Elements of Astronomy*, pp. 427–40 (New York, 1880). In vol. xxxii. (1883) of the *Proceedings American Association Advancement of Science*, W. A. ROGERS has given a full presentation of the German survey of the northern heavens. The occasional bulletins of the Paris astrographic conference contain important researches relating to stellar photography and allied subjects. Recent volumes of *Knowledge* contain numerous papers on the stars and nebulæ, by Mr MAUNDER and others, with fine reproductions of astronomical photographs, especially those of Dr ROBERTS. In the *Journal of the British Astronomical Association* are found very useful lists and abstracts of papers. Occasional bibliographic lists are given in *Bulletin Astronomique*, i.-xv. (1884–98). Recent scientific literature is exhaustively cited in the frequent lists of the *Astrophysical Journal*, vols. i.-viii. (1895–98). — *D. P. T.*

CHAPTER XVIII

TELESCOPES AND HOUSES FOR THEM

KNOWLEDGE of the Sun, Moon and stars gained from the preceding chapters leads naturally to a brief story of the instrument by which astronomers have mainly acquired that knowledge; for before the invention of the telescope, it was impossible that mankind should know very much about the heavenly bodies. The working of a telescope is much affected by its location and the construction of the building in which it is sheltered: a few paragraphs are therefore devoted to these important considerations.

GALILEO (page 12), while not perhaps the original inventor of the telescope, was still undeniably the first to construct one and apply it to the higher purpose of astronomical observation, and he was richly rewarded by the supreme discovery of the satellites of Jupiter. Before his day, not only were no bodies attendant upon the planets known, save our Moon, but many of the planets were themselves unknown; nor was it possible to ascertain the dimensions of those that were known. The accepted truth of the Copernican system had led to the prediction of phases of the planets nearest the Earth and Sun; but final corroboration was lacking till the first telescope achieved this significant revelation. Several catalogues of the stars and Tycho's observations of the planets among them were possible, before the days of telescopes; but so crude were they that their value at the present time is due solely to their antiquity. Although invented early in the 17th century, the telescope was chiefly used as a gazing instrument for a half century; and it was not till 1668 that the mechanical genius of Jean Picard (1620–1682) saw how greatly the telescope might enhance the precision of all astronomical measures of position, if only it were attached to a circle to impart definiteness of alignment. Although Römer

CHRISTIAN HUYGENS (1629–1695)

(page 45) had invented the transit instrument, Picard's invention produced the meridian circle, a modern type of which is shown on page 368; with it may be found the exact place of any heavenly body on the celestial vault with the utmost ease and celerity. However, even the transit or the meridian circle would be greatly restricted in use but for the astronomical clock, the principles governing the pendulum of which had been formulated by Galileo. But here was a delay of nearly a half century, for it was not till 1657 that HUYGENS, famous for his great telescopes and the observations made with them (especially of the planet Saturn), constructed an accurate time-piece by controlling its train with a pendulum.

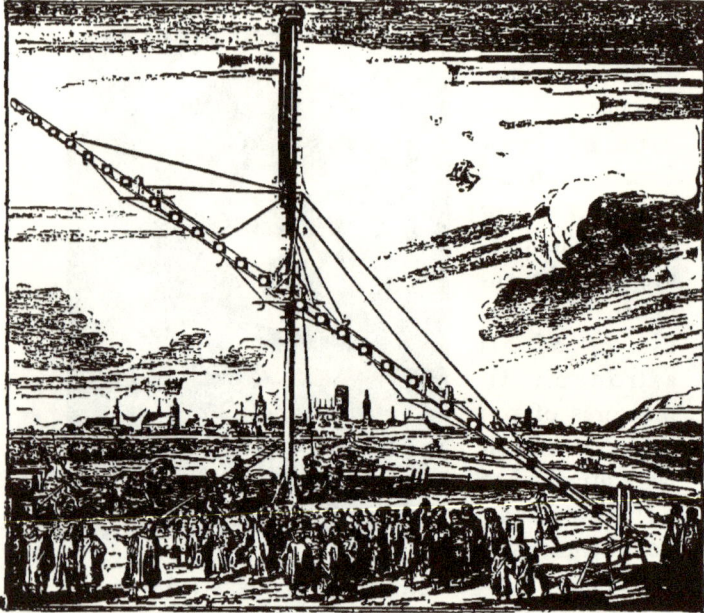

TYPICAL GREAT TELESCOPE OF THE 17TH CENTURY
(*Built and used by* HEVELIUS *at Dantzig*)

In the telescopes of GALILEO and HUYGENS and all their followers for about a century, the image of a remote object was formed by a single lens of double convexity. The glass was full of imperfections; but besides this, the baffling obstacle of the prismatic spectrum surrounded all bright objects with highly colored fringes, so that a limit to the effective size of the telescope seemed to be set by the laws of nature herself. But it was soon found that the disturbing color effects were minimized, if the lens was ground very flat. This, however, necessitated great focal length, and many telescopes were built in the 17th century with but little variation in model from the famous instrument of HEVELIUS (1611–1687) illustrated above; in which object-glass and eye-piece were so far apart that no tube connected them, and they were kept in line by an open framework, strongly trussed. HUYGENS built an aerial telescope, with objective and eye-piece wholly unconnected, the former being mounted in a small swivel tube on top of a high pole, and directed toward the eye-piece by a cord or wire pulled taut by the observer. Definition was exceedingly imperfect except at the centre of the field of view, wind precluded the use of the telescope, and except for a moment at a time it was practically impossible to maintain the object in the field with that steadiness requisite for critical observation. Nevertheless, the long telescope had its day, and one 600 feet in length is said to have been made in Italy, though never successfully used.

A MODERN FIELD-GLASS

(*By* BAUSCH & LOMB. *A ray of light coming down through the objective follows the arrow, is four times reflected in prisms and passes out to the eye, at the arrow-head below*)

GALILEO'S original telescope had but two lenses : a double convex to form an image, and a double concave to examine it with ; and this form of telescope is preserved in the common opera-glass and field-glass of the present day, if only slight magnifying power is desired. Great improvement in optical power has, however, been recently obtained by adjusting two right-angled prisms, properly constructed, in the path of the rays between objective and eye-pieces, whereby the necessary compactness of a hand-glass is secured, and with it the advantages of greater focal length and a larger image at the focus. At the same time the prisms effect that reinversion which is requisite for terrestrial purposes. A magnification of 15 diameters and more is easy.

Sir Isaac Newton (page 93), who was no less a physicist than astronomer, after devoting many years to a study of the imperfections of refracting telescopes, expressed a definitive opinion that a perfect glass of high power was a physical impossibility; and this conclusion, coupled with the unwieldy proportions of the most powerful instruments of that day, forestalled farther attempts to improve the refractor or dioptric telescope. But the genius of Newton led the way by modifications of the reflector, or catoptric telescope, already invented by James Gregory (1638–1675), which forms the image of a distant object by convergence of the rays on reflection from a concave surface of suitable curvature and high polish. In the refractor, the object-glass assembles rays of different colors at different focal points, the violet nearest the glass and the red farthest from it; but in the reflector all rays are focussed at a single point, no matter what their color. Still, as the image is formed directly in front of the mirror itself, the head of the observer, if placed in the ordinary position, would intercept nearly all rays from the object. Gregory had dodged this difficulty by mounting a small secondary concave mirror in the path of the rays, thus throwing them back upon the primary mirror. This he perforated at the centre to allow the doubly reflected rays to reach the eye-piece, which was then screwed into the

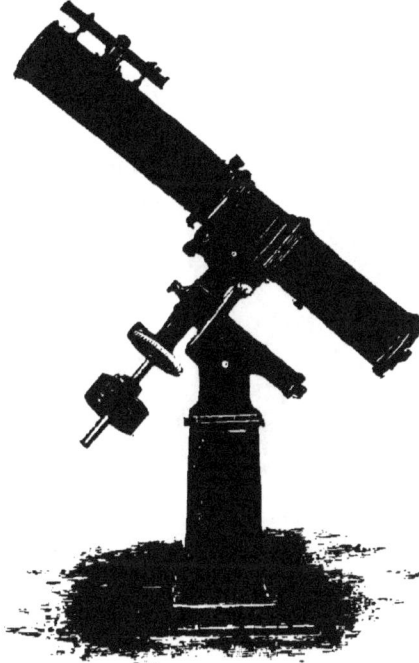

A MODERN REFLECTOR
(*Newtonian type by* BRASHEAR. *Eye-piece near the 'finder' at upper end of tube*)

back of the large reflector. NEWTON improved upon this arrangement and preserved the principal mirror intact by interposing a small diagonal reflector in the path of the rays just before they come to a focus, thereby diverting them to the side of the tube, where

S & T — 21

the eye-piece is set perpendicular to it. The observer then looks into the tube, not at one end, but at one side, at right angles to the direction of its pointing. The Newtonian is the favorite type of construction for the reflector at the present day, especially if moderate in size, as in the preceding illustration of a Newtonian by Mʳ JOHN A. BRASHEAR of Allegheny, who, of all American opticians, has achieved the greatest success in constructing reflectors. A farther improvement of the Newtonian form, by substituting a small totally reflecting prism for the secondary mirror, was carried into effect by HENRY DRAPER.

4-FT. MELBOURNE REFLECTOR
(*Designed and built by* THOMAS GRUBB)

During the past century five builders of reflecting telescopes have stood out pre-eminently: Sir WILLIAM HERSCHEL, whose 4-foot reflector is illustrated on page 13, and who long served a self-apprenticeship by constructing nearly 200 mirrors, one of them 24 inches in diameter which was nearly equal in performance with the 4-foot, and with which his son, Sir JOHN, continued the elder HERSCHEL's researches; Lord ROSSE, whose 'Leviathan,' or 6-foot reflector illustrated on page 247, remains yet unsurpassed in size; WILLIAM LASSELL (1799–1880), an eminent English astronomer, who built two reflectors, duplicates in size of HERSCHEL'S largest instruments, which he used on the island of Malta, 1852–65 ; THOMAS GRUBB, who in 1867 built a 4-foot Gregorian with a silver-on-glass speculum for the government observatory at Melbourne, Australia, and who has a worthy successor in his son, Sir HOWARD, the builder of the

THOMAS GRUBB (1800–1878)

20-inch reflector illustrated on page 292, with which Dr ROBERTS has produced photographs of unusual excellence ; and Dr A. A. COMMON, of Ealing, England, whose first great reflector of 3-foot aperture is now owned by the Lick Observatory, by gift of Mr EDWARD CROSSLEY, and who built in 1889 a 5-foot silver-on-glass reflector, employed to good advantage in photographing the nebulæ. An instrument of like dimensions is now building at the Yerkes Observatory of the University of Chicago.

It will be noticed that all the great reflectors previously mentioned were constructed in Great Britain, and that they achieved extraordinary success in the hands of their builders. A great reflector is, indeed, a delicate instrument, requiring much skill and patience in adjusting, and more than ordinary care in preservation of the specular surfaces from deterioration on exposure to the atmosphere; but by Mr BRASHEAR's formula, a new film of silver is readily deposited upon the glass chemically. Besides this, the great reflector is exceedingly massive, and unwieldy in proportion; and its performance is much influenced by disturbing air-currents, while flexure or sagging of the mirror is a most serious drawback. This is most successfully overcome by making the thickness of the mirror not less than one-sixth its diameter. A few reflectors of great diameter have been built in France: a 4-foot silvered glass reflector at the observatory of Paris by MARTIN, a 39-inch recently completed by the Brothers HENRY for Meudon, and two of 32 inches' diameter at the observatories of Marseilles and Toulouse by FOUCAULT (page 47). Before his time all reflectors were made of speculum metal, an alloy often composed of 126 parts of copper to 59 of tin.

AMASA HOLCOMB, a land surveyor of Southwick, Massachusetts, appears to have been the earliest maker of telescopes in America, having begun in 1826. Although his first instruments were refractors of small size, he afterward made specula as large as 10 inches in diameter, about 30 in all. His favorite mounting was that known as the Herschelian type, in which the speculum is tilted slightly, throwing the image at the side of the tube, instead of centrally. All the light is thereby saved, but a disastrous distortion is introduced. HOLCOMB devised a steady and effective mounting, which received the award of a Franklin Institute medal in 1835. Mʳ E. P. MASON of Yale College, aided by his classmate Professor H. L. SMITH of Hobart College, built in 1838 a reflector of 12 inches aperture, then the largest telescope in America; and it was used conjointly by them in accurate delineation of a few prominent nebulæ.

The signal success of Sir WILLIAM HERSCHEL'S great telescopes, and the giant reflectors of Lord ROSSE inspired many private individuals in America as well as other countries to build instruments of similar design. Among them we mention JOSIAH LYMAN, of Lenox, Massachusetts, whose 9½-inch speculum, with a system of supporting levers to preserve its figure when pointed upon stars in all altitudes, was exhibited before the American Association for the Advancement of Science at the Albany meeting, 1851; and who subsequently figured a 12-inch speculum, now the property of Amherst College Observatory, by gift of his son, the Rev. Dʳ LYMAN of Brooklyn. As early as 1858, HENRY DRAPER (page 298), returning from a visit to Lord ROSSE at Parsonstown, constructed a 15-inch speculum, the perfected result of which was a splendid photograph of the Moon, over 4 feet in diameter. Professor BROOKS, now of Geneva, New York, has built a 9-inch reflector and used it to good advantage in comet work, and Mʳ EDGECOMB of Mystic, Connecticut, has turned out many excellent reflectors. Also we must include an 18-inch mirror ground and figured by Professor SCHAEBERLE, which he used in planetary photography. It was an adapted Cassegrain, or type similar to the Gregorian, except that the secondary mirror is convex instead of concave, thereby producing a larger focal image. Finally, in 1871, Dr DRAPER completed the largest reflector, of 28 inches' diameter, yet constructed in America, which is now part of the equipment of Harvard Observatory.

Still another type of reflector, revived of late in Germany and by Dr COMMON in England, is known as the 'brachy-telescope,' or 'oblique Cassegrain,' a composite of Herschelian and Cassegrain types. The tube is exceptionally short, and the figure of the secondary mirror can be made to neutralize the distortion inevitably produced by tilting the speculum. RANYARD devised a very ingenious type of mounting, completed by Professor WADSWORTH and adaptable to any form of reflector, whereby the advantages of a stationary eyepiece are assured, in all possible pointings of the telescope.

A. C. RANYARD (1845–1894)

Next in order we outline the development of the refracting telescope from the beginning of the 18th century onward. Although NEWTON's experiments had satisfied him that a lens of short focal length could never be made to yield an optically perfect image, EULER (page 96) doubted the accuracy of this conclusion, and suggested that a combination of lenses of glass and water, of suitable curvatures, might produce a composite lens practically free from defects of color — that is, an achromatic objective. In 1733 CHESTER MORE HALL in England was the first to make and use such an objective, by combining a con-

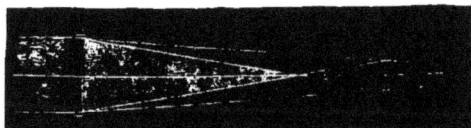

COURSE OF RAYS THROUGH TELESCOPE
(Objective A forms image of object at B)

cave lens of flint glass with the original convex lens of crown glass. Flint glass contains much oxide of lead and is very dense, and its power of dispersing a beam of white light is about double that of an equal prism of crown glass, although the mere refractive powers of equal prisms of the two kinds of glass are the same. If, therefore, a double convex lens of crown glass is capped by a plano-concave or double concave lens of flint-glass, there is a residuum of refraction with the harmful dispersion effectively neutralized. The little diagram above shows in schematic fashion the passage of parallel rays from left to right through such a double objective, also convergence to the focal plane, and subsequent magnification of the image by a small eye-lens, with emergence in parallel pencils suited to human vision.

It was not HALL, however, who obtained the credit for this signal improvement of the telescope, but JOHN

DOLLOND, another English optician, who secured a patent for the invention and reaped most of the benefits therefrom.   No limit to the size of refractors now appeared to be set, save the difficulty of obtaining large disks of glass of the requisite optical purity. This embraces not only high transparency, and freedom from air bubbles and sand-holes, but absolute absence of striæ and waves of irregular density.   So nearly insurmountable were the obstacles in manufacturing flint glass that an achromatic of high excellence as large as 6 inches in diameter was unknown at the beginning of the 19th century.

DOLLOND (1706-1761)

GUINAND (1745–1825), a Swiss watchmaker, by patient experiment finally discovered how to prevent the heavy lead oxide from settling in the molten flint glass; and his methods, in some essentials secret, have been communicated to his successors, MM. ROSETTE & FEIL of Paris, followed by M. MANTOIS, who, with CHANCE Brothers of Birmingham, have made the disks for nearly all the great refractors of the present day.   GUINAND for many years before his death was associated with FRAUNHOFER (page 17) in the manufacture of telescopes at Munich, and together they produced about 1824 an achromatic refractor of nearly 10 inches aperture, with which the elder STRUVE made at Dorpat a celebrated series of

observations of double stars. So critical was FRAUN-
HOFER in his optical work, and so consummate a
mechanician was he that a knowledge of the success
of this great telescope finally penetrated to America,
where at that time a permanent observatory had
scarcely been projected. FRAUNHOFER, too, made
great improvements in the equatorial telescope, by
modifying the old parallactic stand into the German
mounting now univer-
sally met with, and
only improved upon
by the Potsdam
mounting (page 60).

Although DAVID
RITTENHOUSE had es-
tablished a temporary
observatory at Norri-
ton near Philadelphia
for observing the
transit of Venus in
1769, and the elder
BOND (page 240) had
in 1825 a modest pri-
vate observatory at

RITTENHOUSE (1732–1796)

Dorchester, Massachusetts, and Yale College had in
1832 a 5-inch portable telescope of FRAUNHOFER'S
make, the first American observatory, intended for permanent occupation as such, was erected at Chapel Hill,
North Carolina, in the years 1831–32. Few observations were, however, made there; it was soon after
dismantled, and accidentally destroyed by fire about
1838. OLMSTED (page 210) was for a brief period
professor at Chapel Hill. General MITCHEL in 1842
visited Europe in the interest of the astronomical

society of Cincinnati and ordered from MERZ and MÄHLER, of Munich, the successors of FRAUNHOFER, a 12-inch refractor. By his successful tours as a popular lecturer MITCHEL aroused great enthusiasm for astronomical knowledge, and to his forceful and energetic personality may be traced the origin of many American observatories. The great comet of 1843 led directly to the founding of Harvard College Observatory (page 300), and its equipment with a telescope, colossal for that period, of 15 inches aperture, which, with its twin companion at Pulkowa in Russia, formed the culminating labor of the celebrated optical house of Munich.

American telescope builders have always been obliged to order the rough glass for their objectives from foreign makers, and must still do so, the few experiments in optical glass making outside of Europe having proved in the main failures. But in that fine and skilful fashioning of the crude lump into a perfect lens, our countrymen have for the last half century led the world. FITZ of New York made about 30 refractors, between 6 and 13 inches aperture, many of which were subsequently improved by ALVAN CLARK; SPENCER of Canastota, New York, whose chief success was with microscope

ORMSBY MACKNIGHT MITCHEL
(1809–1862)

objectives, undertook telescopes also and completed a 13½-inch glass which the late Dr PETERS of Hamilton College (page 113) made famous by his discovery of 48 small planets; also CLACEY of Boston and BYRNE of New York deserve more than passing mention, not to say many other skilful American opticians who have turned out a host of telescopes of lesser dimension and very satisfactory performance.

Nor have opticians of a high order been lacking in European countries. CAUCHOIX, and the Brothers HENRY in France; STEINHEIL and SCHROEDER in Germany; and COOKE & Sons in England — all have built refracting telescopes of large aperture and excellent definition. The HENRYS have constructed ten of the photographic objectives engaged in the astrographic survey (page 260), their mountings being by GAUTIER; also a 30-inch at M. BISCHOFFSHEIM'S splendid observatory on Mont Gros (page 169), and one slightly larger for the observatory at Meudon, Paris. STEINHEIL'S greatest achievement is an object-glass of 31½ inches diameter for the astrophysical observatory at Potsdam; and COOKE'S, a 25-inch refractor now mounted at Cambridge, England. Within recent years, COOKE has been offering a new photo-visual objective, a triple glass said to be free from the troublesome secondary spectrum of the ordinary double objective, and

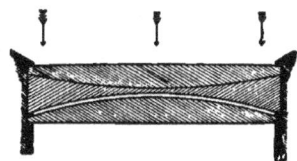

TRIPLE OBJECTIVE (TAYLOR)

(Photographic and visual foci are identical)

dinary double objective, and to converge visual and photographic rays to an identical focus. The curves are the work of Mr H. D. TAYLOR, and the glass employed is known as the new

Jena glass, made from the formulæ of Professor Abbe of the University of Jena, who conducted a course of experiments under the auspices of the German government.

The secondary spectrum gives rise to an effect which becomes especially harmful in very large objectives, surrounding all bright stars and planets with a brilliant bluish luminosity, greatly interfering with the observation of faint adjacent objects like satellites and companion stars. This difficulty is inherent in the glass itself, because dispersion or decomposition by the crown glass cannot be exactly neutralized by the recomposition of the ordinary flint. With the Jena glass, however, Dr Hastings of Yale University has succeeded in producing a double objective practically free from secondary spectrum effect, by careful investigation of the theoretical curvatures, and placing the flint lens on the outside of the combination, so that the light passes through it before reaching the crown.

The curves of the lenses in achromatic objectives have occasioned much research in theoretical optics, and many different constructions have been evolved, among them that of Littrow, with the crown double convex and the flint plano-concave. Gauss (page 99) devised a form in which the crown is plano-convex and the flint concavo-convex. In the type of objective preferred by Clark, the crown is double convex, and the flint double concave, the two lenses being mounted in a long cell with a space between them equalling about $\frac{1}{4}$ their diameter. This construction adds to the weight of the mounted objective, but affords better correction of the aberration, permits easy cleaning of the inner faces, and allows free circulation of air between flint and crown. In all these types the light reaches the crown lens first.

Before passing to a consideration of the remarkable series of telescopes built by the CLARKS, the excellent work of Sir HOWARD GRUBB must be further specified,

27-INCH VIENNA EQUATORIAL
(*Built by Sir* HOWARD GRUBB)

embracing a 27-inch telescope complete (illustrated above), and mounted at the Imperial Observatory of Vienna, one of the best equipped establishments of Europe. Also he has constructed a 26-inch visual and a 28-inch photographic objective for the Royal Observatory at Greenwich, as well as numerous glasses of smaller aperture ; and his mountings are especially commended for rigidity. A type of clock-motion with electric control devised by GRUBB permits the close following of a star in its diurnal motion, with that degree of accuracy necessary to secure circular star-images on a photographic plate during long exposures.

Of all makers of telescopes ALVAN CLARK attained the greatest celebrity. A portrait painter in Boston

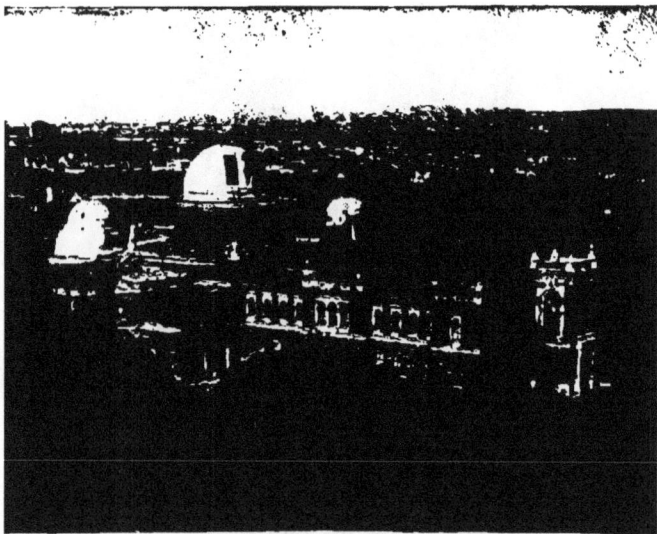

THE IMPERIAL OBSERVATORY AT VIENNA
(*Dr* EDMUND WEISS, *Director*)

in 1844, accident drew his attention to the construction of a small reflector, the success of which led to his making refractors. Two years later he was established in the business of telescope construction with his two sons, ALVAN GRAHAM (page 285) and GEORGE BASSETT. CLARK'S first refractors were introduced to astronomers by DAWES (page 283) who purchased

five of them, and subjected them to the most critical
tests. Above the aperture of 6 inches, the CLARKS
made in all about 75 objectives, with mountings for
the most of them. Their objectives are optically

more perfect than the
atmosphere in which they
can generally be used,
bearing a hundred di-
ameters of magnifying
power for each inch of
aperture, and separating
the components of close
double stars up to the
limit set by DAWES's em-
piric formula. Clark
telescopes have not only
a world-wide fame but a
world-wide distribution.
Between 1860 and 1892
the CLARKS were five
times called upon to

ALVAN CLARK (1804-1887)

construct 'a telescope more powerful than any now
in existence'; and each time the advance was from
one to six inches in excess of the aperture of the then
greatest glass. These were, in order, the 18½-inch now
at Evanston; the 26-inch at Washington, a duplicate
of which was built by them for the University of Virginia; the 30-inch for the Imperial Observatory at
Pulkowa, near Saint Petersburg; the 36-inch for the
Lick Observatory; and the 40-inch, still the largest
refracting telescope in the world, for the Yerkes Observatory of the University of Chicago. Among other
great refractors of their construction are a 23-inch for
Professor YOUNG at Princeton, and a 24-inch for M^r

PERCIVAL LOWELL, now mounted in his private observatory at Flagstaff, Arizona. This glass and the Yerkes objective were the last work of ALVAN GRAHAM CLARK, his constant co-worker being Mʳ CARL LUNDIN, an able optician who has served a quarter-century apprenticeship under all the CLARKS, and now continues the work of the firm at Cambridgeport.

GEORGE B. CLARK (1827–1891)

The Lick 36-inch lenses are mounted in a cast-iron cell faced with inlaid silver where the glasses rest against it. The weight of the objective in its cell is about 700 pounds, and its cost was $53,000. An additional lens, or photographic corrector of 33 inches aperture, was provided at a cost of $13,000. The focal length of the visual object-glass is 56 feet ;

48 feet when the photographic corrector is applied. The Yerkes objective weighs about half a ton, and its crown lens is three inches thick at the centre. The total cost of this telescope, inclusive of its mounting was about $125,000.

ALVAN CLARK was in 1867 awarded the Rumford medal of the American Academy of Arts and Sciences for the perfection of his optical surfaces. The completion of the 30-inch Russian object-glass brought the

YERKES OBSERVATORY OF THE UNIVERSITY OF CHICAGO

(*Located on the shore of Lake Geneva, Williams Bay, Wisconsin. This observatory contains the largest refracting telescope in the world, was 5 years in building, and cost about $500,000, the gift of Mr CHAS. T. YERKES*)

CLARKS the signal honor of the golden medal of the Empire, conferred for the first time by ALEXANDER the Third.

Second in importance only to the objective of a telescope are the mounting by which it is pointed toward the heavenly bodies, and the eyepieces with which the images formed by it are examined. The mountings made by the REPSOLDS of Hamburg are unsurpassed, their largest having been constructed for the 30-inch Clark glass at Pulkowa (page 133).

40-INCH TELESCOPE OF THE YERKES OBSERVATORY

(*Glass by* MANTOIS, *objective by* ALVAN CLARK & *Sons, of Cambridgeport, mounting by*
WARNER & SWASEY *of Cleveland. The tube is about 65 feet long ; and, although the moving parts of this great telescope weigh nearly 15 tons, the whole is easily managed by one
man, through control of several electric motors. The counter-weighted dome-floor rises and
falls 23 feet, on the plan originated by Sir* HOWARD GRUBB. *The photograph shows it near
its lowest position. The dome is 90 feet in diameter, and weighs 140 tons*)

REPSOLD has many foreign rivals who have constructed mountings for smaller instruments. Especially may be mentioned SÉCRETAN of Paris, BAMBERG of Berlin, HEYDE of Dresden, and SALMOIRAGHI in Italy. In America, SAEGMULLER has made fine mountings, one of his largest for a 20-inch glass at Manila. WARNER & SWASEY of Cleveland have met with marked success in the larger mountings. Among their improvements are devices for all necessary manipulations of the telescope from the eye end·; incandescent lamps for illuminating the circles ; wheels and circles for 'setting' in both co-ordinates from the dome-floor; and the adoption of ball-bearings to give ease of motion to the massive steel axes which, in their mountings for the Lick and Yerkes objectives, weigh from 1 to 3½ tons each. The Lick mounting cost $42,000, and the Yerkes about $60,000. The latter is provided with an elaborate system of electric motors, which do the work of clamping and give quick motion to the telescope on either axis as required. Also the dome is turned in either direction, and the floor of the dome elevated or depressed by electric power.

The modern American mountings provoke little unfavorable criticism. In some classes of observation difficulties arise from their lack of rigidity, both as to the axes themselves and the iron-work supporting their bearings. Too great care cannot be expended upon the founding and construction of the pier upon which the mounting rests. In general, our mountings are surpassed in rigidity by those of like dimension from the best English and Irish shops. But if the American telescope is compared with those of foreign makers, in point of subsidiary apparatus for its con-

venient working, we find a constantly growing dispo-
sition on the part of the instrument builder to consult
the wants of the astronomer and to profit by his
suggestions.

A convenient form of mounting known as the
equatorial coudé, or ' elbow equatorial,' was invented
in 1882 by M. LOEWY, now director of the Paris Observatory. As the illustration shows, the instrument

THE EQUATORIAL COUDÉ (DESIGNED BY LOEWY)
(The observer sits in a comfortable room, his telescope being mounted in the open air)

itself is mounted in the open air, while the observer,
as if working with a microscope, sits always in a fixed
position, no matter what the direction of the object he
may be observing. He may therefore avail himself
of comfortable temperatures. Easy and rapid handling, too, are very advantageous, as well as the attach-
ment of cameras and spectroscopes. There are disadvantages, of course, chiefly the loss of light by

reflection from two plane mirrors set parallel at an angle of 45° to the declination and polar axes. The objective is mounted in one side of the upper cube which, with its mirror also, turns round on the declination axis. The long oblique tube is itself the polar axis, at the upper end of which the observer sits, and a powerful clock carries the instrument slowly round to follow the stars in their diurnal motion. First cost of the equatorial coudé exceeds that of the ordinary equatorial, but the expense of a dome is mostly saved, as the coudé is housed under a light rolling structure. This type of instrument, although not yet represented in America, has many examples in the observatories of France. With the larger one at the Paris observatory, built by the brothers HENRY, was taken a remarkable series of lunar photographs, now combined into a complete atlas of the Moon. Two of these are well reproduced in Chapter III (pages 29 and 31.) Definition in the coudé is excellent, and surface deterioration of the silvered mirrors is not troublesome.

Quality of the ocular or eyepiece is a factor of great importance in the performance of an objective. The character, field of view, and power of eyepieces should be carefully suited to the nature of the telescopic work. The ordinary forms are the positive or Ramsden eyepiece (used for micrometers and all kinds of astronomical instruments requiring a reticle), and the negative or Huygenian eyepiece (usually employed for gazing work merely). It is essential that the eyepiece should be achromatic as well as the objective. The Steinheil monocentric eyepiece is a triple glass, achromatic, and composed of two flint menisci of different thicknesses capping a double convex crown on both sides. The performance of this ocular is especially good on stars and planets. Direct observation of the sun requires a polarizing eyepiece, or helioscope, best made in this country by BRASHEAR. It reduces the intense solar light and heat to a degree not harmful to the delicate tissues of the eye.

The micrometer is an accessory of the telescope used in the measurement of small arcs or angles. It is attached in place of the ordinary eyepiece by means of an 'adapter,' or draw-tube. In the field of view are at least three wires, or threads, or spider-lines, as in the illustration adjacent. By turning the micrometer screw the two parallel threads are separated until the space to be measured, as the diameter of a planet, is just embraced between these two lines. The divided head of the micrometer screw and a scale adjacent record the number of whole revolutions and fractional parts ; and this is readily converted into arc, because the arc value of one revolution is easily ascertained by observations upon stars. At night the micrometer lines are rendered luminous by light from a small electric lamp (at the left in the opposite illustration), or an adjustable oil lamp shown below it. A micrometer will usually be provided with a separate equipment or battery of eyepieces, ranging in power from about 9 to 90 for each inch of aperture of the object glass with which they are used. The illustration represents the micrometer of the 36-inch Lick telescope, constructed by SAEGMULLER. For about ten years it has been kept in nearly constant

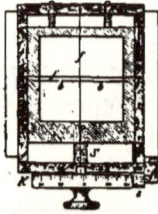

MICROMETER

(*Screw S moves thread bb relatively to stationary horizontal thread f a certain number of revolutions which are read from graduated head K at index s*)

use, notably by Professor BURNHAM in the discovery
and observation of very faint or close double stars.
A similar instrument of unusual proportions was
built by WARNER & SWASEY for the 40-inch Yerkes
telescope.

In considerable part the observations, particularly
of larger arcs, formerly made with the micrometer, are
now supplanted by the greater accuracy attainable

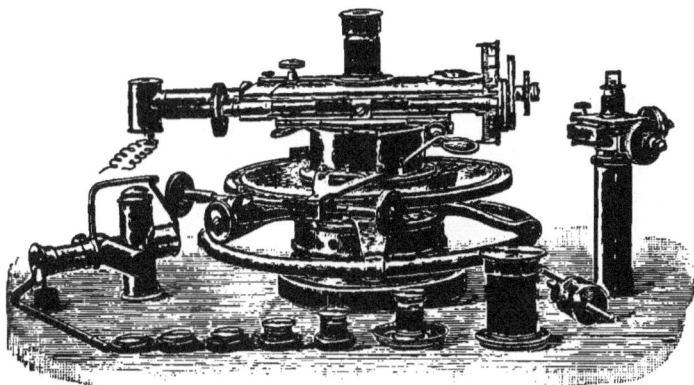

A MODERN MICROMETER
(With battery of eyepieces, oil and electric lamps, etc.)

with the heliometer (page 277) ; but more especially
the photographic plate and measuring engines are re-
placing both. As early as 1850 the younger BOND
obtained with the 15-inch Harvard telescope, fine
photographic impressions of Vega, and later of the
double star Castor, showing an elongate disk. This was
the beginning of stellar photography. BARTLETT, at
West Point, and CAMPBELL, in New York, secured
good pictures of the annular eclipse of 26th May,

1854, the first celestial phenomenon photographically observed. By measuring his plates of Mizar, BOND in 1857 proved photography capable of results equally precise with direct observation of stellar images. The year following, RUTHERFURD of New York undertook celestial photography at his private observatory with an objective of $11\frac{1}{4}$ inches aperture figured by himself. Plates prepared by the old wet collodion process were used, and exposures of 5 to 10 seconds brought out the belts of Jupiter, and even Saturn's ring. In 1863, RUTHERFURD began the construction of the first objective figured solely for photographic rays, and he obtained the required actinic correction by means of his neat

RUTHERFURD (1816–1892)

adaptation of the spectroscope in examining the chromatic condition of an objective. So great was the improvement that an exposure of one second gave an image of Castor, which required ten seconds in the uncorrected telescope. RUTHERFURD later demonstrated the feasibility of converting any visual objective into a nearly equivalent photographic telescope by mounting in front of it a meniscus of flint glass. In 1869 he constructed the first instrument of this character, 13 inches in aperture, and with it were taken, in

August, 1871, the first photographs of the sun, showing the minute granulations of its surface. His pictures of sunspots have not yet been surpassed (page 52), and his lunar photographs compare favorably with recent work at Paris and Mount Hamilton.

THE 40-FOOT HORIZONTAL PHOTO-HELIOGRAPH
(*As mounted at the author's eclipse station in Japan, 1887*)

The transits of Venus, in 1874 and 1882, offered exceptional opportunities for the application of photography, and a type of instrument known as a 'horizontal photo-heliograph,' originated by the elder WINLOCK in 1869, was brought into service by the

American Commission, whose work was directed in the main by Professor NEWCOMB. The CLARKS built eleven of these instruments, eight for the Government, and one each for the Princeton, Harvard, and Lick observatories. They have photographically corrected objectives 5 inches in diameter, and of 40 feet focal length; are mounted in the meridian, and the sun's rays are thrown constantly through the objective by 7-inch plane mirrors, suitably mounted and driven by clock-work. In 1888, BRASHEAR built a similar instrument for the Imperial Observatory of Japan at Tokyo. Besides the large scale of the original image, the sun appearing about 4⅜ inches in diameter on the plate, the stationary image and plate-holder allow the photographer to avail himself of any required number of assistants, all of whom work within the dark room, which thus replaces the moving camera as ordinarily employed in celestial photography. The preceding illustration represents one of the American photo-heliographs as mounted by the writer at Shirakawa, Japan, for the total eclipse of 1887. On the right are mirror and objective, and in the background the photographic house, the 40-foot tube between them being housed from the direct rays of the sun. With similar instruments Professor SCHAEBERLE in 1893, in Chili, and Professor CAMPBELL in 1898, in India, obtained very fine pictures of the sun's corona, — not, however, by using the mirror, but by rigidly fixing the objective high in the air, and following the sun's motion with a slowly-moving plate-holder.

As it is not our present purpose to trace the history of astronomical photography, but only to outline salient departures, we mention but a few additional developments. Most interesting of all is the train of circumstances leading up to our modern methods of determining star places by photographic means.

On the early morning of the 8th September, 1882, Mr FINLAY, first assistant at the Observatory of the Cape of Good Hope, discovered and first observed a bright comet in Hydra (page 200). Several photographers in South Africa obtained photographic impressions of the new comet within a month following, by the use of ordinary apparatus only; and on reporting this fact to Dr GILL, he at once began to take photographs of the comet, with the assistance of Mr ALLIS of Mowbray, whose only suitable lens was a Ross doublet, but 2½ inches in diameter, and of 11 inches focal length. They not only succeeded in taking fine pictures of the comet, but their plates on development showed hundreds of stars, well defined over an area so large as to suggest at once the practicability of employing similar and more powerful means for the construction of star maps. Admiral MOUCHEZ, late director of the Paris Observatory, endorsed the views of Dr GILL, and his encouragement of the Brothers HENRY led them to devote their attention to the construction of suitable lenses. As Dr GILL well says, 'The brilliant results which Messrs. HENRY soon attained are still fresh in the minds of astronomers, and mark an epoch in the history of astronomy in the nineteenth century.' One of these is the remarkable photograph of the Pleiades, already shown on page 237. Meanwhile Dr GILL himself went zealously forward with the work of star charting by means of a 6-inch Dallmeyer lens, the photographic work being done by Mr WOODS at the charges of the Government Grant Fund of the Royal Society.

The plates have all been measured by Professor

KAPTEYN with a parallactic apparatus of his own devising, dependent upon the following principle : hold up against the sky a photographic negative of the

PLATE-MEASURING ENGINE
(*Designed and built by the* REPSOLDS)

same region, taking care to keep it perpendicular to the line of vision through the center of the plate ; then if it be removed from the eye to a distance equal to the focal length of the lens, every star can be exactly covered by its dark image on the negative. Then by substituting for the eye an instrument suitable for measuring stellar positions in the sky, their positions are quickly read off from the plate itself. This ingenious apparatus is quite different from the ordinary plate-measuring engine, shown opposite, with which stellar photographs are usually converted into catalogue positions. This instrument measures merely plane co-ordinates, while that of Professor KAPTEYN measures spherical co-ordinates, as the geometer designates them. The completed catalogue of stars of the southern firmament, the joint research of Dr GILL and Professor KAPTEYN, a work of enormous magnitude, is now appearing in the *Annals of the Cape Observatory*. As a result of Dr GILL's initiative came also the Astrographic Congress (p. 260), with its comprehensive plans, and more than a dozen photographic telescopes at work for years in both hemispheres — all as a direct issue from one comet and a few nearly accidental photographs of it taken in remotest Africa.

Meanwhile, but quite independently, stellar photography had been making rapid progress in America. Professor PICKERING, employing at first very small photographic instruments of the ordinary type, extended his researches by means of an 8-inch Voigtländer lens, refigured by CLARK, and since known as the Bache telescope (page 299). The star charts taken with this instrument in Cambridge and Peru, and with its duplicate, the Draper telescope, have yielded results of high importance ; not only in the determination of stellar magnitudes, but, by affording the means of studying certain regions of the sky at many different epochs, they have brought to light many new stars. But their use as spectroscopes has secured the greatest contribution to stellar astronomy

VARIABLE NEBULA SURROUNDING ETA CARINAE (ARGÛS)
(*Photographed with the 13-inch Boyden telescope*)

(pages 297–301). By mounting prisms in front of the objectives, on the plan first tried by FRAUNHOFER in 1823, all stars in the field impress their spectra, sometimes as many as 1,000, on a single plate.

The requisite width of spectrum is given by mounting the prism with its edge east and west, and suitably varying the clock-motion from the true sidereal rate, according to the degree of dispersion employed, as well as the color and magnitude of the stars in the photographic field. These researches have culminated in the construction and use of the Bruce telescope, already described and illustrated (pages 263–265), and which has fully met the successes predicted for it. Its enormous power is readily inferred from the fact that on a single plate 14 X 17 inches were counted no less than 400,000 stars. This extraordinary result is a necessary consequence of the great aperture of the lens with a relatively short focal length, a combination which photographers designate by the term 'quick-acting.' The rapidity of lenses of the doublet type may be inferred on comparison of the 8-inch Bache telescope of 44 inches focal length with the standard 13-inch astrographic telescope of 134 inches focus; the former requires but an hour's exposure to record a star of the 14.7 magnitude, while the latter consumes three hours in impressing the same image.

During the progress of this work Professor PICKERING devised a novel form of double objective which is equally efficient for both photographic and optical work. This long-sought result is obtained by figuring the two faces of the crown lens very different in curvature, and modifying the cell so that the flint lens may be moved toward the focal plane when desired. For visual observations the two lenses are in contact, the more convex surface of the crown lying next to the concave face of the flint. By reversing the crown lens

HALE'S TYPE OF SPECTRO-HELIOGRAPH

(*Constructed by* BRASHEAR)

and separating the flint three inches from it, the combination becomes an objective perfectly corrected for
the actinic rays — or nearly so. The fact, however,
seems to be that there is a trifling sacrifice of quality,
in both the optical and the photographic combination,
for the sake of this convenient and inexpensive union

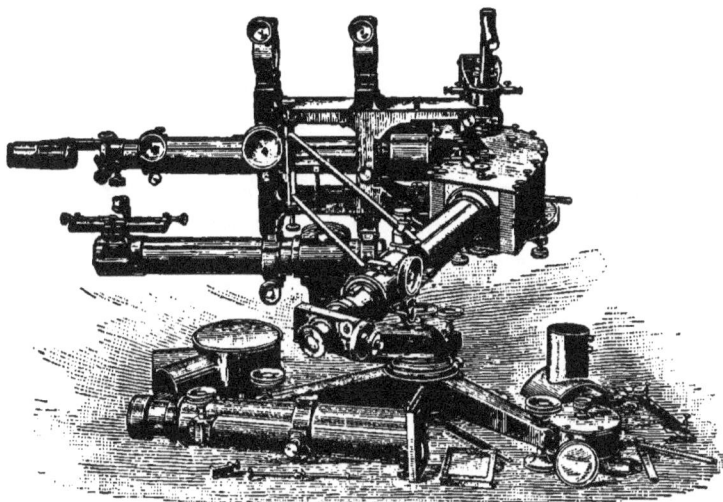

UNIVERSAL SPECTROSCOPE BY BRASHEAR

*(Arranged for visual work. Light enters the spectroscope at the extreme
left, passes through the prisms or is reflected from the grating in the
flat-topped box at the right, and comes to the eye which is placed at the
eyepiece on the micrometer in the middle of the engraving. Below it are
the tube and plateholder, which are substituted when spectra are to be
photographed)*
(By special permission of the American Book Company)

of two objectives in one. A 13-inch telescope of this
pattern was made by the CLARKS, with which the fine
photograph of η Carinæ was taken (page 350), and a
construction practically identical was independently

invented by Sir GEORGE STOKES. The 28-inch Greenwich lens figured by Sir HOWARD GRUBB is an example of this type.

The reflectors built by Mr BRASHEAR have been described earlier in this chapter. But his mechanical skill and untiring energy have enabled him to play a most important rôle in the recent progress of physical astronomy. His optical surfaces exhibit the last order of precision and finish; and many are the physicists and astronomers whose work has been facilitated by his deft constructions. Principally by spectroscopes (page 353) has his reputation been enhanced, and he has built the largest instruments of this character ever constructed, among them the Lick spectroscope (page 288), and one for the Yerkes telescope. Also Professor HALE's lesser spectro-heliograph, illustrated on page 352, is Mr BRASHEAR's workmanship, and the work accomplished with it has already been described on pages 63–65.

Likewise, Mr BRASHEAR has achieved great success in figuring objectives, both optical and photographic. His largest object-glass is of 18 inches aperture, and it acquired instant fame from its use by Mr PERCIVAL LOWELL and his assistants in observing Mars, at Flagstaff, Arizona, during the opposition of 1894. This glass is now mounted at the Flower Observatory of the University of Pennsylvania. Photographic objectives of all sizes have come from Mr BRASHEAR's competent hands — the 6-inch Willard lens (ordinary portrait), re-figured for Professor BARNARD, with which, at the Lick Observatory, he obtained his unsurpassed photographs of the Milky Way; an 8-inch for Professor TERAO, of Tokyo; a 10-inch for the Yerkes Observatory; and a pair of 16-inch doublets for Dr WOLF, of Heidelberg, used to excellent advantage in the photographic discovery of small planets (page 115). The adjacent illustration is reproduced from that part of a photographic plate on which was discovered a small planet ; it shows as a faint elongate trail, indicating the amount of its motion during the exposure of two hours. Perhaps the best form of mounting for short-focus cameras of this description is that shown in the reproduction on p. 357, from a design by HEYDE of Dresden, for the observatory at Moscow. It readily admits of the necessarily long exposures without reversal from one side of the pier to the other.

As the telescope and its accessories become very large and massive, the disadvantages of the Fraunhofer or German mounting become more and more pronounced, on account of the overhang of the declination axis, and its flexure,

DISCOVERY OF A SMALL PLANET

(by photography. The planet is the long object at the centre)

as well as the hindrance of the pier itself, preventing very long exposures or periods of observation, except by reversal of the telescope to the opposite side of the pier. So there is a present tendency to revert to the old type of parallactic stand first invented by RÖMER and now universally known as the English mounting, in which the polar axis is supported at its upper end by a second bearing and pier, and the axis

itself is a double fork with the telescope swung between
its prongs.   Of this type is the old equatorial mounting at Greenwich devised by Sir GEORGE AIRY.   The
chief disadvantage is that the sky close around the pole
and underneath it is inaccessible ; but this is more than
counterbalanced by the cardinal advantage of having
the entire space beneath the polar axis free and clear,
so that a celestial body can be followed from one
horizon to the other without reversal of the instrument.
Several of the astrographic telescopes were mounted
in this fashion; also in 1898 the reflector of one
metre aperture at the observatory at Meudon.   The
clock-motion is imparted with great steadiness, and
the general design of the English mounting permits a
very satisfactory solution of the engineering problems
that arise in the construction of great telescopes.
This type of mounting received the mature approval
of ALVAN GRAHAM CLARK.

With Mr BRASHEAR is associated Dr HASTINGS of
Yale University, who calculated the curvatures of all
the objectives fashioned by the former in recent years.
Dr HASTINGS is himself a practical optician as well as
theoretical physicist, having made many objectives
with his own hands.   The largest has an aperture of
9.4 inches, and is mounted at the Johns Hopkins Observatory.   Contrary to the usual practice of opticians, no
examination of the quality of image formed by the
lens was made until the work upon it was finished ; and
on first trial the objective was found so nearly satisfactory that only twenty minutes were required to render it sensibly perfect.   In this achromatic glass, with
the flint lens in advance of the crown, with a space
of 0^{in}.1 between them, a variation of this latter distance will effectively correct the objective for changes

of temperature. The curves of Mr BRASHEAR'S objectives being based on the formulæ of Dr HASTINGS there is a close agreement of the actual focal length

A MODERN CAMERA FOR STELLAR PHOTOGRAPHY
(*With equatorial mounting designed by* HEYDE)

with the calculated value, which is often a marked advantage.

The effect of focal length upon the performance of a lens is excellently shown by the wonderful photographs of the Milky Way recently obtained by Professor BARNARD with a magic-lantern lens of 1½ inches diameter, and 5½ inches focus. In a few minutes' exposure it photographs what the ordinary quick-acting portrait lens requires several hours to show. The scale is of course very small, and the cloud-forms of the Galaxy are so compressed that they act, not as an aggregation of individual stars, but as a surface. The earth-lit portion of the Moon was well depicted by exposure of a single second; the brighter cloud-forms of the Milky Way appeared in ten to fifteen minutes; and a great wing-like nebula involving ν Scorpii was discovered by Professor BARNARD with this simple lantern lens.

The recent application of photography to the trails of meteors is very suggestive. Lenses of short focus and a large field of view are best. Professor BARNARD, on the 13th November 1893, obtained the opposite photograph of a meteor which shot across the field while he was making an exposure more than two hours in duration in order to secure the faint comet below (comet IV, 1893). But a still more remarkable delineation has been secured : on the 18th June 1897, when the Bache telescope at Arequipa was photographing spectra of the stars in the constellation Telescopium, a bright meteor flashed across the field, and the lines of its spectrum were therefore registered upon the plate — the first spectrum of a meteor ever recorded by photography. There were six lines, and their intensity varied in different parts of the photograph, showing a change in character of the meteor's light as its image trailed across the plate. The lines were mostly due to hydrogen, resembling stars that have bright line spectra. Thus

ETEOR-TRAIL PHOTOGRAPHED AT THE LICK OBSERVATORY, 13 November 1893

(By Professor BARNARD *with 6-inch Willard lens. Exposure* 2^h 5^m *upon stars and the faint object below, comet* IV 1893)

the chance spectrum of this meteor will, Professor PICKERING hopes, aid in ascertaining the conditions of temperature and pressure in such stars. The expected mid-November meteoric showers of 1899 and 1900 will undoubtedly yield further and similar results. These, with trails from Dr ELKIN's multiple cameras (page 212) will probably provide data for a more accurate orbit of the Leonids. More than thirty were photographed in November 1898.

An arrangement of multiple cameras for observation of total solar eclipses was first worked out by the writer in 1889, for the eclipse of the 22d December in West Africa. . In all, 23 instruments, chiefly photographic, were attached to a massive polar axis, and pointed parallel to each other, following accurately on the eclipsed Sun. The engraving opposite illustrates many of them ; also in the foreground are the pneumatic contrivances by which exposing shutters, plate-holders, and all other moving devices for eclipse observation were operated automatically. The control was effected by a perforated strip of paper, similar to the music sheets now commonly used in automatic organs. Each perforation in the eclipse. sheet represented, not a musical note, but a mechanical movement of some particular device. By the liberality of Mr D. WILLIS JAMES of New York, a trustee of Amherst College, a similar battery of instruments was duly installed in Yezo by the Amherst Eclipse Expedition in 1896. One of the numerous devices was constructed with reference to securing on a single plate the imprint of a few coronal streamers complete in their entire length. This had never been done, because exposure long enough to give the outer corona burns out the film where the

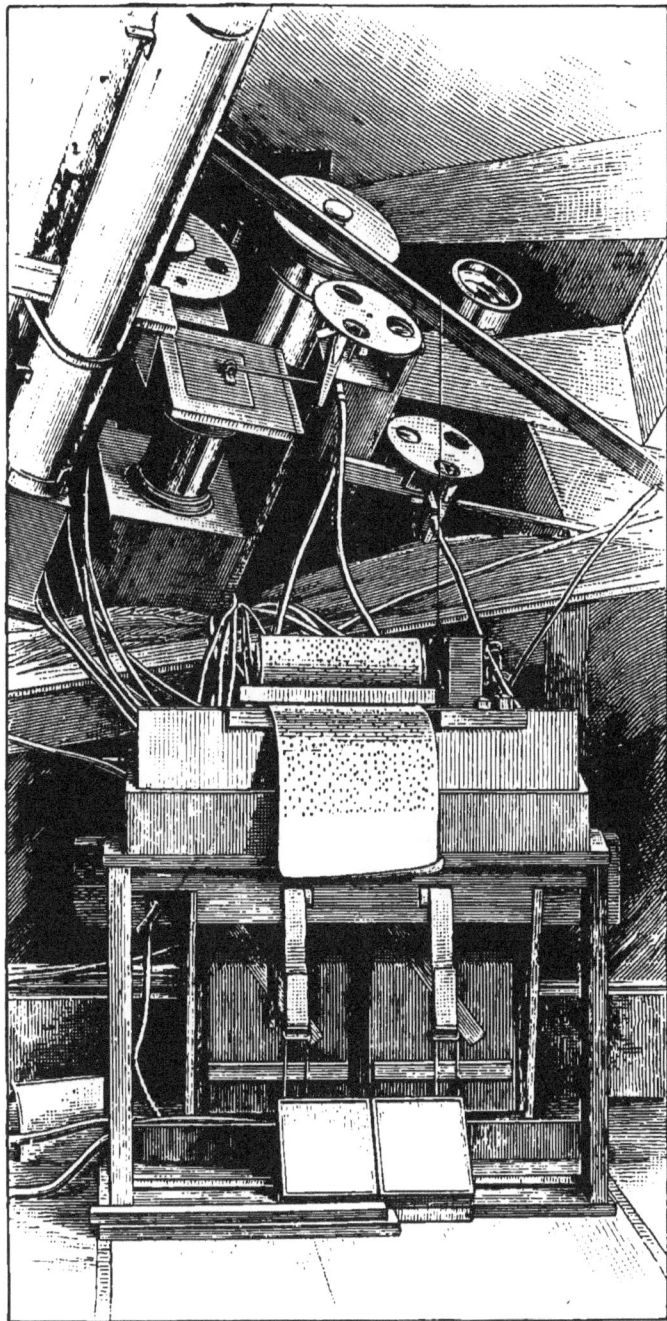

THE PNEUMATIC COMMUTATOR AND PHOTOGRAPHIC BATTERY OF
ECLIPSE INSTRUMENTS (TODD)

(As mounted at Cape Ledo, Africa, for the total eclipse of December 1889)

AUTOMATIC DEVICE FOR PHOTOGRAPHING THE SOLAR CORONA IN RINGS (TODD)

(*Operates by electricity. Designed and built on board Mr JAMES's yacht 'Coronet,' June 1896*)

bright inner corona would otherwise be. In the device illustrated opposite, three concentric rings and a
central disk are removed from the photographic field
successively, permitting long exposure for the outer
and short for the inner corona.

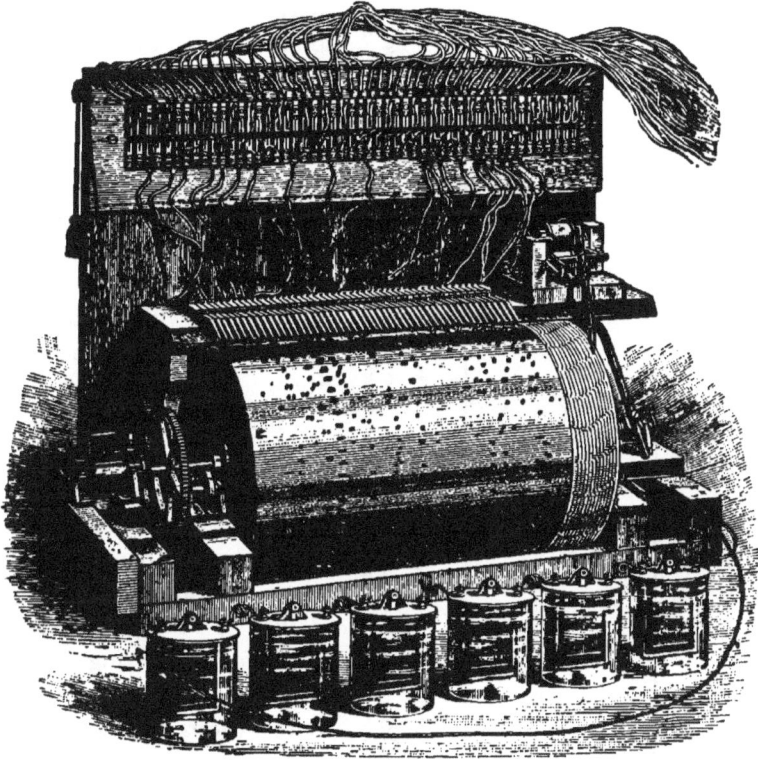

THE ELECTRIC COMMUTATOR (TODD)
(*Amherst Eclipse Expedition to Japan*, 1896)

Instead of pneumatic operation, the necessary movements, about 500 in all, were secured at the critical
instants by the control of an electric commutator

(preceding page). The pneumatic and electric systems both worked perfectly; but clouds at both stations unfortunately obscured the corona. The ease with which a number of delicate pieces of photographic apparatus can be arranged, the certainty with which they can be operated; the large number of photographs obtainable by this means for subsequent study, the dispensing with manual movements during totality when every nerve is at high tension — all point to the desirability of repeating the experiment during the coming eclipses of 1900 and 1901. Also the novel device of Mʳ BURCKHALTER, for obtaining the bright inner and faint outer corona on the same plate by means of a suitably shaped and swiftly whirling vane, revolving in the photographic field during exposure, was successful during the India eclipse of 1898, and is worthy of permanent

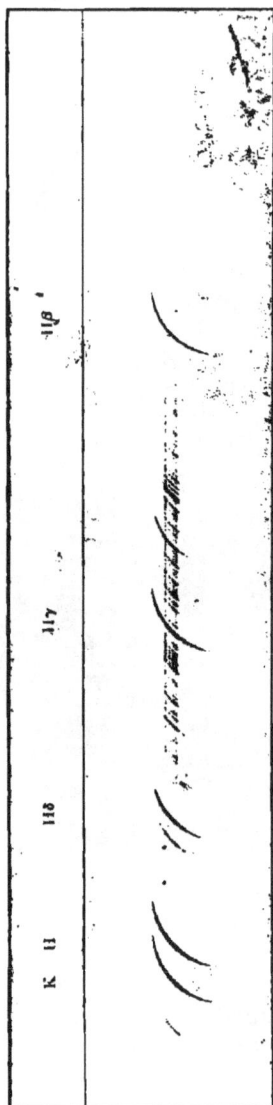

FLASH-SPECTRUM NEAR TOTALITY IN INDIA, 22d January 1898

(From a photograph by Professor NAEGAMVELA. The dark crescents are due to bright lines in the spectrum of the reversing-layer. The two spots between H and Hδ are due to a hydrogen protuberance)

adoption in eclipse programmes of the future. But even more important is the photography of the flash spectrum, near beginning and end of the total phase. The unobscured limb of the Sun is then a very slender crescent; and by using the objective prism, or prismatic camera, the developed image on the plate comes out as in the opposite photograph — not a single crescent, but a succession of crescents whose position and size enables us to say to what substances the light owes its origin, and to study the depth of the reversing layer all around the Sun's edge. The total eclipses of 1896 and 1898 indicate that the depth of this stratum is about 700 miles.

The advances of astronomy by the aid of photography are too numerous for further treatment here. In 1840 the Moon was first photographed, in 1850 a star, in 1854 a solar eclipse, in 1872 the spectrum of a star, in 1880 a nebula, in 1881 a comet, in 1897 the spectrum of a meteor, and in 1898 a stellar occultation by the Moon. All these epoch-making photographs were made in America. Nearly every branch of astronomical science has been advanced by the instrumentality of photography, so universal are its applications, and so persistently have they been put to the successful test.

Before passing to the construction of observatories and suitable sites for them, one or two paragraphs will be devoted to the instruments of the old astronomy, or astronomy of precision, as it is often termed. Most important of all is the meridian circle, the joint invention of Römer and Picard, before whose day the position of a star or planet on the celestial sphere was observed by means of a quadrant. Hevelius conducted a long and valuable series of observations with an instrument of this sort, illustrated adjacent; but they were unfortunately destroyed by fire at Dantzig, the scene of his labors. Two persons were neces-

HEVELIUS AND HIS CONSORT OBSERVING
(The instrument is a quadrant of the 17th century, before telescopes were adapted to measurement of angles)

sary to make a complete observation, and the star's direction was fixed, not by the telescope, but by sights or pinules. A surprising degree of accuracy was reached by carefully repeated measures.

From the middle of the 17th century onward, we must omit the intervening forms of arc-measuring instruments, down to the end of the 19th century, when we find astronomers almost universally employing the meridian circle. An excellent type by the best modern maker is shown on the following page, by courtesy of Professor PAYNE, editor of *Popular Astronomy*, an American monthly that provides much information about all kinds of astronomical apparatus, including telescopes, their manufacture, and use. As its name implies, the meridian circle revolves in a vertical plane north and south ; and its construction demands a house with a sliding or removable roof, with shutters opening below it, as at the extreme left in the illustration. A clear strip of sky is thereby accessible for observation in every part, and unobstructedly free from the zenith down to the north and south horizons. The exterior of a meridian room is shown in the left-hand wing of Dr ROBERTS's observatory, page 290.

Within a meridian room the telescope, whether a transit instrument or a meridian circle, surmounts two stable piers, between which it turns round on bearings called Y's. Circles adjacent to the telescope, and on either side of it are rigidly attached to the horizontal axis ; and they, with both telescope and axis turn round as a unit, the axis of the telescope describing the plane of the meridian. The circles enable any known star to be found, by means of its distance from the zenith. Then the time of its passing the wires in the reticle is observed by a clock or chronometer, usually with the aid of a chronograph which electrically records the precise instants at which the transits take place. If a star's position on the celes-

THE REPSOLD MERIDIAN CIRCLE OF CARLETON COLLEGE OBSERVATORY, MINNESOTA

(A modern observing room, showing collimators on the right and left, reversing carriage, and other accessories for meridian observations)

tial vault is already known, one complete meridian observation provides the means of finding the latitude of the observatory, and the local time ; or conversely, if these elements have previously been found, a complete observation enables us to assign the star's exact geometric position on the celestial sphere ; in other words, we determine its right ascension and declina-

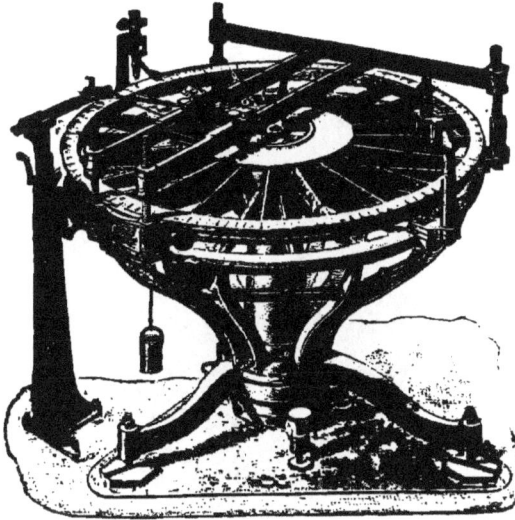

AUTOMATIC DIVIDING-ENGINE
(For graduating circles. Built by SÉCRETAN *of Paris)*

tion, as technically called. Meridian instruments of high precision and excellent workmanship are constructed in America as well as in Europe. BUFF & BERGER of Boston, SAEGMULLER of Washington, and WARNER & SWASEY of Cleveland, are the chief makers. Accurate division of their circles is the chief difficulty of construction, but this end is now attained with all necessary precision by means of graduating engines, similar to the one in the last illustration. Errors of a whole second of arc in the position of a line on the face of a silver circle are now uncommon in the best modern instruments. This means that of all the lines automatically engraved on a fine circle 20 inches in diameter, one is rarely misplaced by so much as the half of $\frac{1}{10,000}$ of an inch.

ZACH (1754–1832)

But few observers are able to assign the absolute time when a star is actually crossing a line in the field of view: most will record the instant after the star crosses, while a few will always set down the instant before that event. Any difference between the absolute and observed times is termed personal equation; and the value of it is susceptible of exact ascertainment in several ways. The best observer is not one whose personal equation is least, but the one whose equation remains most nearly constant. Avoidance of the effect of personal equation is practically attained by the photo-chronograph, an instrument devised by Father FARGIS of Georgetown College for recording transits photographically. The telescope is set upon the star as if for an ordinary visual observation ; but the eye-piece is removed and a tiny sensitive plate inserted in its stead. In front of it is a little strip of metal, or occulting bar, upon which the star's image trails in crossing the field ; and this bar is connected with an armature, a circuit through which is sent by the clock at the beginning of every second, as for the ordinary chronograph. This throws the bar aside for an instant, so that the developed plate shows a succession of black dots, or instantaneous and equidistant impressions of the star. Holding a lantern for a few seconds in front of the objective will impress the transit lines upon the plate, in their correct relation to the stellar dots ; and the fraction of a second when the star was crossing a wire can then be found by measurement, practically free from personal equation.

THE ALMUCANTAR
(*Devised by* CHANDLER)

The first astronomer who discussed the inherent and unavoidable deficiencies of instruments and their effects upon observations was the versatile VON ZACH, in his *Correspondance Astronomique* in 1819. In our day a determination of the errors of an instrument gives vastly more trouble than the subsequent observations and the reduction or calculation of them. In the almucantar as used by Dr CHANDLER, many of the troublesome errors of the meridian circle are skilfully eluded. The flotage principle is adopted for the supports of the telescope ; and while the small errors of the meridian circle must be found from observations on less than half its circle of reference, the almucantar has the great advantage of permanent visibility of its entire circle of reference, which is the small circle parallel to the horizon and passing through the north pole of the heavens, often called an almucantar.

So perfect is the construction of the marine chronometer at the present day that all observatories now use this convenient instrument in recording and carrying on the time. Its portability, and ease of regulation to a rate nearly invariable with changes of temperature, enforce its use on expeditions where astronomical clocks would be impracticable. The chronometer has received but few cardinal modifications of design since the days of JOHN HARRISON, who early in the 18th

HARRISON (1693–1776)

century began his great invention, in competition for a prize of $100,000 offered by the British government in 1714, for the ' discovery of a method of finding the longitude at sea within 30 miles.' HARRISON'S chronometer was taken twice to Jamaica, and on other voyages, actually finding the longitude with a precision even greater than the margin stipulated, and in 1765 he was paid the reward, — but only half of it, because the prize was divided between him and the brilliant TOBIAS MAYER, of Göttingen, another competitor, who had succeeded in bringing the lunar tables to such a state of perfection that the longitude of a ship could

MAYER (1723-1762)

be found from observations of the Moon as accurately as by the chronometric method. MAYER invented also the principle of the repeating circle, for eliminating the effect of errors of division by repeating the measure of an angle on different parts of the graduated circle. HARRISON'S chronometer was in the latter part of the century improved by both ARNOLD and EARNSHAW; and is to-day essentially in the form in which they left it. It is not our purpose here to enter upon an explanation of the method of using the chronometer at sea. That will be found exemplified to the fullest detail in the *American Practical Navigator* of NATHANIEL BOWDITCH, whose fame rests also upon his admirable translation of the *Mécanique Céleste* of LA PLACE (page 246), which he supplemented by useful annotations. It is sufficient to remark here that the chronometer simply carries Greenwich time; the difference between which and the ship's local time gives the longitude.

BOWDITCH (1773-1838)

The observatories amply illustrated in the preceding chapters embody many excellent features, also some constructive faults. Astronomers a half century and more in the past were accustomed to build their observatories high up from the ground, a construction far from desirable and always to be avoided if possible. Better is it to choose an original site as elevated as convenient, and then build the observatory upon it only a single story high. Architectural effect must often be sacrificed to scientific utility. An attempt to combine both met with some measure of success in the instructional observatory on the following page, built from the writer's specifications in 1886. The instruments possess great stability, and all practical observers appreciate the absence of long stairways between dome and transit room. A better form of dome wall is shown in Dʳ ROBERTS's observatory (page 290), the corners of the square dome-room affording a welcome convenience for observing-chairs and other accessories. This is the form adopted at Cambridge (page 300), which exemplifies the proper construction of a great observatory, in marked contrast to the main building of the Naval Observatory at Washing-

SMITH COLLEGE OBSERVATORY, NORTHAMPTON

(*Miss* MARY E. BYRD, *Director*)

ton (page 101). Telescopes mounted in individual domes, surrounded by low trees and heavy turf, meet with minimum interference from the radiating heat of early evening. A quiet atmosphere must be secured at any sacrifice. Massive walls of brick and stone are far from conducing to this end; and the observer whose instrument is mounted therein must frequently wait several hours for them to cool down before beginning his work on summer evenings. The best type of observatory construction is that which utilizes a minimum of material, so that very little solar heat can be stored in its walls during the day. Local disturbance of the air in early evening is then but slight. Louvers, as employed by Dr GILL in his heliometer house at the Cape, or better still, ivy-grown walls, as in the observatory at Oxford University, shown below, contribute effectively to this desirable end.

Often circumstances enforce the building of an observatory surmounting a dwelling house, or a large

OXFORD UNIVERSITY OBSERVATORY
(*Professor* H. H. TURNER, *Director*)

school building. Mr BRENNER'S observatory opposite is an example of this construction; and the observatories built on top of the new high schools at Springfield and Northampton, Massachusetts, according to plans furnished by the writer, show the best that can be done under restricted conditions. It is often difficult to get a telescope so mounted to perform its best: violent tremors of whatever sort, communicated directly to it, are fatal to delicate observation. Also high winds are very unfavorable; and the effect of warmer air from the building below, not to

MANORA OBSERVATORY, LUSSINPICOLO, ISTRIA
(*Mr* LEO BRENNER, *Director*)

say from adjacent chimneys, is often troublesome. This is best avoided by entering the dome, not by a stairway from underneath, but from an exterior gallery in the open air, usually easy to arrange, if specified in due season. But if the telescope is small, a portable mounting carried out and in as required, and used upon a firm foundation, will usually yield the most satisfactory vision, and at very little expense or inconvenience. At the end of this chapter will be found a list of manuals for the use and

3-INCH PORTABLE TELESCOPE
(*With altazimuth mounting*)

care of such instruments.   The earliest and most successful of these is by the Rev. T. W. WEBB, late Pre-
bendary of Hereford Cathedral, and entitled *Celestial
Objects for Common Telescopes,* which in 1893 passed
to a fifth edition under the able revision of the Rev.
Mr ESPIN.   Its closely stocked pages comprise a handy
book for the practical observer which few professional
astronomers can afford to despise.

THOMAS WILLIAM WEBB (1807-1885)

No less important than optical perfection of a telescope is excellence of the site where it is to be located
and used.   The greater freedom from cloud the better; the horizon should be unobstructed, especially to
the south and west; and gravel foundations for the
instruments are worth seeking for.   But in addition to

all this, optical quietude and transparency of the atmosphere is of the utmost significance. That a perfect telescope may perform perfectly, it must be located in a perfect atmosphere. Currents of warm air are continually rising from the earth, and colder air is coming down to take its place. By removing the eye-piece of the telescope these atmospheric waves become plainly visible ; and when examining a star or planet, their effect appears as blurring, distortion or quivering of the image as seen through these shifting air strata of variant temperatures and consequently variant densities. At Mr LOWELL's observatory in Arizona, his careful assistants have made a thorough study of these atmospheric waves, and their influence upon the conditions of telescopic vision. If the air is much disturbed, only the lower magnifying powers can be used, and much of the expected advantage of a great telescope is unavailable.

All hindrances of atmosphere are found best avoided in arid regions of our globe, at elevations of 3,000 to 10,000 feet above sea level. A dry atmosphere is a prime essential for steady vision. On the American continent several observatories are now maintained at mountain elevation : the Boyden Observatory, Arequipa, Peru (page 382), at 8,000 feet ; the Lowell Observatory, Flagstaff, Arizona, at 7,000 feet ; and the Lick Observatory, Mount Hamilton, California (page 122), at 4,000 feet. Higher mountains have as yet been but partially investigated, the personal infelicities of occupying them seeming to counterbalance the gain afforded by greater elevation. But it is an easy assertion that the great telescope of the future, if it is to do the work for which it is best adapted, must be located at a mountain elevation of

10,000 to 15,000 feet, in a region where steadiness of atmosphere has previously been assured by critical and protracted exploration with that sole condition in view. Whether on the lofty table-lands of Asia, or those of South America, is yet undecided.

Of the three greatest benefactors of American astronomy, now deceased, two bequeathed their fortune largely to provide for great telescopes at mountain elevation. Both scientific trusts have been so administered as to assure very great scientific good. JAMES LICK, in 1874, by a deed of trust, left the sum of about $700,000 for the building of a telescope more powerful than any then in existence, and for equipping and maintaining an observatory in conjunction with it.

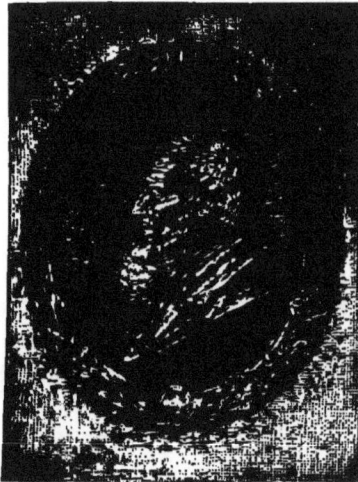

JAMES LICK (1796–1876)

Fourteen years were consumed in preliminaries, and in building the observatory, completing and installing the instruments. The subsequent history of the Lick Observatory is more widely known than that of any other scientific institution in the world. Could Mr LICK have known the advantages to science and the general diffusion of astronomical knowledge accruing from this bequest, he would probably have endowed this single object with his entire estate, exceeding $4,000,000.

Uriah A. Boyden, who acquired a snug fortune of about $230,000 from his inventions and improvements in turbine water-wheels, bequeathed it all to three trustees to 'establish and maintain, in conjunction with others, an astronomical observatory on some mountain peak.' Harvard College Observatory having received an unrestricted bequest of $100,000 from Robert Treat Paine, the two were consolidated in

URIAH ATHERTON BOYDEN (1804-1879)

1886, and the search for a suitable peak began. In our own country, lofty elevations in Colorado and California were investigated, with unsatisfactory conclusions, after which the field of search was transferred to South America. The Chilean desert of Atacama was tested, and later a temporary station was

established at Chosica in Peru, followed by the permanent observatory now maintained at Arequipa in the same country. Two photographic telescopes are constantly in use at this remote station, and a powerful refractor could advantageously be added to the working equipment.

A third great benefactor of astronomy died in New Haven in 1889 — ELIAS LOOMIS, whose fortune,

ELIAS LOOMIS (1811–1889)

exceeding $300,000, he left to Yale Observatory, reserving its income exclusively for the employment of astronomers who should make observations and calculate them, and for their subsequent publication. Professor LOOMIS was widely known, not only as a lecturer and teacher at Yale, but as a writer of highly successful text-books on mathematics and astronomy. Also he was a keen investigator in meteorology and terrestrial physics, his magnetic charts of the United States being the first published.

Most of the great telescopes of the world have in their turn signalized their extraordinary power by some important astronomical discovery — or at least some significant piece of research which could not have been done so well with a lesser instrument. To specify in part: HERSCHEL'S reflector first revealed the planet Uranus; Lord ROSSE'S ' Leviathan,' the spiral nebulæ; the 15-inch Cambridge telescope, Saturn's dusky ring; the 18½-inch Chicago refractor, the companion of Sirius; the Washington 26-inch refractor the satellites of Mars; the 30-inch Russian objective, the nebulosities of the Pleiades; and, last of all, the 36-inch Lick telescope brought to light a fifth satellite of Jupiter. At the time of these discoveries, each of the great telescopes was almost the only instrument powerful enough to have made the discovery. With such a record, are we not safe in predicting further advances with larger telescopes still? Two American opticians of wide experience and high competency stand ready to undertake them, if only the funds are forthcoming; and as yet there is no indication that a refracting telescope of five, or even six, feet aperture would fail of a gratifying success to its projector.

According to ALVAN GRAHAM CLARK, a six-foot objective would not necessitate a combined thickness of more than six inches; and he also says, ' Who knows how soon glass still more transparent may be at hand, considering the steady improvement made in this line, and the fact that the present disks are infinitely superior to the early ones? ' The obstacles to

overcome in mounting and using such a glass disappear when we reflect that as yet the astronomer
has but just begun to invoke the fertile resources of
the modern engineer. The younger CLARK further
wrote, in 1893, this remarkable union of fact and
prophecy : 'The increase in size of even our present
great refractors is not a possibility but a fact, and
with this will come large acquisitions to our present
stock of knowledge. The new astronomy, as well as
the old, demands more power. Problems wait for
their solution, and theories to be substantiated or dis-
proved. The horizon of science has been greatly
broadened within the last few years, but out upon the
border-land I see the glimmer of new lights that wait
for their interpretation, and the great telescopes of
the future must be their interpreters.'

EX·LIBRIS

JOHANNIS COVCH ADAMS

TELESCOPES (HISTORY, MOUNTING, AND USE)

GRANT, *History of Physical Astronomy* (London 1852).

GRUBB, T., 'Mountings,' *Report Brit. Assoc. Adv. Sci.* 1857, p. 195.

HERSCHEL, J. F. W., Article 'Telescope,' 8th ed. *Encyclopædia Britannica*, vol. xxi. (1860), 117.

WOLF, R., *Geschichte der Astronomie* (Munich 1877).

POGGENDORFF, *Geschichte der Physik* (Leipzig 1879).

KNIGHT, E. H., 'Telescope,' *American Mechanical Dictionary*, vol. iii. (Boston 1881).

PICKERING, 'Large Telescopes,' *Proc. Am. Acad. Arts and Sciences*, xvi. (1881), 364.

BURNHAM, DENNING, YOUNG, 'Small *vs.* Large Telescopes,' *Sid. Mess.* iv. (1885), 193, 259; v. (1886), 1.

SERVUS, *Die Geschichte des Fernrohrs* (Berlin 1886). Bibliog.

HOLDEN, 'Lick Telescope,' *Publ. Lick Obs.* i. (1887).

YOUNG, 'Great Telescopes,' *The Forum*, iv. (1887), 78.

GILL, 'Telescope,' *Encyclopædia Britannica*, 9 ed. xxiii. (Edinburgh 1888).

TODD, 'Telescopic Work in U. S.,' *Encyclopædia Britannica*, xxiii. (Philadelphia 1888).

COMMON, 'Telescopes,' *Nature*, xlii. (1890), 183.

CHAMBERS, *Descriptive and Practical Astronomy* (Oxford 1890).

PLUMMER, 'Pulkowa Refractor,' *Nature*, xlii. (1890), 204.

WOLF, R., *Handbuch der Astronomie, ihrer Geschichte und Litteratur* (Zurich 1890–92).

SWIFT, 'Telescopes,' APPLETON'S *Annual Cyclopædia* (1890–1898).

RANYARD, 'Penetrating Power,' *Knowledge*, xiv. (1891), 154.

HASTINGS, 'History of Telescope,' *The Sidereal Messenger*, x. (1891), 335.

KNIGHT, W. H., 'Telescopes in the U. S.,' *The Sidereal Messenger*, x. (1891), 393.

VOGEL, NEWCOMB-ENGELMANN'S *Populäre Astronomie* (Leipzig 1892).

CLARK, 'Possibilities of the Telescope,' *North American Review*, clvi. (1893), 48.

CLARK, 'Future Great Telescopes,' *Astronomy and Astrophysics*, xii. (1893), 673.

WARNER, 'Engineering Problems of Large Refractors,' *Astronomy and Astrophysics*, xii. (1893), 695.

GRUBB, H., 'Great Telescopes,' *Knowledge*, xvii. (1894), 98.
BARNARD, 'Great Telescopes,' *Pop. Astron.* ii. (1895), 245.
NEWCOMB, 'Telescopes,' JOHNSON'S *Univ. Cycl.* viii. (1895).
DENNING, 'Great Telescopes,' *Nature*, lii. (1895), 232.
GERLAND, VALENTINER'S *Handwörterbuch der Astronomie* (Breslau 1895–98).
BARNARD, 'Large and Small Telescopes,' *M. N. Roy. Astron. Soc.* lvi. (1896), 163.
HALE, 'Yerkes Telescope,' *Astrophys. Jour.* vi. (1897), 37.
LEWIS, HOLLIS, 'Large Refractors,' *The Observatory*, xxi. (1898), 239, 270.
FOWLER, 'Concise Knowledge Library,' *Astronomy*, part ii. (New York 1898).

GLASS, OBJECTIVES, AND TESTING

RUTHERFURD, 'Spectroscope in Testing Objective,' *Am. Jour. Sci.* lxxxix. (1865), 307.
GUNDLACH, 'Quadruple Objectives,' *Astr. Nachr.* xci. (1878), 177.
HASTINGS, 'Achromatic Objective,' *J. H. Univ. Circ.* ii. (1882), No. 19.
SMITH, H. L., 'Short-focus,' *Sidereal Messenger*, i. (1883), 239.
GRUBB, H., *Trans. Roy. Dubl. Soc.* i. (1883), 1.
HASTINGS, 'Figuring Lenses,' *Sidereal Messenger*, i. 244; ii. 39 (1883–84).
PENDLEBURY, *Lenses and Systems of Lenses* (London 1884).
GRUBB, H., 'Objectives and Mirrors: Preparation and Testing,' *Proc. Roy. Inst.* xi. (1886), 413.
YOUNG, 'Lick Objective,' *Boston Jour. Chem.* (October 1886).
HASTINGS, 'Achromatic Objective,' *Am. Jour. Sci.* cxxxvii. (1889), 291.
CONROY, 'Reflection and Transmission of Light by Glass,' *Phil. Trans.* clxxx. (1889), 245.
KEELER, 'Lick Objective,' *Publ. Astron. Soc. Pacific*, ii. (1890), 160.
COOKE, *On the Adjustment and Testing of Telescopic Objectives*, (York 1891).
FOWLER, 'Testing Objectives,' *Nature*, xlv. (1892), 204.
MAUNDER, 'Adjustment,' *Jour. Brit. Astron. Assoc.* ii. (1892), 219.
TAYLOR, 'New Achromatic Objective,' *M. N. Roy. Astron. Soc.* liv. (1894), 328.

BRASHEAR, 'Optical Glass,' *Pop. Astron.* i. (1894), 221, 241, 291, 447.

BRASHEAR, 'Adjustment and Care of Objectives,' *Pop. Astron.* ii. (1894), 9, 57.

VOGEL, 'Absorption by Objective,' *Astrophys. Jour.* v. (1897), 75.

BRASHEAR, 'Glass Manufacture,' *Popular Astron.* vi. (1898), 104.

REFLECTORS AND THEIR MOUNTINGS

HERSCHEL, W., ' 40-ft. Reflector,' *Phil. Trans. for* 1795, 347.

ROSSE, ' 6-ft. Reflector,' *Philosophical Transactions* 1840, p.503; also cli. (1861), 681.

MASON, SMITH, ' 12-in. Reflector,' *Trans. Am. Phil. Soc.* vii. (1841), 165.

LASSELL, 'Polishing Specula,' *Mem. Roy. Astr. Soc.* xviii. (1850), 1.

DRAPER, ' Construction of Reflector,' *Smithson. Contr. Knowl.* xiv. (1865).

LASSELL, ' 4-ft. Reflector,' *Mem. Roy. Ast. Soc.* xxxvi. (1867).

GRUBB, T., ' Melbourne Reflector,' *Phil. Trans. Roy. Soc.* clix. (1869), 127.

THORNTHWAITE, *Hints on Reflecting and Refracting Telescopes* (London 1877).

COMMON, ' Mounting for 3-ft. Reflector,' *Mem. Roy. Astr. Soc.* xlvi. (1881), 173.

BRASHEAR, ' Silvering a Reflector,' *Brit. Jour. Photo. Alm.* 1888.

JONES, G. S., ' Figuring a Mirror,' *Sid. Mess.* ix. (1890), 353.

COMMON, ' 5-ft. Reflector,' *Mem. Roy. Astron. Soc.* l. (1892), 113.

COMMON, 'Mounting for Reflector,' *M. N. Roy. Astron. Soc.* liii. (1893), 19.

CLERKE, *The Herschels and Modern Astronomy* (London 1895).

CALVER, *Hints on Silvered-glass Reflecting Telescopes* (London 1895).

SCHAEBERLE, 'Cassegrain,' *Pub. Ast. Soc. Pacific,* vii. (1895), 185.

STONEY, ' Supporting Large Specula,' *M. N. Roy. Astron. Soc.* lvi. (1896), 454.

HALE, ' Reflectors *vs.* Refractors,' *Astrophysical Journal,* v. (1897), 119.

WADSWORTH, ' Coudé Reflectors,' *Astrophysical Journal,* v. (1897), 132.

GAUTIER, ' Sidérostat à lunette de 60^m de foyer et 1^m.25 d'ouverture,' *Annuaire Bur. Long.* 1899.

SPECTROSCOPY AND SPECTROSCOPES

SCHUSTER, 'Modern Spectroscopy,' *Proc. Roy. Inst.* ix. (1881), 493.'

LOCKYER, 'Spectroscopic Apparatus and Methods,' *The Chemistry of the Sun* (London and N. Y. 1887).

MAUNDER, in CHAMBERS'S *Descriptive Astronomy*, ii. (Oxford 1890).

KEELER, 'Elementary Principles of Spectroscopes,' *The Sidereal Messenger*, x. (1891), 433.

DESLANDRES, *Comptes Rendus*, cxiii. (1891), 308; cxiv. (1892), 276.

FROST, 'Potsdam Spectrograph,' *Astronomy and Astro-Physics*, xii. (1893), 150.

HALE, 'Spectroheliograph,' *Astronomy and Astro-Physics*, xii. (1893), 241.

KEELER, 'Spectroscope,' *Pop. Astron.* i., ii. (1893–94).

FROST, SCHEINER, *Treatise on Astronomical Spectroscopy* (Boston and London 1894). Bibliography.

REED, 'Spectroscope,' *Pop. Astron.* ii. (1895).

YOUNG, 'Spectroscope and Adjustments,' *The Sun*, p. 351 (New York 1896); *Pop. Astron.* v. (1897), 318.

WADSWORTH, 'Resolving Power,' *Astro. Jour.* vi. (1897), 27.

MICHELSON, 'Echelon Spectroscope,' *Astrophys. Jour.* viii. (1898), 37.

CAMPBELL, 'New Spectrograph,' *Astrophys. Jour.* viii. (1898), 123.

POOR, 'Concave Grating,' *M. N. Roy. Astron. Soc.* lviii. (1898), 291.

The books and papers on celestial spectroscopy are exhaustively cited in Dr ALFRED TUCKERMAN'S 'Index to the Literature of the Spectroscope,' *Smithsonian Miscellaneous Collections*, No. 658 (1888), pp. 8, 66.

GENERAL TREATISES, SPECIAL INSTRUMENTS, AND NEW METHODS

HANSEN, *Die Theorie des Aequatoreals* (Leipzig 1855).

SHADWELL, *Management of Chronometers* (London 1861).

KARL, *Principien der Astronomischen Instrumentenkunde* (Leipzig 1863).

NEWCOMB, 'Theory of Meridian Circle,' *Washington Observations* 1865, Appendix i.

KAISER, 'Double-image Micrometer,' *Leyden Obs.* iii. (1872).

DAVIS, C. H., 'Ratings of Chronometers,' *Washington Observations* 1875, App. iii.

GILL, 'Heliometer,' *Dun Echt Obs. Publications*, ii. (1877).

SEELIGER, *Theorie des Heliometers* (Leipzig 1877).

LOCKYER, *Stargazing : Past and Present* (London 1878).

GODFRAY, *A Treatise on Astronomy* (London 1880).

TODD, 'Orbit-sweeper,' *Proc. Am. Acad.* xv. (1880), 270.

HARKNESS, 'Flexure of Meridian Instruments,' *Wash. Obs.* 1882, App. iii.

CLARK, L., *Treatise on the Transit Instrument* (London 1882).

V. KONKOLY, *Praktische Anleitung zur Anstellung Astron. Beobachtungen* (Brunswick 1883).

HARRINGTON, 'Tools of the Astronomer,' *Sidereal Messenger*, ii. (1883–84).

BRASHEAR, 'Optical Surfaces,' *Proc. Am. Assoc. Adv. Sci.* xxxiii. (1884), 255.

LOEWY, 'Equatorial Coudé,' *Bulletin Astron.* i. (1884), 265.

RADAU, 'Heliostats,' *Bulletin Astron.* i. (1884), 153.

NEWCOMB, *Recent Improvements in Astron. Instruments* (Washington 1884).

HOUGH, 'Printing Chronograph,' *Sidereal Mess.* v. (1886), 161.

CHANDLER, 'Almucantar,' *Annals Harv. Coll. Obs.* xvii. (1887).

GRUBB, H., 'Electric Control,' *M. N. R. A. S.* xlviii. (1888), 352.

CHANDLER, 'Square Bar Micrometer,' *Mem. Am. Acad.* xi. (1888), 158.

LANGLEY, YOUNG, PICKERING, 'Wedge Photometer,' *Mem. Am. Acad.* xi. (1888), 301.

SANFORD, 'Personal Equation,' *Am. Jour. Psychology*, ii. (1889), 3, 271, 403. Bibliography.

MICHELSON, 'Interference Methods,' *Phil. Mag.* xxx.(1890), 1.

ROGERS, J. A., 'Correction of Sextants,' *Smithsonian Misc. Coll.* No. 764 (1890).

HAGEN, *The Photochronograph* (Georgetown Coll. Obs. 1891).

CLERKE, 'Instruments and Methods,' *Hist. of Astron. during XIX Century* (London 1893).

HUGGINS, M. L., 'Astrolabe,' *Pop. Astron.* ii. (1895), 199.

BECKER, E., HERZ, 'Micrometer, Personal Equation, Transit Instrument,' VALENTINER'S *Handwörterbuch der Astronomie* (Breslau 1895–98).

STONEY, 'Siderostat,' *M. N. R. Ast. Soc.* lvi. (1896), 456.

HAMY, 'Temperature Effect,' *Bull. Astr.* xiii. (1896), 178.

TURNER, 'Cœlostat,' *M. N. R. A. S.* lvi. (1896), 408.

MÜLLER, *Die Photometrie der Gestirne* (Leipzig 1897). Bibliography.

UPDEGRAFF, 'Flexure,' *Trans. Acad. Sci. St Louis* iii. (1897), No. 11.

TODD, *A New Astronomy*, Chapter ix. (New York 1897).

LIPPMANN, 'Cœlostat,' *Bulletin Astronomique*, xiv. (1897), 369.

The literature of astronomical instruments by the able writers of the *Encyclopædia Britannica* is summarized in BALDWIN's *Systematic Readings in E. B.*, p. 94 (Chicago and New York), 1897. Many novel devices of service to astronomer and astrophysicist are described and figured in the Berlin monthly begun in 1881, entitled *Zeitschrift für Instrumentenkunde.*

OBSERVATORIES

STRUVE, F. G. W., *Description de l'Observatoire de Poulkova* (St Petersburg 1845).

MAIN, *Encycl. Brit.* 8th ed. iii. (1853), 816.

LOOMIS, *Recent Progress of Astronomy, especially in U. S.* (New York 1856).

ANDRÉ, RAYET, ANGOT, *L'Astronomie Pratique et les Observatoires en Europe et en Amérique* (Paris 1874–78).

PERROTIN, *Visite à divers Observatoires d'Europe* (Paris 1881).

NEWCOMB, *No. Am. Rev.* cxxxiii. (1881), 196; *Nature*, xxvi. (1882), 326.

DREYER, 'Observatories,' *Encycl. Brit.*, 9 ed. xvii. (1884), 708.

PERROTIN, *Annales de l'Observatoire de Nice*, i. (Paris 1885).

WINTERHALTER, 'European Observatories,' *Washington Obs.* 1885, App. i.

TODD, 'Observatories in America,' *Nation*, xlvii. (1888), 25.

YOUNG, 'European Observatories,' *Scribner's Magazine*, iv. (1888), 82.

LOVE, 'First College Obs. in U. S.,' *Sid. Mess.* vii. (1888), 417.

CLERKE, 'Cape Observatory,' *Contemp. Review*, lv. (1889), 380.

LANCASTER, *Liste Générale des Observatoires*, etc. (Brussels 1890).

CHAMBERS, *Descriptive and Practical Astronomy* (Oxford 1890).

SWIFT, 'New Observatories,' APPLETON's *Annual Cyclopædias* (1890–98).

GILL, 'Work in Modern Observatory,' *Proc. Royal Inst.* xiii. (1891), 402.

UPTON, 'Ancient and Modern Observatories,' *The Sidereal Messenger*, x. (1891), 481.

PROCTOR, RANYARD, *Old and New Astronomy* (London and New York 1892).

ALGUÉ, ' Observatory at Manila,' *Astron. and Astro-Phys.* xiii. (1894), 85.

PAYNE, ' Free Public Observatories,' *Astron. and Astro-Phys.* xiii. (1894), 110.

LYNN, ' Observatories a Century Ago,' *Knowledge*, xviii. (1895), 86.

NEWCOMB, ' Observatories,' JOHNSON's *Univ. Cycl.* vi. (1895).

VALENTINER'S *Handwörterbuch der Astronomie* (Breslau 1895 –1899).

STONEY, WADSWORTH, ' Astrophysical Observatory,' *Astrophys. Jour.* iv. (1896), 238.

JANSSEN, ' *Annales de l'Observatoire d'Astronomie Physique,*' i. (Paris 1896).

JACOBY, ' Cape Observatory,' *Pop. Astron.* iii. (1896), 217.

STONEY, ' Future Astrophys. Observatory,' *M. N. Roy. Astron. Soc.* lvi. (1896), 452.

HALE, ' Yerkes Observatory,' *Astrophys. Jour.* v., vi. (1897).

HALE, ' Aim of the Yerkes Observatory,' *Astrophys. Jour.* vi. (1897), 310.

PAYNE, ' Yerkes Observatory,' *Pop. Astron.* v. (1897), 115.

NEWCOMB, ' Aspects of American Astronomy,' *Pop. Astron.* v. (1897), 351.

MEYERS *Konversations-Lexikon*, Bd. xvi., p. 418 (Leipzig and Vienna 1897).

WITT, ' Meudon Observatory,' *Himmel und Erde*, x. (1898), 529.

COLIN, ' End of Tananarivo Observatory,' *The Observatory*, xxi. (1898), 305.

MEYER, *Das Weltgebäude* (Leipzig and Vienna 1898).

For detailed descriptions of observatories, and instruments and the work done with them, consult the volumes of their annals, especially Pulkowa, Greenwich, Paris, Washington, Cambridge (U. S.), Nice, Mount Hamilton (Lick), Dunsink, and Leyden. For reports of observatories, refer to the *Monthly Notices Royal Astronomical Society* for February each year, and to the *Vierteljahrsschrift der Astronomischen Gesellschaft*, published at Leipzig. The publications of observatories of all countries are excellently summarized by Dr COPELAND, in the elaborate *Catalogue of the Crawford Library of the Royal Observatory*, pp. 325–45 (Edinburgh 1890).

CELESTIAL PHOTOGRAPHY

BOND, G. P., 'Stellar Photography,' *Astron. Nachrichten*, xlvii. (1858) ; xlviii. (1858) ; xlix. (1859).

DE LA RUE, 'Celestial Photography in England,' *Report Brit. Assoc. Adv. Sci.* 1859, p. 130.

DRAPER, 'Reflector for Photography,' *Smithson. Contr. Knowl.* xiv. (1865).

RUTHERFURD, 'Astr. Photography,' *Am. Jour. Sci.* lxxxix. (1865), 304.

NEWCOMB, 'Photography of Precision,' *Papers U. S. Transit of Venus Commission*, i. (Washington 1872).

HARKNESS, 'Photoheliograph,' *Memoirs Roy. Astr. Soc.* xliii. (1877), 129.

GOULD, 'Celestial Photography,' *The Observatory*, ii. (1879), 13.

PRITCHARD, 'Photography of Precision,' *Mem. Roy. Astr. Soc.* xlvii. (1883), 1.

COMMON, 'Telescopes for Astr. Photog.,' *Nature*, xxxi. (1884–5), 38, 270.

COMMON, 'Photography as Aid to Astronomy,' *Proc. Roy. Institution*, xi. (1885), 367.

WINTERHALTER, 'Astrophotographic Congress,' *Washington Obs.* (1885, App. i.).

HENRY, 'Astr. Photography,' *Nature*, xxxiv. (1886), 35.

STRUVE, O., 'Die Photographie im Dienste der Astronomie,' *Bull. Acad. Sci. St Pétersbourg*, xxx. (1886), N^r 4.

RAYET, 'History,' *Bulletin Astronomique*, iv. (1887), 165.

COMMON, 'Astron. Photography,' *Nineteenth Century*, xxi. (1887), 227.

MOUCHEZ, 'Astrographic Survey,' *Annuaire Bur. Long.* 1887.

MOUCHEZ, *La Photographie Astronomique* (Paris 1887).

GILL, 'Applications of Photography in Astronomy,' *Proc. Royal Institution*, xii. (1887), 158.

v. KONKOLY, *Praktische Anleitung zur Himmelsphotographie* (Halle 1887).

————, 'Astr. Photography,' *Sid. Mess.* vii. (1888), 138, 181 ; *Edinburgh Review*, clxvii. (1888), 23.

PICKERING, E. C., 'Stellar Photography,' *Mem. Am. Acad.* xi. (1888), 179.

CHAMBERS, 'Astron. Photography,' *Descriptive Astronomy*, ii. (Oxford 1890).

SCHOOLING, *Westminster Review*, cxxxvi. (1891), 303.

RUSSELL, *The Star Camera* (Sydney 1891).

394   *Stars and Telescopes*

BROTHERS, *Photography: its History, Processes, Apparatus, and Materials* (London 1892).

FRAISSINET, 'Paris Astrographic Observatory,' *La Nature*, 1893.

TAYLOR, 'Photographic Objectives,' *M. N. Roy. Astron. Soc.* liii. (1893), 359.

TURNER, LEWIS, 'Astrographic Chart,' *The Observatory*, xvi. (1893), 407; *Astron. and Astro-phys.* xiii. (1894), 20.

ROBERTS, 'Progress in Photography,' *Proc. Am. Phil. Soc.* xxxii. (1894), 97.

RUSSELL, 'Progress,' *Pop. Astron.* ii. (1894–95).

PICKERING, W. H., 'Astr. Photog.,' *Ann. Harv. Coll. Obs.* xxxii. (1895).

BARNARD, 'Astronomical Photography,' *The Photographic Times*, xxvii. (1895), 65.

ROBERTS, 'Reflector and Portrait Lens,' *Jour. Brit. Astr. Assoc.* vi. (1896), 332.

GILL, KAPTEYN, 'Stellar Photography,' *Annals Cape Observatory*, iii. (London 1896).

SCHAEBERLE, 'Planetary Photography,' *Pop. Astron.* iii. (1896), 280.

TURNER, 'Photo. Transit Circle,' *M. N. Roy. Astron. Soc.* lvii. (1897), 349.

WADSWORTH, 'Planetary Photography,' *The Observatory*, xx. (1897).

PICKERING, E. C., 'Bruce Photographic Telescope,' *Pop. Astron.* iv. (1897), 531.

DESLANDRES, *Spécimens de Photographies Astronomiques* (Paris 1897).

SCHEINER, *Die Photographie der Gestirne* (Leipzig 1897). Bibliography.

LOEWY, PUISEUX, 'Lunar Photography,' *Annuaire Bur. Long.* 1898, A.

SCHWEIGER-LERCHENFELD, *Atlas der Himmelskunde* (Vienna 1898).

BARNARD, 'Development of Photography,' *Proc. Am. Assoc. Adv. Science*, xlvii. (Salem 1898); *Pop. Ast.* vi. (1898), 425.

In HOUZEAU'S *Vade-Mecum de l'Astronome*, p. 339 (Brussels 1882), the earlier scientific papers on astronomical photography are catalogued. All the preceding bibliographies need to be supplemented by the classified lists of *The Astrophysical Journal*, i.–ix. (1895–99).

MOUNTAIN OBSERVATORIES

SMYTH, C. P., 'Teneriffe,' *Phil. Trans. Roy. Soc.* cxlviii. (1858), 465.

SMYTH, C. P., *Teneriffe, an Astronomer's Experiment* (London 1858).

LANGLEY, 'Aetna,' *Am. Jour. Science*, xvii. (1879), 259; xx. (1880), 33.

LANGLEY, 'Mount Whitney Expedition,' *Nature*, xxvi. (1882), 314.

PICKERING, 'Mountain Observatories,' *Sid. Mess.* ii. (1883), 105.

COPELAND, 'Experiments in the Andes,' *Copernicus*, iii. (1884), 193.

————, 'Mountain Observatories,' *Edinburgh Review*, clx. (1884), 351 ; *Pop. Sci. Monthly*, xxvi. (1885), 388.

TODD, 'Lick Observatory,' *Science*, vi. (1885), 181, 186.

HOLDEN, *Handbook of the Lick Observatory* (San Francisco 1888).

RANYARD, *Knowledge*, xii. (1889), 125.

JANSSEN, 'Mt. Blanc Observatory,' *Annuaire Bur. Long.* (1894–1899). Also *Rept. Smithsonian Institution* 1894, p. 237.

HOLDEN, 'Mountain Observatories,' *Smithsonian Misc. Collections*, 1035 (1896). Bibliography.

DOUGLASS, 'Atmosphere, Telescope, and Observer,' *Pop. Ast.* v. (1897), 64.

SEE, 'Air Waves and Currents,' *Popular Astron.* v. (1898), 479.

SEE, 'Mountain Observatories,' *Popular Astron.* vi. (1898), 65.

DOUGLASS, 'Scales of Seeing,' *Popular Astron.* vi. (1898), 193.

Popular literature of the telescope, celestial photography, and the spectroscope has experienced an enormous growth, particularly in recent years. A nearly complete bibliography of titles may be found on reference to POOLE's *Indexes*, at the pages given below ; supplemented, of course, by the annual volumes for 1897, 1898 : —

Volume of Index	Observatories	Photography	Spectroscopes	Telescopes
I. (1800–81)	p. 936	p. 1003	p. 1233	p. 1291
II. (1882–86)	319	340	413	433
III. (1887–91)	309	330	403	423
IV. (1892–96)	410	440	540	567

PRACTICAL ASTRONOMY AND MAKING OBSERVATIONS

PEARSON, *Introduction to Practical Astronomy* (London 1824–29). With Appendix and Plates.

MASON, *Introduction to Practical Astronomy* (New York 1841).

LOOMIS, *Introduction to Practical Astronomy* (New York 1855).

CHAUVENET, *Manual of Spherical and Practical Astronomy* (Philadelphia 1863).

PROCTOR, *Half-hours with the Telescope* (New York 1873).

SAWITSCH, PETERS, *Abriss der Praktischen Astronomie* (Leipzig 1879).

SMYTH, CHAMBERS, *A Cycle of Celestial Objects* (Oxford 1881).

SOUCHON, *Traité d'Astronomie Pratique* (Paris 1883).

DOOLITTLE, *Treatise on Practical Astronomy as applied to Geodesy and Navigation* (New York 1885).

NOBLE, *Hours with a Three-inch Telescope* (New York 1887).

SERVISS, *Astronomy with an Opera Glass* (New York 1888).

OLIVER, *Astronomy for Amateurs* (London and New York 1888).

DENNING, *Telescopic Work for Starlight Evenings* (London 1891).

CAMPBELL, *Handbook of Practical Astronomy* (Ann Arbor 1891).

GREENE, *Introduction to Spherical and Practical Astronomy* (Boston 1891).

COLAS, *Celestial Handbook* (Chicago 1892).

WEBB, ESPIN, *Celestial Objects for Common Telescopes* (London 1893–94).

GIBSON, *The Amateur Telescopist's Handbook* (New York 1894).

COMSTOCK, 'Practical Astronomy,' *Bull. Univ. Wisconsin*, i. (1895), No. 3.

FOWLER, *Popular Telescopic Astronomy* (London 1896).

MEE, *Observational Astronomy* (Cardiff and London 1897).

BRENNER, *Handbuch für Amateur Astronomen* (Leipzig 1898).

Supplementary to the foregoing handbooks are the annual 'Companion' to *The Observatory;* also the monthly notes in *Popular Astronomy*, and in the *Publications of the Astronomical Society of the Pacific*, together with ephemerides in the *Annuaire du Bureau des Longitudes.*

THE SMALL PLANETS BETWEEN MARS AND JUPITER

NUM-BER	NAME	DATE OF DISCOVERY	DISCOVERER AND HIS NUMBER	PLACE OF DISCOVERY
1	Ceres	1 Jan. 1801	Piazzi	Palermo
2	Pallas	28 Mar. 1802	Olbers $_1$	Bremen
3	Juno	1 Sept. 1804	Harding	Lilienthal
4	Vesta	29 Mar. 1807	Olbers $_2$	Bremen
5	Astræa	8 Dec. 1845	Hencke $_1$	Driessen
6	Hebe	1 July 1847	Hencke $_2$	Driessen
7	Iris	13 Aug. 1847	Hind $_1$	London
8	Flora	18 Oct. 1847	Hind $_2$	London
9	Metis	25 Apr. 1848	Graham	Markree
10	Hygeia	12 Apr. 1849	De Gasparis $_1$	Naples
11	Parthenope	11 May 1850	De Gasparis $_2$	Naples
12	Victoria	13 Sept. 1850	Hind $_3$	London
13	Egeria	2 Nov. 1850	De Gasparis $_3$	Naples
14	Irene	19 May 1851	Hind $_4$	London
15	Eunomia	29 July 1851	De Gasparis $_4$	Naples
16	Psyche	17 Mar. 1852	De Gasparis $_5$	Naples
17	Thetis	17 Apr. 1852	Luther $_1$	Bilk
18	Melpomene	24 June 1852	Hind $_5$	London
19	Fortuna	22 Aug. 1852	Hind $_6$	London
20	Massalia	19 Sept. 1852	De Gasparis $_6$	Naples
21	Lutetia	15 Nov. 1852	Goldschmidt $_1$	Paris
22	Calliope	16 Nov. 1852	Hind $_7$	London
23	Thalia	15 Dec. 1852	Hind $_8$	London
24	Themis	5 Apr. 1853	De Gasparis $_7$	Naples
25	Phocæa	6 Apr. 1853	Chacornac $_1$	Marseilles
26	Proserpine	5 May 1853	Luther $_2$	Bilk
27	Euterpe	8 Nov. 1853	Hind $_9$	London
28	Bellona	1 Mar. 1854	Luther $_3$	Bilk
29	Amphitrite	1 Mar. 1854	Marth	London
30	Urania	22 July 1854	Hind $_{10}$	London
31	Euphrosyne	1 Sept. 1854	Ferguson $_1$	Washington
32	Pomona	26 Oct. 1854	Goldschmidt $_2$	Paris
33	Polyhymnia	28 Oct. 1854	Chacornac $_2$	Paris
34	Circe	6 Apr. 1855	Chacornac $_3$	Paris
35	Leucothea	19 Apr. 1855	Luther $_4$	Bilk

NUM-BER	NAME	DATE OF DISCOVERY	DISCOVERER AND HIS NUMBER	PLACE OF DISCOVERY
36	Atalanta	5 Oct. 1855	Goldschmidt $_3$	Paris
37	Fides	5 Oct. 1855	Luther $_5$	Bilk
38	Leda	12 Jan. 1856	Chacornac $_4$	Paris
39	Lætitia	8 Feb. 1856	Chacornac $_5$	Paris
40	Harmonia	31 Mar. 1856	Goldschmidt $_4$	Paris
41	Daphne	22 May 1856	Goldschmidt $_5$	Paris
42	Isis	23 May 1856	Pogson $_1$	Oxford
43	Ariadne	15 Apr. 1857	Pogson $_2$	Oxford
44	Nysa	27 May 1857	Goldschmidt $_6$	Paris
45	Eugenia	27 June 1857	Goldschmidt $_7$	Paris
46	Hestia	16 Aug. 1857	Pogson $_3$	Oxford
47	Aglaia	15 Sept. 1857	Luther $_6$	Bilk
48	Doris	19 Sept. 1857	Goldschmidt $_8$	Paris
49	Pales	19 Sept. 1857	Goldschmidt $_9$	Paris
50	Virginia	4 Oct. 1857	Ferguson $_2$	Washington
51	Nemausa	22 Jan. 1858	Laurent	Nismes
52	Europa	4 Feb. 1858	Goldschmidt $_{10}$	Paris
53	Calypso	4 Apr. 1858	Luther $_7$	Bilk
54	Alexandra	10 Sept. 1858	Goldschmidt $_{11}$	Paris
55	Pandora	10 Sept. 1858	Searle	Albany
56	Melete	9 Sept. 1857	Goldschmidt $_{12}$	Paris
57	Mnemosyne	22 Sept. 1859	Luther $_8$	Bilk
58	Concordia	24 Mar. 1860	Luther $_9$	Bilk
59	Olympia	12 Sept. 1860	Chacornac $_6$	Paris
60	Echo	15 Sept. 1860	Ferguson $_3$	Washington
61	Danaë	9 Sept. 1860	Goldschmidt $_{13}$	Châtillon
62	Erato	14 Sept. 1860	Foerster & Lesser	Berlin
63	Ausonia	10 Feb. 1861	De Gasparis $_8$	Naples
64	Angelina	4 Mar. 1861	Tempel $_1$	Marseilles
65	Maximiliana	8 Mar. 1861	Tempel $_2$	Marseilles
66	Maia	9 Apr. 1861	Tuttle $_1$	Cambridge, U. S.
67	Asia	17 Apr. 1861	Pogson $_4$	Madras
68	Leto	29 Apr. 1861	Luther $_{10}$	Bilk
69	Hesperia	29 Apr. 1861	Schiaparelli	Milan
70	Panopæa	5 May 1861	Goldschmidt $_{14}$	Paris
71	Niobe	13 Aug. 1861	Luther $_{11}$	Bilk
72	Feronia	29 May 1861	Peters $_1$ & Safford	Clinton
73	Clytia	7 Apr. 1862	Tuttle $_2$	Cambridge, U. S.
74	Galatea	29 Aug. 1862	Tempel $_3$	Marseilles
75	Eurydice	22 Sept. 1862	Peters $_2$	Clinton

NUM-BER	NAME	DATE OF DISCOVERY	DISCOVERER AND HIS NUMBER	PLACE OF DISCOVERY
76	Freia	21 Oct. 1862	D'Arrest	Copenhagen
77	Frigga	12 Nov. 1862	Peters $_8$	Clinton
78	Diana	15 Mar. 1863	Luther $_{12}$	Bilk
79	Eurynome	14 Sept 1863	Watson $_1$	Ann Arbor
80	Sappho	2 May 1864	Pogson $_5$	Madras
81	Terpsichore	30 Sept. 1864	Tempel $_4$	Marseilles
82	Alcmene	27 Nov. 1864	Luther $_{13}$	Bilk
83	Beatrix	26 Apr. 1865	De Gasparis $_9$	Naples
84	Clio	25 Aug. 1865	Luther $_{14}$	Bilk
85	Io	19 Sept. 1865	Peters $_4$	Clinton
86	Semele	4 Jan. 1866	Tietjen	Berlin
87	Sylvia	16 May 1866	Pogson $_6$	Madras
88	Thisbe	15 June 1866	Peters $_5$	Clinton
89	Julia	6 Aug. 1866	Stephan	Marseilles
90	Antiope	1 Oct. 1866	Luther $_{15}$	Bilk
91	Ægina	4 Nov. 1866	Borrelly $_1$	Marseilles
92	Undina	7 July 1867	Peters $_6$	Clinton
93	Minerva	24 Aug. 1867	Watson $_2$	Ann Arbor
94	Aurora	6 Sept. 1867	Watson $_3$	Ann Arbor
95	Arethusa	23 Nov. 1867	Luther $_{16}$	Bilk
96	Ægle	17 Feb. 1868	Coggia $_1$	Marseilles
97	Clotho	17 Feb. 1868	Tempel $_5$	Marseilles
98	Ianthe	18 Apr. 1868	Peters $_7$	Clinton
99	Dike	28 May 1868	Borrelly $_2$	Marseilles
100	Hecate	11 July 1868	Watson $_4$	Ann Arbor
101	Helena	15 Aug. 1868	Watson $_5$	Ann Arbor
102	Miriam	22 Aug. 1868	Peters $_8$	Clinton
103	Hera	7 Sept. 1868	Watson $_6$	Ann Arbor
104	Clymene	13 Sept. 1868	Watson $_7$	Ann Arbor
105	Artemis	16 Sept. 1868	Watson $_8$	Ann Arbor
106	Dione	10 Oct. 1868	Watson $_9$	Ann Arbor
107	Camilla	17 Nov. 1868	Pogson $_7$	Madras
108	Hecuba	2 Apr. 1869	Luther $_{17}$	Bilk
109	Felicitas	9 Oct. 1869	Peters $_9$	Clinton
110	Lydia	19 Apr. 1870	Borrelly $_3$	Marseilles
111	Ate	14 Aug. 1870	Peters $_{10}$	Clinton
112	Iphigenia	19 Sept. 1870	Peters $_{11}$	Clinton
113	Amalthea	12 Mar. 1871	Luther $_{18}$	Bilk
114	Cassandra	23 July 1871	Peters $_{12}$	Clinton
115	Thyra	6 Aug. 1871	Watson $_{10}$	Ann Arbor

NUM-BER	NAME	DATE OF DISCOVERY	DISCOVERER AND HIS NUMBER	PLACE OF DISCOVERY
116	Sirona	8 Sept. 1871	Peters $_{13}$	Clinton
117	Lomia	12 Sept. 1871	Borrelly $_4$	Marseilles
118	Peitho	15 Mar. 1872	Luther $_{19}$	Bilk
119	Althæa	3 Apr. 1872	Watson $_{11}$	Ann Arbor
120	Lachesis	10 Apr. 1872	Borrelly $_5$	Marseilles
121	Hermione	12 May 1872	Watson $_{12}$	Ann Arbor
122	Gerda	31 July 1872	Peters $_{14}$	Clinton
123	Brunhilda	31 July 1872	Peters $_{15}$	Clinton
124	Alceste	23 Aug. 1872	Peters $_{16}$	Clinton
125	Liberatrix	11 Sept. 1872	Pros. Henry $_1$	Paris
126	Velleda	5 Nov. 1872	Paul Henry $_1$	Paris
127	Johanna	5 Nov. 1872	Pros. Henry $_2$	Paris
128	Nemesis	25 Nov. 1872	Watson $_{13}$	Ann Arbor
129	Antigone	5 Feb. 1873	Peters $_{17}$	Clinton
130	Electra	17 Feb. 1873	Peters $_{18}$	Clinton
131	Vala	24 May 1873	Peters $_{19}$	Clinton
132	Æthra	13 June 1873	Watson $_{14}$	Ann Arbor
133	Cyrene	16 Aug. 1873	Watson $_{15}$	Ann Arbor
134	Sophrosyne	27 Sept. 1873	Luther $_{20}$	Bilk
135	Hertha	18 Feb. 1874	Peters $_{20}$	Clinton
136	Austria	18 Mar. 1874	Palisa $_1$	Pola
137	Melibœa	21 Apr. 1874	Palisa $_2$	Pola
138	Tolosa	19 May 1874	Perrotin $_1$	Toulouse
139	Juewa	10 Oct. 1874	Watson $_{16}$	Pekin
140	Siwa	13 Oct. 1874	Palisa $_3$	Pola
141	Lumen	13 Jan. 1875	Paul Henry $_2$	Paris
142	Polana	28 Jan. 1875	Palisa $_4$	Pola
143	Adria	23 Feb. 1875	Palisa $_5$	Pola
144	Vibilia	3 June 1875	Peters $_{21}$	Clinton
145	Adeona	3 June 1875	Peters $_{22}$	Clinton
146	Lucina	8 June 1875	Borrelly $_6$	Marseilles
147	Protogeneia	10 July 1875	Schulhof	Vienna
148	Gallia	7 Aug. 1875	Pros. Henry $_3$	Paris
149	Medusa	21 Sept. 1875	Perrotin $_2$	Toulouse
150	Nuwa	18 Oct. 1875	Watson $_{17}$	Ann Arbor
151	Abundantia	1 Nov. 1875	Palisa $_6$	Pola
152	Atala	2 Nov. 1875	Paul Henry $_3$	Paris
153	Hilda	2 Nov. 1875	Palisa $_7$	Pola
154	Bertha	4 Nov. 1875	Pros. Henry $_4$	Paris
155	Scylla	8 Nov. 1875	Palisa $_8$	Pola

NUM-BER	NAME	DATE OF DISCOVERY	DISCOVERER AND HIS NUMBER	PLACE OF DISCOVERY
156	Xantippe	22 Nov. 1875	Palisa $_9$	Pola
157	Dejanira	1 Dec. 1875	Borrelly $_7$	Marseilles
158	Coronis	4 Jan. 1876	Knorre $_1$	Berlin
159	Æmilia	26 Jan. 1876	Paul Henry $_4$	Paris
160	Una	20 Feb. 1876	Peters $_{23}$	Clinton
161	Athor	19 Apr. 1876	Watson $_{18}$	Ann Arbor.
162	Laurentia	21 Apr. 1876	Pros. Henry $_5$	Paris
163	Erigone	26 Apr. 1876	Perrotin $_3$	Toulouse
164	Eva	12 July 1876	Paul Henry $_5$	Paris
165	Loreley	9 Aug. 1876	Peters $_{24}$	Clinton
166	Rhodope	15 Aug. 1876	Peters $_{25}$	Clinton
167	Urda	28 Aug. 1876	Peters $_{26}$	Clinton
168	Sibylla	27 Sept. 1876	Watson $_{19}$	Ann Arbor
169	Zelia	28 Sept. 1876	Pros. Henry $_6$	Paris
170	Maria	10 Jan. 1877	Perrotin $_4$	Toulouse
171	Ophelia	13 Jan. 1877	Borrelly $_8$	Marseilles
172	Baucis	5 Feb. 1877	Borrelly $_9$	Marseilles
173	Ino	1 Aug. 1877	Borrelly $_{10}$	Marseilles
174	Phædra	2 Sept. 1877	Watson $_{20}$	Ann Arbor
175	Andromache	1 Oct. 1877	Watson $_{21}$	Ann Arbor
176	Idunna	14 Oct. 1877	Peters $_{27}$	Clinton
177	Irma	5 Nov. 1877	Paul Henry $_6$	Paris
178	Belisana	6 Nov. 1877	Palisa $_{10}$	Pola
179	Clytemnestra	11 Nov. 1877	Watson $_{22}$	Ann Arbor
180	Garumna	29 Jan. 1878	Perrotin $_5$	Toulouse
181	Eucharis	2 Feb. 1878	Cottenot	Marseilles
182	Elsa	7 Feb. 1878	Palisa $_{11}$	Pola
183	Istria	8 Feb. 1878	Palisa $_{12}$	Pola
184	Deiopeia	28 Feb. 1878	Palisa $_{13}$	Pola
185	Eunice	1 Mar. 1878	Peters $_{28}$	Clinton
186	Celuta	6 Apr. 1878	Pros. Henry $_7$	Paris
187	Lamberta	11 Apr. 1878	Coggia $_2$	Marseilles
188	Menippe	18 June 1878	Peters $_{29}$	Clinton
189	Phthia	9 Sept. 1878	Peters $_{30}$	Clinton
190	Ismene	22 Sept. 1878	Peters $_{31}$	Clinton
191	Kolga	30 Sept. 1878	Peters $_{32}$	Clinton
192	Nausicaä	17 Feb. 1879	Palisa $_{14}$	Pola
193	Ambrosia	28 Feb. 1879	Coggia $_3$	Marseilles
194	Procne	21 Mar. 1879	Peters $_{33}$	Clinton
195	Eurykleia	22 Apr. 1879	Palisa $_{15}$	Pola

NUM-BER	NAME	DATE OF DISCOVERY	DISCOVERER AND HIS NUMBER	PLACE OF DISCOVERY
196	Philomela	14 May 1879	Peters $_{34}$	Clinton
197	Arete	21 May 1879	Palisa $_{16}$	Pola
198	Ampella	13 June 1879	Borrelly $_{11}$	Marseilles
199	Byblis	9 July 1879	Peters $_{35}$	Clinton
200	Dynamene	27 July 1879	Peters $_{36}$	Clinton
201	Penelope	7 Aug. 1879	Palisa $_{17}$	Pola
202	Chryseis	11 Sept. 1879	Peters $_{37}$	Clinton
203	Pompeia	25 Sept. 1879	Peters $_{38}$	Clinton
204	Callisto	8 Oct. 1879	Palisa $_{18}$	Pola
205	Martha	13 Oct. 1879	Palisa $_{19}$	Pola
206	Hersilia	13 Oct. 1879	Peters $_{39}$	Clinton
207	Hedda	17 Oct. 1879	Palisa $_{20}$	Pola
208	Lachrymosa	21 Oct. 1879	Palisa $_{21}$	Pola
209	Dido	22 Oct. 1879	Peters $_{40}$	Clinton
210	Isabella	12 Nov. 1879	Palisa $_{22}$	Pola
211	Isolda	10 Dec. 1879	Palisa $_{23}$	Pola
212	Medea	6 Feb. 1880	Palisa $_{24}$	Pola
213	Lilæa	16 Feb. 1880	Peters $_{41}$	Clinton
214	Aschera	26 Feb. 1880	Palisa $_{25}$	Pola
215	Œnone	7 Apr. 1880	Knorre $_{2}$	Berlin
216	Cleopatra	10 Apr. 1880	Palisa $_{26}$	Pola
217	Eudora	30 Aug. 1880	Coggia $_{4}$	Marseilles
218	Bianca	4 Sept. 1880	Palisa $_{27}$	Pola
219	Thusnelda	30 Sept. 1880	Palisa $_{28}$	Pola
220	Stephania	19 May 1881	Palisa $_{29}$	Vienna
221	Eos	18 Jan. 1882	Palisa $_{30}$	Vienna
222	Lucia	9 Feb. 1882	Palisa $_{31}$	Vienna
223	Rosa	9 Mar. 1882	Palisa $_{32}$	Vienna
224	Oceana	30 Mar. 1882	Palisa $_{33}$	Vienna
225	Henrietta	19 Apr. 1882	Palisa $_{34}$	Vienna
226	Weringia	19 July 1882	Palisa $_{35}$	Vienna
227	Philosophia	12 Aug. 1882	Paul Henry $_{7}$	Paris
228	Agathe	19 Aug. 1882	Palisa $_{36}$	Vienna
229	Adelinda	22 Aug. 1882	Palisa $_{37}$	Vienna
230	Athamantis	3 Sept. 1882	De Ball	Bothkamp
231	Vindobona	10 Sept. 1882	Palisa $_{38}$	Vienna
232	Russia	31 Jan. 1883	Palisa $_{39}$	Vienna
233	Asterope	11 May 1883	Borrelly $_{12}$	Marseilles
234	Barbara	12 Aug. 1883	Peters $_{42}$	Clinton
235	Carolina	28 Nov. 1883	Palisa $_{40}$	Vienna

NUM- BER	NAME	DATE OF DISCOVERY	DISCOVERER AND HIS NUMBER	PLACE OF DISCOVERY
236	Honoria	26 Apr. 1884	Palisa 41	Vienna
237	Cœlestina	27 June 1884	Palisa 42	Vienna
238	Hypatia	1 July 1884	Knorre 3	Berlin
239	Adrastea	18 Aug. 1884	Palisa 43	Vienna
240	Vanadis	27 Aug. 1884	Borrelly 18	Marseilles
241	Germania	12 Sept. 1884	Luther 21	Düsseldorf
242	Kriemhild	22 Sept. 1884	Palisa 44	Vienna
243	Ida	29 Sept. 1884	Palisa 45	Vienna
244	Sita	14 Oct. 1884	Palisa 46	Vienna
245	Vera	6 Feb. 1885	Pogson 8	Madras
246	Asporina	6 Mar. 1885	Borrelly 14	Marseilles
247	Eukrate	14 Mar. 1885	Luther 22	Düsseldorf
248	Lameia	5 June 1885	Palisa 47	Vienna
249	Ilse	16 Aug. 1885	Peters 43	Clinton
250	Bettina	3 Sept. 1885	Palisa 48	Vienna
251	Sophia	4 Oct. 1885	Palisa 49	Vienna
252	Clementina	11 Oct. 1885	Perrotin 6	Nice
253	Matilda	12 Nov. 1885	Palisa 50	Vienna
254	Augusta	31 Mar. 1886	Palisa 51	Vienna
255	Oppavia	31 Mar. 1886	Palisa 52	Vienna
256	Walpurga	3 Apr. 1886	Palisa 53	Vienna
257	Silesia	5 Apr. 1886	Palisa 54	Vienna
258	Tyche	4 May 1886	Luther 23	Düsseldorf
259	Altheia	28 June 1886	Peters 44	Clinton
260	Huberta	3 Oct. 1886	Palisa 55	Vienna
261	Prymno	31 Oct. 1886	Peters 45	Clinton
262	Valda	3 Nov. 1886	Palisa 56	Vienna
263	Dresda	3 Nov. 1886	Palisa 57	Vienna
264	Libussa	17 Dec. 1886	Peters 46	Clinton
265	Anna	25 Feb. 1887	Palisa 58	Vienna
266	Aline	17 May 1887	Palisa 59	Vienna
267	Tirza	27 May 1887	Charlois 1	Nice
268	Adorea	9 June 1887	Borrelly 15	Marseilles
269	Justitia	21 Sept. 1887	Palisa 60	Vienna
270	Anahita	8 Oct. 1887	Peters 47	Clinton
271	Penthesilea	13 Oct. 1887	Knorre 4	Berlin
272	Antonia	4 Feb. 1888	Charlois 2	Nice
273	Atropos	8 Mar. 1888	Palisa 61	Vienna
274	Philagoria	3 Apr. 1888	Palisa 62	Vienna
275	Sapientia	15 Apr. 1888	Palisa 63	Vienna

NUM-BER	NAME	DATE OF DISCOVERY	DISCOVERER AND HIS NUMBER	PLACE OF DISCOVERY
276	Adelheid	17 Apr. 1888	Palisa $_{64}$	Vienna
277	Elvira	3 May 1888	Charlois $_3$	Nice
278	Paulina	16 May 1888	Palisa $_{65}$	Vienna
279	Thule	25 Oct. 1888	Palisa $_{66}$	Vienna
280	Philia	29 Oct. 1888	Palisa $_{67}$	Vienna
281	Lucretia	31 Oct. 1888	Palisa $_{68}$	Vienna
282	Clorinda	28 Jan. 1889	Charlois $_4$	Nice
283	Emma	8 Feb. 1889	Charlois $_5$	Nice
284	Amelia	29 May 1889	Charlois $_6$	Nice
285	Regina	3 Aug. 1889	Charlois $_7$	Nice
286	Iclea	3 Aug. 1889	Palisa $_{69}$	Vienna
287	Nephthys	25 Aug. 1889	Peters $_{48}$	Clinton
288	Glauca	20 Feb. 1890	Luther $_{24}$	Düsseldorf
289	Nenetta	10 Mar. 1890	Charlois $_8$	Nice
290	Bruna	20 Mar. 1890	Palisa $_{70}$	Vienna
291	Alice	25 Apr. 1890	Palisa $_{71}$	Vienna
292	Ludovica	25 Apr. 1890	Palisa $_{72}$	Vienna
293	Brasilia	20 May 1890	Charlois $_9$	Nice
294	Felicia	15 July 1890	Charlois $_{10}$	Nice
295	Theresia	17 Aug. 1890	Palisa $_{73}$	Vienna
296	Phaëthusa	19 Aug. 1890	Charlois $_{11}$	Nice
297	Cecilia	9 Sept. 1890	Charlois $_{12}$	Nice
298	Baptistina	9 Sept. 1890	Charlois $_{13}$	Nice
299	Thora	6 Oct. 1890	Palisa $_{74}$	Vienna
300	Geraldina	3 Oct. 1890	Charlois $_{14}$	Nice
301	Bavaria	16 Nov. 1890	Palisa $_{75}$	Vienna
302	Clarissa	14 Nov. 1890	Charlois $_{15}$	Nice
303	Josephina	12 Feb. 1891	Millosevich $_1$	Rome
304	Olga	14 Feb. 1891	Palisa $_{76}$	Vienna
305	Gordonia	16 Feb. 1891	Charlois $_{16}$	Nice
306	Unitas	1 Mar. 1891	Millosevich $_2$	Rome
307	Nike	5 Mar. 1891	Charlois $_{17}$	Nice
308	Polyxo	31 Mar. 1891	Borrelly $_{16}$	Marseilles
309	Fraternitas	6 Apr. 1891	Palisa $_{77}$	Vienna
310	Margarita	16 May 1891	Charlois $_{18}$	Nice
311	Claudia	11 June 1891	Charlois $_{19}$	Nice
312	Pierretta	28 Aug. 1891	Charlois $_{20}$	Nice
313	Chaldæa	30 Aug. 1891	Palisa $_{78}$	Vienna
314	Rosalia	1 Sept. 1891	Charlois $_{21}$	Nice
315	Constantia	4 Sept. 1891	Palisa $_{79}$	Vienna

NUM-BER	NAME	DATE OF DISCOVERY	DISCOVERER AND HIS NUMBER	PLACE OF DISCOVERY
316	Goberta	8 Sept. 1891	Charlois $_{22}$	Nice
317	Roxana	11 Sept. 1891	Charlois $_{23}$	Nice
318	Magdalena	24 Sept. 1891	Charlois $_{24}$	Nice
319	Leona	8 Oct. 1891	Charlois $_{25}$	Nice
320	Katharina	11 Oct. 1891	Palisa $_{80}$	Vienna
321	Florentina	15 Oct. 1891	Palisa $_{81}$	Vienna
322	Phaeo	27 Nov. 1891	Borrelly $_{17}$	Marseilles
323	Brucia	20 Dec. 1891	Wolf$_1$	Heidelberg
324[1]	Bamberga	25 Feb. 1892	Palisa $_{82}$	Vienna
325	Heidelberga	4 Mar. 1892	Wolf $_2$	Heidelberg
326	Tamara	19 Mar. 1892	Palisa $_{83}$	Vienna
327	Columbia	22 Mar. 1892	Charlois $_{26}$	Nice
328	Gudrun	18 Mar. 1892	Wolf $_3$	Heidelberg
329	Svea	21 Mar. 1892	Wolf $_4$	Heidelberg
330	Adalberta	18 Mar. 1892	Wolf $_5$	Heidelberg
331	Etheridgea	1 Apr. 1892	Charlois $_{27}$	Nice
332	Siri	19 Mar. 1892	Wolf $_6$	Heidelberg
333	Badenia	22 Aug. 1892	Wolf $_7$	Heidelberg
334	Chicago	23 Aug. 1892	Wolf $_8$	Heidelberg
335	Roberta	1 Sept. 1892	Staus $_1$	Heidelberg
336	Lacadiera	19 Sept. 1892	Charlois $_{28}$	Nice
337	Devosa	22 Sept. 1892	Charlois $_{29}$	Nice
338	Boudrosa	25 Sept. 1892	Charlois $_{30}$	Nice
339	Dorothea	25 Sept. 1892	Wolf $_9$	Heidelberg
340	Eduarda	25 Sept. 1892	Wolf $_{10}$	Heidelberg
341	California	25 Sept. 1892	Wolf $_{11}$	Heidelberg
342	Endymion	17 Oct. 1892	Wolf $_{12}$	Heidelberg
343	Ostara	15 Nov. 1892	Wolf $_{13}$	Heidelberg
344	Desiderata	15 Nov. 1892	Charlois $_{81}$	Nice
345	Tercidina	23 Nov. 1892	Charlois $_{82}$	Nice
346	Hermentaria	25 Nov. 1892	Charlois $_{83}$	Nice
347	Pariana	28 Nov. 1892	Charlois $_{84}$	Nice
348	May	28 Nov. 1892	Charlois $_{85}$	Nice
349	Dembowska	9 Dec. 1892	Charlois $_{86}$	Nice
350	Ornamenta	14 Dec. 1892	Charlois $_{87}$	Nice
351	Yrsa	16 Dec. 1892	Wolf $_{14}$	Heidelberg
352	Gisela	12 Jan. 1893	Wolf $_{15}$	Heidelberg
353	1893 *F*	16 Jan. 1893	Wolf $_{16}$	Heidelberg
354	Eleonora	17 Jan. 1893	Charlois $_{88}$	Nice
355	1893 *E*	20 Jan. 1893	Charlois $_{89}$	Nice

[1] Nearly all the discoveries following 324 have been made photographically.

NUM-BER	NAME	DATE OF DISCOVERY	DISCOVERER AND HIS NUMBER	PLACE OF DISCOVERY
356	1893 G	21 Jan. 1893	Charlois $_{40}$	Nice
357	1893 J	11 Feb. 1893	Charlois $_{41}$	Nice
358	1893 K	8 Mar. 1893	Charlois $_{42}$	Nice
359	1893 M	10 Mar. 1893	Charlois $_{43}$	Nice
360	1893 N	11 Mar. 1893	Charlois $_{44}$	Nice
361	1893 P	11 Mar. 1893	Charlois $_{45}$	Nice
362	1893 R	12 Mar. 1893	Charlois $_{46}$	Nice
363	1893 S	17 Mar. 1893	Charlois $_{47}$	Nice
364	1893 T'	19 Mar. 1893	Charlois $_{48}$	Nice
365	1893 V	21 Mar. 1893	Charlois $_{49}$	Nice
366	1893 W	21 Mar. 1893	Charlois $_{50}$	Nice
367	1893 AA	19 May 1893	Charlois $_{51}$	Nice
368	1893 AB	19 May 1893	Charlois $_{52}$	Nice
369	Aëria	4 July 1893	Borrelly $_{18}$	Marseilles
370	1893 AC	14 July 1893	Charlois $_{53}$	Nice
371	1893 AD	16 July 1893	Charlois $_{54}$	Nice
372	1893 AH	19 Aug. 1893	Charlois $_{55}$	Nice
373	1893 AJ	14 Sept. 1893	Charlois $_{56}$	Nice
374	1893 AK	18 Sept. 1893	Charlois $_{57}$	Nice
375	1893 AL	18 Sept. 1893	Charlois $_{58}$	Nice
376	1893 AM	18 Sept. 1893	Charlois $_{59}$	Nice
377	1893 AN	20 Sept. 1893	Charlois $_{60}$	Nice
378	1893 AP	6 Dec. 1893	Charlois $_{61}$	Nice
379	1894 AQ	8 Jan. 1894	Charlois $_{62}$	Nice
380	1894 AR	8 Jan. 1894	Charlois $_{63}$	Nice
381	1894 AS	10 Jan. 1894	Charlois $_{64}$	Nice
382	1894 AT	29 Jan. 1894	Charlois $_{65}$	Nice
383	1894 AU	29 Jan. 1894	Charlois $_{66}$	Nice
384	Burdigala	11 Feb. 1894	Courty $_1$	Bordeaux
385	Ilmatar	1 Mar. 1894	Wolf $_{17}$	Heidelberg
386	1894 AY	1 Mar. 1894	Wolf $_{18}$	Heidelberg
387	1894 AZ	5 Mar. 1894	Courty $_2$	Bordeaux
388	1894 BA	7 Mar. 1894	Charlois $_{67}$	Nice
389	1894 BB	8 Mar. 1894	Charlois $_{68}$	Nice
390	1894 BC	24 Mar. 1894	Bigourdan $_1$	Paris
391	Ingeborg	1 Nov. 1894	Wolf $_{19}$	Heidelberg
392	Wilhelmina	4 Nov. 1894	Wolf $_{20}$	Heidelberg
393	1894 BG	4 Nov. 1894	Wolf $_{21}$	Heidelberg
394	1894 BH	19 Nov. 1894	Borrelly $_{19}$	Marseilles
395	1894 BK	30 Nov. 1894	Charlois $_{69}$	Nice

NUM-BER	NAME	DATE OF DISCOVERY	DISCOVERER AND HIS NUMBER	PLACE OF DISCOVERY
396	1894 BL	1 Dec. 1894	Charlois $_{70}$	Nice
397	1894 BM	19 Dec. 1894	Charlois $_{71}$	Nice
398	1894 BN	28 Dec. 1894	Charlois $_{72}$	Nice
399	1895 BP	23 Feb. 1895	Wolf $_{22}$	Heidelberg
400	1895 BU	15 Mar. 1895	Charlois $_{73}$	Nice
401	Ottilia	16 Mar. 1895	Wolf $_{23}$	Heidelberg
402	1895 BW	21 Mar. 1895	Charlois $_{74}$	Nice
403	1895 BX	18 May 1895	Charlois $_{75}$	Nice
404	1895 BY	20 June 1895	Charlois $_{76}$	Nice
405	1895 BZ	23 July 1895	Charlois $_{77}$	Nice
406	1895 CB	23 Aug. 1895	Charlois $_{78}$	Nice
407	1895 CC	13 Oct. 1895	Wolf $_{24}$	Heidelberg
408	1895 CD	13 Oct. 1895	Wolf $_{25}$	Heidelberg
409	1895 CE	9 Dec. 1895	Charlois $_{79}$	Nice
410	1896 CH	7 Jan. 1896	Charlois $_{80}$	Nice
411	1896 CJ	7 Jan. 1896	Charlois $_{81}$	Nice
412	Elisabetha	7 Jan. 1896	Wolf $_{26}$	Heidelberg
413	Edburga	7 Jan. 1896	Wolf $_{27}$	Heidelberg
414	1896 CN	16 Jan. 1896	Charlois $_{82}$	Nice
415	1896 CO	7 Feb. 1896	Wolf $_{28}$	Heidelberg
416	Vaticana	4 May 1896	Charlois $_{83}$	Nice
417	1896 CT	6 May 1896	Wolf $_{29}$	Heidelberg
418	1896 CV	3 Sept. 1896	Wolf $_{30}$	Heidelberg
419	1896 CW	3 Sept. 1896	Wolf $_{31}$	Heidelberg
420	Bertholda	3 Sept. 1896	Wolf $_{32}$	Heidelberg
421	Zähringia	3 Sept. 1896	Wolf $_{33}$	Heidelberg
422	Berolina	8 Oct. 1896	Witt $_{1}$	Berlin
423	1896 DB	7 Dec. 1896	Charlois $_{84}$	Nice
424	1896 DF	31 Dec. 1896	Charlois $_{85}$	Nice
425	1896 DC	28 Dec. 1896	Charlois $_{86}$	Nice
426	1897 DH	25 Aug. 1897	Charlois $_{87}$	Nice
427	1897 DJ	27 Aug. 1897	Charlois $_{88}$	Nice
428	Monachia	18 Nov. 1897	Villiger $_{1}$	Munich
429	1897 DL	23 Nov. 1897	Charlois $_{89}$	Nice
430	1897 DM	18 Dec. 1897	Charlois $_{90}$	Nice
431	1897 DN	18 Dec. 1897	Charlois $_{91}$	Nice
432	1897 DO	18 Dec. 1897	Charlois $_{92}$	Nice
433	Eros	13 Aug. 1898	Witt $_{2}$	Berlin
434	Hungaria	11 Sept. 1898	Wolf $_{34}$	Heidelberg
435	1898 DS	11 Sept. 1898	* Wolf & S $_{35}$	Heidelberg

NUM- BER	NAME	DATE OF DISCOVERY	DISCOVERER AND HIS NUMBER	PLACE OF DISCOVERY
436	1898 DT	13 Sept. 1898	* Wolf & S $_{86}$	Heidelberg
437	1898 DU	8 Nov. 1898	Charlois $_{93}$	Nice
438	1898 DV	6 Nov. 1898	Wolf & S $_{87}$	Königstuhl
439	1898 DW	6 Nov. 1898	* Wolf & V $_{38}$	Königstuhl
440	1898 DX	6 Nov. 1898	Wolf & V $_{39}$	Königstuhl
441	1898 DY	13 Nov. 1898	Wolf & V $_{40}$	Königstuhl
442	1898 DZ	19 Nov. 1898	Wolf & V $_{41}$	Königstuhl
443	1898 EA	19 Nov. 1898	Wolf & S $_{42}$	Königstuhl
444	1898 EB	13 Oct. 1898	Coddington $_1$	Mt Hamilton
445	1898 EC	14 Oct. 1898	Coddington $_2$	Mt Hamilton
446	1898 ED	8 Dec. 1898	Charlois $_{94}$	Nice
447	1899 EE	15 Feb. 1899	Wolf & S $_{43}$	Königstuhl
448	1899 EF	17 Feb. 1899	Wolf & S $_{44}$	Königstuhl
449				
450				

Planet (433) Eros possesses exceptional interest because of the large eccentricity of its orbit, and its mean distance of only 1.46, that of Mars being 1.52. But it is not proper to describe the path of Eros as lying within that of Mars, because the perihelion points of the two orbits lie on opposite sides of the Sun. Only about half of Eros's path, therefore, is within that of Mars : but so eccentric is the orbit that it approaches within 14 million miles of our own path round the Sun. When Eros comes to opposition near the time of perihelion, as must have been the case in 1894, its stellar magnitude is about the 7th, or almost within naked-eye visibility. Professor PICKERING and Mrs FLEMING have re-discovered the planet on about 20 Harvard plates exposed near this time, greatly enhancing the accuracy of the orbit calculated by Dr CHANDLER. Not until 1924 does so favorable an opposition again occur; but in December 1900 it will approach us within 31 million miles, or about four million miles nearer than Mars ever does. This opposition will give a new and very accurate value of the solar parallax, because of the precision with which a diskless object of this sort can be measured. The opposition of 1924 will provide a value of the Sun's distance far exceeding in accuracy that from all other methods now available. Probably Eros is less than 20 miles in diameter, and its close approach to the Earth every 30 years will introduce perturbations of its motion of great interest to gravitational astronomers. It consumes 1¾ years in journeying once completely round the Sun; and its variation between such wide extremes of distance suggests many important problems to the astrophysicist.

* S. or V. signifies that Dr WOLF was assisted by SCHWASSMANN or VILLIGER in making the discovery.

INDEX

www.ingramcontent.com/pod-product-compliance
Lightning Source LLC
Chambersburg PA
CBHW020905210326
41598CB00018B/1774